SCIENTIFIC NIHILISM

SUNY Series in Philosophy
George R. Lucas, Jr., editor

Scientific Nihilism

On the Loss and Recovery of Physical Explanation

Daniel Athearn

STATE UNIVERSITY OF NEW YORK PRESS

Published by
State University of New York Press, Albany

For information, address State University of New York Press,
State University Plaza, Albany, N.Y., 12246

Production by Cathleen Collins
Marketing by Nancy Farrell

Library of Congress Cataloging in Publication Data

Athearn, Daniel, 1951–
    Scientific nihilism: On the loss and recovery of physical explanation /
Daniel Athearn.
        p.   cm. — (SUNY series in philosophy)
    Includes bibliographical references and index.
    ISBN 0-7914-1807-3 (alk. paper).—ISBN 0-7914-1808-1 (pbk.   :
alk. paper)
    1. Causality (Physics)   2. Physics—Philosophy.   3. Science—
Philosophy.   I. Title.   II. Series.
QC6.4C3A86   1994
530'.01—dc20   93-12869
                                                                CIP

10   9   8   7   6   5   4   3   2   1

Contents

Acknowledgments   ix

PART ONE

Current Outlooks in the Shadow
of the Technicalization of Physics

Introduction   1

1 The Causalist Quest in Physical Science   15
   Early Philosophical Turbulence   16
   Advanced Causalist Physics   21
   The New Era of Physics and the Reign of the Cult of Surfaces 29

2 "Law Explanation"   57
   Foundations of Logico-Empiricism   57
   Instrumentalism and Antirealism: Breaking the Link   61
   The "Covering Law" Model   67
   Why Theory of Explanation?   81

3 Philosophy and the Structure of Causation   85
   The Positivist Conception of Science   87
   Schlick's Humean Arguments   90
   Positivists on Indeterminism   105
   Hobart's Defense of Hume   112
   Is "Productionism" Anthropomorphism?   114
   Conclusions   115

4 Causal Realist Projects, I   119
   Composition of Causes   121
   Ontology of Latent Properties   126

Contents

The Success of Critical Realism		138
Causality and Forces		146
Conclusions		150
5	Causal Realist Projects, II	153
	The Theory of Transmission	155
	Explanation as a Theme: A Mark of Basic Acausalism	166
	"Probabilistic" Process	169
	Conclusion	172

PART TWO

Physical Ontology

Introduction		175
6	Radiation and Causality	191
	Enigmatic Physical Activity	193
	Empty Space Events?	201
	Causal Contour	209
	The Transition in Interaction	214
	Conclusion	223
7	Time, Space, and Genetic Structure	225
	Whitehead on Time and "Process"	226
	A Sense-Making Application	253
	The Ontological Shift	264
	Conclusions and Ramifications	272
8	Interatomic Reality	279
	Realism and Quantum Interpretation	279
	Richness of the Concept of Causality	286
	Quantum Mysteries	289
	Propagation and Spatial Representation	299
	Explanation and Causal Composition	302
	Summary and Transition	310
9	Absolute Causal Reference	313
	History of the Problem	314
	Whitehead's Solution	317
	Flat Transition and Structured Transition	324
	Conclusion	327

10 Velocity C and Emergent Extension 329
Relativity: Theories, Models, Explanations 329
The Nature of the Problem 334
Furnishing the Physical Context 337
Whitehead's Theory 343
Conclusions 364

Notes 367

Index 381

Acknowledgments

THE COMPLETION OF THIS WORK owes something to each of the following: Catherine Wilson and Cheyney Ryan for their assistance with major portions of Part One; the whole philosophy faculty at the University of Oregon for invaluable instruction in thinking; my friend Douglas Donkel for stimulating the theme of "identity and difference"; all the friends and acquaintances that have provided discussion and served as a fruitful testing ground for the ideas; Hannah Harrison, for her many beneficial suggestions; my parents, who nurtured "perfectly free inquiry" (Earl Morse Wilbur); and George R. Lucas, Jr., for inestimable encouragement and suggestions.

PART ONE

Current Outlooks in the Shadow
of the Technicalization of Physics

Introduction

THIS BOOK is a work of philosophy, intended for anyone who is interested and is willing to think it through, even though it is principally concerned with some questions of physical science which are naturally assumed to be the business of specialists. This might bring the phrase "philosophy of science" to mind, but this heading carries just the wrong associations. Though Part One is largely an examination of ideas in philosophy of science, the purpose is to develop a critique of the overall course of this discipline from an unconventional point of view.

"Philosophy of science" in current usage means *theory of science,* a form of epistemology or theory of knowledge, which is a traditional project of philosophy. Theory of science does not *engage in* science (here meaning compartmentalized research in the natural and social sciences), but thinks about the results of science as it finds them. A basic characteristic of its recent tradition is that across different positions and perspectives on the issues, and even across widely varying approaches, a common assumption is maintained: that contemporary physics is paradigmatic of achieving science and therefore can be regarded as a principal source of examples when pursuing questions about the nature of scientific knowledge in general. A typical framework of questions within this orientation is as

1

follows: What are the attributes of a theory in science (physics) that lead to its general acceptance? Is such a theory then "true"? Is the success of a particular theory explained in any part by its situation in the history of the science? What is the correct response to suggestions that modern mathematical theories are not achievements in the explanation of natural phenomena, but only instruments of technical advance? When physicists refer to entities seemingly "unobservable" and purely "theoretical," should these references be construed realistically?

The present work has completely different guiding aims and assumptions. For one thing, it is crucially interested in demonstrating that today's physics exemplifies only certain aspects of scientific inquiry. More generally, this work does not adhere to the theme of "the nature of science" at all; indeed, insofar as it seeks to develop an investigation into certain physical phenomena directly, it can only understand itself *as* science—though it should be emphasized that "science" here breaks the bounds of its contemporary connotation of specialized research disciplines cloistered to some extent by technical terminology and advanced levels of specialized knowledge and training, and takes on a historically expanded meaning as systematic truth-seeking inquiry sharing common ground with philosophy. However one prefers to classify this investigation, it considers a diversity of issues having some direct bearing on central questions of physical explanation. The subjects of these questions comprise a multifarious natural domain roughly encompassed by physical forces of attraction and repulsion (gravitational, electric, magnetic) and "radiant energy" or radiation. In the procedure that evolves for the direct investigation of these diverse interrelated natural phenomena they are treated as a category and thematized in their underlying unity, with focus on electromagnetic radiation such as ordinary light and on the general physical referent of "field" understood naively as some causal factor in the vicinity of an object that generates specific effects on other objects. The project as a whole brings together historical, philosophical, and scientific dimensions.

But what business can philosophy have in the study of these ubiquitous aspects of nature—forces and radiation—which roughly comprise the natural subject matter and resources of the science we know as "physics"? This science is marked by a high degree of specialization in abstruse technical discourse and high-level mathematics. Is there some legitimate role for philosophy here, or even a means and mode of access for the general philosopher? What could I mean by "investigation"? Consider that until sometime in the last century the basic scientific search for an understanding of this rough grouping of varied phenomena fell under the heading of "natural philosophy" (only subsequently did "philosophy" and "science" come to mean entirely separate things). This indicates the general possibility of philosophy and natural science coinciding, though so far perhaps all examples are anachronisms. More concretely, the term "natural philosophy" marks the fact that the quest for scientific understanding in this area was not at that time considered an inherently technical pursuit involving the mastery of specialized procedures of theory construction and application that can at best be conveyed to the lay public through analogies and parables.

This book does not employ mathematical symbols or rely on technical terminology of physics in discussing this physical category. Because of this, readers who are trained physicists might find it an odd and unlikely landscape of "science": "will it not in the end amount to a philosopher's verbal adventure, at best of 'poetic' significance?" But for the same reason the book should be of special value to nonphysicist philosophers, authored as it is by one of their own. Unfortunately its basic premise—that "natural philosophy" in the old sense can be resurrected—is likely to be received with incredulity by at least some philosophers and at least some physicists. There is no point in withholding, evading, or veiling the fact that at some point the inquiry will presume to constructively engage some physical questions which as a matter of historical fact are no longer pursued by science, questions which according to a consensus of authorities can only lead nowhere! Natural philosophy groped its way in basic questions about how light propagates and about the origin of forces acting between objects. Despite a general impression that "we have the answers now," the truth is that this particular exertion of primitive wonder in physical science became radically stymied and ceased altogether. The present project seeks to restore this old-fashioned quest for physical or causal explanations (under a sweeping transformation) to a vital and productive condition, rescuing it from its current status as a complete anachronism that was only justified during a more innocent stage of science. Again, the conventional limitation of the term "science" to specialized and technical research is here abandoned.

Is this not an arrogant intrusion onto the turf of physics? It is no such intrusion, simply because this science that lies currently in oblivion but which I intend to revive in a new form is distinguished from today's physics by a fundamental difference in mode of treatment of the physical domain in question. Historically, the methods and results of theoretical science concerning light and the various forces or "fields" can be divided into two distinct aspects:

(1) The seemingly antiquated attempts toward correct *narrative causal explanations* of the facts as they are revealed by observation and experimentation. These explanations had consisted of hypotheses about unseen physical occurrences in apparently empty space (or passing through objects) that might generate particular phenomena or groups of phenomena. Light propagation, for instance, was for a time explained as vibration in a medium of mysterious substance present everywhere, and electricity was once explained as fluids flowing around and through charged or conducting material. In the case of forces of attraction and repulsion such attempts have been so little successful that scientists have widely adopted the stand-in supposition that objects act on other objects "at a distance," opting out of the question of the mediating process or background activity that would seem to be required of a satisfactory narrative physical explanation.

(2) The accumulation and organization of *empirical knowledge,* covering facts about the observed phenomena themselves apart from any inquiry into their causes. This includes the projects of cataloguing and correlating the multifarious electric, magnetic, and gravitational phenomena, most notably the inventive process by which "regularities," patterns of interdependence among measured quantities occurring in experimental research, are formulated mathematically. The whole essential content of what is today called "physical theory," a phrase generally limited to physics by tacit convention, consists of mathematical models of correlations disclosed by technically advanced measurement, in short, of laws. The function of laws is to *predict* in various senses, for example, as a procedure for deriving a certain measurable quantity from given quantitative conditions, or by suggesting a mathematical symmetry

leading to the discovery of a new entity or phenomenon. The results of this aspect of scientific inquiry as they come to be adopted and applied merely express in formal models the quantitative correlations that are found in observation and measurement in the process of exploring and formulating the detail in the phenomena to an increasing degree of precision and completeness.

The fundamental distinction underlying and orienting this inquiry is that between narrative explanations and organized technical-empirical knowledge (especially quantitative laws). These two distinct modes of physical theory are hereafter referred to as "aspect (1)" and "aspect (2)" inquiry, respectively. The basic aim of this book is to show that a missing dimension of narrative physical explanation in the domain of physics is worth pursuing and can be successful.

There has been wide divergence among scientists, among eras of science, and among departments of science as to the status of narrative (historical or causal) explanation within the general aims of science. Chapter 1 relates the history in the course of which an intuitive commitment to the mode of science I have labeled aspect (1) was for centuries sustained by an eminent tradition of physicists, and then was abandoned by physics in the twentieth century—truly an extraordinary development, though it hardly seems extraordinary any longer except from a curiously naive point of view. Even before this official abandonment, powerful currents in philosophy and physics had fostered the view that the quest for laws of nature is the true method and direction of science, a view that ignores or overlooks the fact that the scope of laws and prediction has limits even within the natural sciences. Whatever the general rationales for this empiricist concept of science, it received a potent affirmation when theoretical physics in fact shifted entirely away from the traditional attempts to explain light and forces by means of causal accounts of goings-on in the background of objects and intervening space. (Talk of "particles" and "waves" by physicists today sounds like narrative physical explanation, but actually it is merely useful modeling terminology for one or another special, isolated context, that is, light may be *treated as* a wave or a particle depending on the context; the situation is confusing, but these modeling strategies do not answer the question as to what light propagation actually *is*, but rather *stand in place of* such an answer in

conventional discourse.) The upshot is that the very pursuit of
"natural philosophy" is now treated as an innocent error marking
the progress of science, as similar in a way to the theory of heat
involving the concentration and flow of "caloric fluid," or the
theory that combustion is the release of "phlogiston" from objects.
This historical shift in the aims of physics can be described even less
dramatically by saying that causal questions about phenomena of
electricity, magnetism, and light all finally settled into place alongside the corresponding question about gravitation, which had already for the most part been treated according to Newton's dictum,
hypotheses non fingo, "I feign [or spin, or frame] no hypotheses"
(regarding generative processes). Newton in fact did expend considerable effort at framing a causal hypothesis about gravity,
though his confidence in the result was little or nil; but physicists in
our time, and conspiring with them the most influential thinkers in
philosophy of science, have consigned aspect (1) to the museum. In
sciences other than fundamental physics and physical cosmology,
however, nothing threatens the vitality, status, or legitimacy of narrative causal explanation.[1]

The title and overall aim of this book can now be explained on
the basis of this distinction between aspects of inquiry in the history
of physics, in light of the prevailing view that aspect (1) or "natural
philosophy" is an extinct species of science.

The book seeks to revive and fundamentally reorient natural
philosophy in the old sense. In this philosophical and scientific
project the assumption is sustained that detailed causal explanations for the natural domain in question—and by "causal explanations" I mean narrative accounts identifying and comprehending
the physical processes (e.g., light "waves") that generate observed
phenomena, the essential content of aspect (1) inquiry—can be
fruitfully pursued in spite of well-solidified contemporary convictions to the contrary. In the course of the investigation it is shown
that the actual problems of physical explanation are not *technical*
matters of physics at all, and that inquiry into the problems and the
possibility of their resolution in a new understanding can proceed
indefinitely without any prior specialization in aspects of physics
which to the nonphysicist are a closed book. The actual proposals of
physical explanation are the subjects of Part Two. Part One is a
necessary preparatory phase of the inquiry, a process of demonstrat-

ing the legitimacy of the project against the antipathetic intellectual atmosphere of the present by unveiling the historical sources of this antipathy, showing especially how the dominant ideas in philosophy of science have gone wrong. This entails a critical examination of doctrines and ways of thinking that pervade the contemporary scientific and public culture of ideas. In the process of this therapy of historical criticism Part Two is prepared for by a consideration of the basic relevant concepts of explanation and causality.

The overall argument for the feasibility of actually reviving narrative causal explanation of radiation and forces is based on developing the possibility that the general concept of a causal process as employed in physical explanations can in the course of concrete applications be developed beyond the framework of the "classical" or mechanistic causal models of previous centuries, going on the supposition that it is only the latter *form* of causal explanation—roughly, models involving local motion of matter—whose possibilities have been surpassed by advanced experiments. This derives important guidance from Alfred North Whitehead's idiosyncratic project of laying new foundations for physical theory. Renewed and reoriented natural philosophy is given the name "physical ontology," for reasons that become fully apparent in Part Two.

The term "ontology" is often regarded as a synonym for "metaphysics." On the standard assumption that "physics" will from now on be a name for a purely aspect (2) science, physical ontology (despite its name) will be seen as "metaphysics" in the highly peculiar sense that it seeks to extend inquiry *in this area* beyond empirical formulas and expedient models, even though it has nothing directly to do with the series of synoptic systems of the world in the history of philosophy. The name "*physical* ontology" directly repudiates the label "metaphysics," and it also conveys in more than one way that the inquiry so named is not nonscience. I do not subscribe to the clean partitioning of philosophy from science that is widely assumed today, and as I have indicated I seek in the course of the book to present physical ontology as an example of science and philosophy coinciding, or if one prefers, as a "luminous bridge" between them, as Heidegger said.[2] The present use of "ontology" corresponds *approximately* to its occasional use by physicists who engage in unorthodox explanatory interpretation of quantum theory and quantum phenomena. Entirely apart from this comparison, there is one recent and unambiguous precedent for my use of the term in the field of philosophy, discussed in Chapter 7.

The philosophical outlook I am calling "scientific nihilism" is the main topic of Part One. This is a specially adapted term that has certain connections with other uses of the word "nihilism" by philosophers, though these connections are not directly discussed in this book; it does not, of course, refer to a destructive political movement or antimoral doctrine. "Scientific nihilism" designates an outlook or broadly shared intellectual attitude built around the core dogma, based on decisions in physics itself, that today it is no longer feasible to engage in the search for narrative causal explanations in fundamental physics, and that therefore the legitimate course of scientific theory in this area can be presumed to be confined from now on to aspect (2). This conclusion and its generalization into a concept of science can be called the philosophically and scientifically determinative marks of the twentieth century. Seldom do scientists or philosophers of science bother to state this core doctrine explicitly, though one can find it stated in one form or another in elementary science textbooks and popular expositions. Discussions that link physics and philosophy are permeated with tacit acceptance of the "obvious" truth that in the illustrious case of physics the aim of narrative explanation has proven thoroughly naive, whatever is the case with other sciences. It is not only that certain physical explanations come to be seen as unobtainable; rather, for this domain an entire inner impetus of physical science, the very "feeling" that science ought to disclose the generative factors manifested in the phenomena, slips away as the whole attention of the science is shifted to other concerns.

Acausalism about fundamental physical science is only the core of scientific nihilism, around which have accumulated layers of assumptions and methodological doctrines that dominate a scientific/academic community in which philosophy of science has an influence second only to physics itself (which is no indicator of the overall status of *philosophy*). Taken as a whole this institutional/ dogmatic structure is antithetical to the basic claims of physical ontology; indeed the two outlooks are radically alien to one another. Philosophy of science as theory of science has been largely animated by the project of following out the apparent acausalist implications of twentieth-century physics—essentially a philo-

sophical *reaction* as if to the decree of an unique authority, partly
due to the purely emotional factor of the prestige of "hard" mathematical physics with its technical accomplishments. A useful mark
of scientific nihilism is the notion that *real* scientific knowledge
and/or scientific explanation is exemplified by aspect (2), following
the paradigm of current physics, with aspect (1) implicitly suppressed in its status. (Sciences still characterized by narrative explanation, such as biology and geology, tend to be devalued by this
outlook as "inexact," "merely descriptive," and so on.) It can be
found wherever the call for "empirically adequate" theories is
sounded, whereas physical ontology presumes the contrary, that
science seeks the true and correct explanation in the given case. As I
explain in Chapter 1, aspects of scientific nihilism range from philosophical doctrines to cultural conditions.

"Nihilism" is a strong term for something that enjoys a bland,
half-conscious general acceptance, and which moreover is hardly
seditious but rather has the most authoritative sources. But to the
extent that one can recover a degree of innocent expectation with
respect to physical science its implicit claims should seem extreme
and even bewildering: "The propagation of light *can never* be physically comprehended in a causal-narrative sense." As Chapter 1
relates, the attempts (which have been officially abandoned) to
determine narrative causal explanations for fundamental natural
phenomena belonged to the basic scientific commitments of some
of the greatest physicists of past centuries, however sparse or nil
their successes in such attempts. Suppose it is true that science has
been *forced to give up on* the quest for these explanations (whether
because of difficulties present in the phenomena or on methodological or epistemological grounds). If all possibilities for this pursuit have really permanently and finally collapsed, then the present
situation in the history of physics is one for which even our fashionable relativism about scientific "paradigms" has not quite prepared
us. To appreciate it one must be careful about assumptions regarding what it is that contemporary physical theory achieves, in light
of the difference of the two aspects. If abandonment of aspect (1)
were inescapable, this would mean, beneath all veils and disguises,
that natural science is forced to acknowledge one of two versions of
scientific nihilism (or some vague mixture of the two): (a) that such
commonplace facts as electromagnetic propagation and whatever

physical processes generate forces are all *permanently impenetrable* to the detailed causal-narrative comprehension that science eventually achieves for other aspects of the natural world; or (b) the position, which is actually a centerpiece of philosophical controversy, that such "unobservable" causal processes have no physical reality apart from "their" observable and quantifiable "effects." One might complain that the term "nihilism" befits the startling and strange antirealist conclusion (b), but not so much the more familiar notion of a realm entirely outside "ordinary concepts," which does not go so far as to deny the reality of this "realm." But these are roughly equivalent versions of nihilism from the point of view of physical ontology, since the "other reality" of (a) would bar access by science both to itself and to its connections with the phenomena it would explain (regardless of the successes of empirical models and formulas), and is therefore precisely nothing to a science that seeks to disclose what is. A common way of evading these nihilist options is to suppose that laws somehow *constitute* explanation and/or understanding, which if feasible would allow physics to be a science of exemplary *explanatory* success in spite of the collapse of narrative physical explanations. Chapter 2 is devoted to exposing the failings of this popular strategy of adopting a special concept of "scientific explanation." There are a variety of obscuring rationalizations, but the scientific nihilist doctrine uncovered is that there are no real physical explanations for light and other basic physical phenomena.

A broad intellectual culture has become versed in nihilist conclusions (in a number of guises), tending to favor them as marks of sophistication; and who can argue with theoretical physicists? Contributing to this general complaisance has been the fact that the influential discipline of philosophy of science is already caught up in the question of whether the familiar "particle" of physicists (this is one technical term for radiation, generally recognized as a term of utility and not of description) is an actual physical entity opaque to any actual description, or merely a useful construct of science. This cultivated displacement of the actual physical problems is accompanied by a general chorus of educated pronouncements about the limitations and essential subjectivity of human knowledge, with weighty intellectual traditions behind them. But the best way to counteract jadedness to nihilism is to turn away from epistemology, naively supposing that its importance to physical science is at best ambiguous, and back to the physical matters. One thing to note initially is that the entities or processes referred to in different contexts as particles, radiation, or fields underlie all physical actuality: their activity is inherent to space empty of matter, and they also comprise in different configurations the fundamental structure of matter itself. A physical explanation of phenomena of the field, assuming such is possible, would be an ultimately basic, background kind of physical explanation; that it would necessarily be *ontological*, which I can here explain as "involving a transformation in elemental physical concepts," is evidenced, I submit, in the array of unresolved physical enigmas that present themselves to physics and cosmology today (whatever the reaction to them). The status of such prospects is certainly important for the nonspecialized science-minded inquirer or philosopher of nature, since this physical domain includes familiar causal processes pervading the world, and not only arcane "theoretical entities" known only through the specialized occupation of physics; moreover, one class of mysteries concerns a quite general cosmological understanding. First and foremost, the questions I seek to revive are basic problems in the understanding of ordinary physical phenomena, rather than special mysteries of a remote subobservational world.

I do not wish to portray those to whom I attribute a nihilist point of view in one context or another as necessarily disposed toward attacks on traditional notions of physical reality. Even a reputation for "realism" can take shape just where the founding decisions of scientific nihilism are guiding the theoretical method. When it came to the discussion of quantum theory, the gentle Einstein famously exemplified the "realist" physicist who will not be satisfied until his criteria of physical comprehensibility are met; but his own approach in formulating the theory of relativity had a nihilist impact equal to that of quantum theory. As discussed in the first chapter, Einstein's approach to theory, with the success and reknown it achieved, directly inspired three monumental nihilist doctrines about the nature of science: P. W. Bridgman's operationalism, Karl Popper's doctrine of falsifiability, and Thomas Kuhn's conclusions regarding progress in science. For each of these philosophers, the scientific status of a theory is (to some degree inexplicitly) made to depend entirely on its workability in technical "problem solving" and in guiding the progress of experiments. The reason a fascination with the example of Einstein contributes to this conclusion is that the special and general theories of relativity carry out, for a certain sphere of problems, a *confinement of theory within aspect (2).* (From the point of view adopted here this means they are neither correct nor incorrect explanations.) This decision of method, due to the great success and prestige of its products (when carried out with sufficient genius), is adopted as a model for a concept of science in general. A powerful ally in these decisions was Copenhagenist quantum physics, which consciously renounced the quest for causal understanding of the strange "microcosmic" reality and embraced the predictive formula as the complete content of theory. Because the nihilist outlook as it is represented in the abovementioned philosophers of science involves determinations as to what should be considered a correct concept of true and genuine science (albeit based on a "one-sided diet of examples," as the Wittgensteinians say), it earns the appellation "scientific." I wish also to exploit this word for its tone of intellectual rigor, as a sardonic comment.

The history of scientific nihilism has the same narrative structure as Nietzsche's conception of the nihilism of values. Nietzsche's basic account if I understand him is this: It begins to dawn on European civilization that certain traditional icons of moral and metaphysical value have become "untenable"; the general reaction to this incipient realization, building beneath the surface of what is said and thought, is the judgment that "all is meaningless," "all is vain"; but with an enlarged perspective and a deeper discernment the historical moment is seen as the preparation of a transition toward a future "revaluation of values," imparting value back into values and also transforming the whole character of valuative thinking.[3] What I call scientific nihilism takes place as follows: A certain centuries-old scientific quest for narrative physical explanation had relied on a mechanistic conceptual framework in the absence of an alternative, and had only fragmentary and transitory successes; when it finally became clear that this framework is not applicable to this particular quest, the conclusion took hold that the *quest itself* had been scientifically outdated. A body of literature accreted around this doctrine while remaining fixated on physics and empirical theory as an emerging paradigm of science. But as I see it, this whole development in the history of physical science is actually only a transitional stage in which the natural and traditional quest for causes has gone into dormancy awaiting a conceptual transformation. "Nihilism" in my usage means the *slipping into oblivion* of basic *questions* of physical science, just when—indeed because—these questions are revealing themselves as pointing beyond the traditional conceptual foundations of physical explanations.

"Causalism" and "acausalism" are more suitable terms for the historical study comprising Part One than "ontology" and "nihilism." Besides the fact that "ontology" is not completely explained until Part Two, the latter terms are mainly appropriate for recent outlooks in philosophy of science and recent decisions concerning the possibilities of theoretical physics, since before the present century it was not at all obvious that an *ontological* crisis was on the horizon for physics, that is, it could not then be predicted that this science would be forced to adopt one of two courses, that of undertaking a shift in its foundational conception of physical being (ontology), or that of effectively abandoning a certain fundamental category of physical explanations (scientific nihilism).

Part One overall is a critical examination of the sources and supporting arguments of traditional neo-Humean (positivist and logico-empiricist) philosophy of science, the articulation of scientific nihilism by professional philosophers. Beginning with Chapter 2, two key doctrines of this tradition are examined: (a) that laws or systems of laws can themselves comprise explanations of physical phenomena, or at least that genuine scientific knowledge, whether explanatory or not, consists of laws (this is allied with the hazardous tendency to view mathematical models as descriptions); and (b) the traditional analysis of the causal relation derived from David Hume. Regarding (b), a major pillar of neo-Humean doctrine has been the persistent difficulty philosophers have had with causation: under standard assumptions the special connection that is supposed to distinguish causal from noncausal sequences of events radically eludes reflection. A complete resolution of this problem as it pertains to physical causality is offered in the course of the book. The last two chapters in Part One discuss major contemporary "causal realist" philosophers of science who achieve important clarities in the process of sustaining counternihilist intuitions, and the discussion shows how they nevertheless remain without an auspicious orientation with respect to the pursuit of constructive physical ontology because of the pervasive influence of scientific nihilist doctrine.

Part Two contains among other things an answer to the objection that the aim of a causal ontology for physics can never succeed because of certain utterly intractable impasses for causal explanation that have been brought to light by recent experiment and analysis. To answer this complaint I propose a background physical hypothesis that contextualizes these extraordinary discoveries, dissipating their appearance of permanent impassability to causal understanding and to some extent concretely explaining them. The result is a theory of radiation developing in the context of a narrative theory of the causal background of physical space. There are no immediately accessible answers, whatever expertise the reader may have.

1

The Causalist Quest in Physical Science

ONE OF THE PREPARATORY TASKS is to beckon contemporary philosophy and scientific thinking back toward a healthy concern and involvement with natural science in a certain domain, after a period of deepening alienation. By "natural science" I mean a general and unfolding understanding of the natural world attained through a history of inquiry whose legacy is broadly accessible to educated people, not a collection of specialized disciplines and departments. My urging is that natural science needs to assert itself *philosophically* in questions related to physics, and, as if that were not enough, that it needs to proceed down an uncharted path. Its eventual course has to be one of *ontological* research, capable of resurrecting questions directed toward a first and most basic understanding of a profound matter: the nature of the causal activity that comprises physical space whether empty or occupied by objects. As such it has to move directly against the major current of our time regarding such questions and regarding conceptions of physical science in general, because this current is dominated by the influence of *a characteristic unique to physics in its present era*, namely, a hegemony of what I have designated aspect (2) (see Introduction), whereas physical ontology is primarily a concern with aspect (1) on its own. The current aspect-hegemony is reflected in and reinforced by the fundamental positions of influential philosophers of science.

The purpose of the brief revisionist history of physical science presented here is to orient the revived natural philosophy historically and scientifically, or in other words, to recover a sense for the questions whose consignment to oblivion I have termed scientific nihilism. This preparatory process requires complete autonomy from the recent nihilist consensus, which arose from conceiving theoretical science in general on the model of contemporary theories of physics. This autonomous point of view is achieved by emphasizing the history of aspect (1), its erstwhile vitality and its current entombment. Specifically, the argument aims to show that the new project of explanation, correctly oriented, will neither employ nor emulate the theoretical and technical/mathematical procedures and methodologies of today's physics, because this science as we know it has effectively confined the whole of constructive theorizing to mathematical formulas (often viewed, mistakenly I believe, as encoded or symbolic explanatory descriptions of which one might venture an unofficial narrative and inherently speculative interpretation) and convenient modeling language. Theory thus confined has its own rationale and an open-ended applicability but cannot itself lead to causal explanation, let alone furnish an ontology. Fortunately, as I will try to show, intuitions about the meaning of "physical explanation" that can be emulated by the natural philosopher in a time of scientific nihilism are on display in some powerful figures from the history of physics. The chapter traces this history with the spotlight on these intuitions, but it begins with a singular giant of science who stands at the opposite pole from these causalist physicists.

Early Philosophical Turbulence

The story begins with Galileo Galilei (1564–1642), the great forefather of modern acausalist and mathematical physics. One of the forces that shaped his thought was a love of mathematics, and another was a reaction against the Scholastic physicists in the late Aristotelian tradition of his time. The Scholastics, and even Galileo himself at first, sought to understand, for example, the acceleration of a falling body by an ad hoc conjecture about its immediate cause. The trajectory of a thrown object was thought to be explained as the result of a continually impressed force that for some reason diminishes over time.[1] But Galileo developed a radically different scientific approach. Concerning free fall, his new procedure was to attend closely to the phenomenon itself, assume the simplest proportions among its measures, and produce a mathematical analysis or "definition" of acceleration. By this means, and by applying a mathematical theory of proportions, he predicted with dazzling accuracy a truth that is striking to a naive expectation: that the rate of acceleration in free-fall (in a vacuum) is unaffected by weight. The upshot was that he provided science with the modern concept of *inertia*. It is widely thought that "Galilean science" is a synonym for "modern science," that is, Western science roughly since Galileo's time, but this chapter should help to show that this is a mistaken notion stemming from twentieth-century prejudices.

The kind of success in the furtherance of science achieved by Galileo's revolutionary methods can produce the strong suggestion that to conjecture about the hidden causes for physical phenomena in general is simply the wrong way for science to proceed. Indeed, Galileo was led by the fruitfulness of his mathematical approach, which no doubt bolstered his reaction against the Scholastic-Aristotelian approach, to think that in general it was vain to seek for the "true and internal essences of natural substances" and that science should "content [itself] with a knowledge of some of their properties."[2] Given this attitude, it is not surprising that Galileo was not inclined to speculate about occurrences in the surroundings of objects that would account, for example, for electric and magnetic forces.

Pierre Gassendi (1592–1655) was a follower of Galileo, and understood well the objections to speculations beyond the sphere of what can be confirmed by observation. But after duly expressing qualms and asking the reader to indulge him some "uncertain conjectures" and "murky babblings," he felt justified in pursuing serious reflections on questions about the possible causes of magnetism, static electricity, gravity, and light propagation, sketching "a theory that seems closer to the truth than the others."[3] One of his written works echoes Galileo by attacking the possibility of "Aristotelian science," and this work has given Gassendi a reputation as a strict empiricist, even though this is contradicted by his serious aspect (1) conjectures. "Aristotelian science" meant science guided by the expectation that the "necessary" causes of things could be determined by logical demonstration, so that his criticism does not speak against the possibility of causal conjectures that are not demonstratively certain, but may have some degree of plausibility and some claim to truth. Also, there is in this same work a passage that, whether intentionally or not, movingly defends the calling of a speculative natural philosophy:

> Therefore, I conclude that whatever certainty there is in mathematics is related to appearances, and in no way related to genuine causes or the inner natures of things. However, I must add that with the help of mathematics I can become certain that, for instance, the earth is round; this can be made manifest through the eclipses of the moon and the varying height of the poles. But why is the earth round? what is its true nature? is it animate or not? and if it has a soul, what kind does it have? what functions does it perform? what properties is it alloted? why does it lie motionless at the center, or if it moves, what impels it? The same questions are to be asked about the sun and the other stars, and likewise about sound, which is the subject of music, and light, the subject of optics, and so forth. Truly, the moment you pass beyond things that are apparent, or fall under the province of the senses and experience, in order to inquire about deeper matters, both mathematics and all other branches of knowledge become completely shrouded in darkness.[4]

Regarding one of the matters that fall under the province of speculation, that of the ultimate nature of the matter-space relation, Gassendi did have a disputant in his contemporary René Descartes (1596–1650). Gassendi thought that the ultimate structure of things had to consist of atoms moving around in a void, whereas Descartes considered the void an impossibility, and thought that substance must be present everywhere, a "plenum." Today a physicist views these theories of fundamental physical structure founded on the contrast between empty and occupied space as too simple and naive for advanced knowledge; in modern "field" physics there is an implicit, background recognition of the existence of *empty space events*. However, there is no basic physical account or narrative description of such events, despite the thorough utility of a merely convenient technical terminology.

To try to determine where Descartes stood on the question whether Galileo had formed the correct conception of physics will end in confusion. He sought to build up all knowledge from self-evident certainties and rules of inference, forging a mathematical/deductive approach into methodological precepts for developing a system of the world, a system free from the unfounded hypotheses he thought were characteristic of philosophies of the past. This is a stance fully ambivalent between a Galilean anticonjectural method and an Aristotelian quest for encompassing knowledge. Against Galileo and Gassendi, Descartes was opposed to the overriding emphasis upon what "appears to the senses," and proposed elaborate mechanistic explanations of magnetism and light in terms of invisible particles and processes of the plenum; however, he capitulated in a way to the empiricist proscription on knowledge in a remark displaying vacillation as to whether these explanations were to be taken seriously as candidates for the truth.[5] Gassendi was more thoughtful on this point, and merely disclaimed any undue expectation of certainty or adequacy for his conjectures.

The explicit historical discussion on the proper aims and expectations of physics that had been sparked by Galileo's work died out in this confused condition. Meanwhile, in early research into magnetism and electricity, explanatory conjectures were developed and refined along with experimental progress as if the quest for a causal understanding of the phenomena unquestionably belonged to the basic mission of science.

Next our interest comes to rest on Isaac Newton (1642–1727). Due to the dominant prejudices of today, Newton is often treated as if he were a pillar of acausalist physics, the man who formulated complete laws of motion and gravitation and, when it came to the question of the cause of gravity, said *"hypotheses non fingo."* This Latin phrase, whose suggestion depends quite a bit on the chosen translation, is understood on this view to encapsulate the thought that it is illegitimate for science to engage in speculations about the causes of forces, or about "unobservable" causes in general. But this is a misrepresentation of Newton's views. We know from his letters to Boyle and Bentley and from other writings that he had no doubt there was something, whether "material or immaterial," *mediating across the space between objects* that would explain forces, and that he tried to account for the force of gravity by the action of a pervasive material medium but was never satisfied with the results. There occurs in the third letter to Bentley a passage whose forceful causalism brought it fame even in our era:

> That gravity should be innate, inherent and essential to matter, so that one body may act upon another at a distance through a vacuum, without the mediation of anything else, by and through which their action and force may be conveyed from one to another, is to me so great an absurdity, that I believe no man who has in philosophical matters a competent faculty of thinking, can ever fall into it.[6]

As discussed further below (page 29), biographical material indicates that the reason Newton did not publish his causal conjectures about gravity was that he could not conceive a satisfactory mechanism, not that he did not condone such conjectures philosophically, that is, from the standpoint of a doctrine or theory of the true scientific method. Also, Newton devised the corpuscular theory of light, which dominated optics for the subsequent century and a half. Though on the surface narrative theories of light are more tractable than speculations about gravity, especially for this innocent era, both are conjectures beyond the conceivable province of direct observation.

In a wide-ranging debate by correspondence with Leibniz, the Newtonian metaphysician Samuel Clarke (1675–1729) said the following:

> That one body should attract another without any intermediate means, is indeed not a miracle, but a contradiction: for 'tis supposing something to act where it is not. But the means by which two bodies attract each other, may be invisible and intangible, and of a different nature from mechanism; and yet, acting regularly and constantly, may well be called natural; being much less wonderful than animal-motion, which is yet never called a miracle.
>
> If the word *natural forces*, means here mechanical; then all animals, and even men, are as mere machines as a clock. But if the word does not mean, mechanical forces; then gravitation may be effected by regular and natural powers, though they be not mechanical.[7]

Note that Clarke is not suggesting a teleological explanation, as if to say that that gravitation might be effected by something having the purposefulness of animal and human behavior; indeed if it were anything of this sort, then for the theistic worldview it might have to result somehow from God's volition, which would make it miraculous, that is, outside the scope of natural science, and would also seem to require that it be to some degree unpredictable and changeable, rather than ever-present and constant. Clarke merely urges an openness to the possibility that the "intermediate means" may be neither teleological nor mechanical (meaning roughly: involving matter in motion). This remarkably open suggestion, which stops short of any positive conception of the "intermediate means," anticipates the course that narrative causal explanation is finally challenged by the evidence to follow (assuming its aims remain valid) in the twentieth century.

To summarize: Galileo began an early historical reflection concerning the basic goals and prospects of physical science by throwing his weight behind mathematical analysis of the empirical properties of things (at least for nonastronomical matters) in reaction against Scholastic physics. But despite the influence of Galileo's breakthrough, the procedure of speculating beyond the observations for the specific causes of what is observed was far from vanquished; there was even the phenomenon of Gassendi, who generally promoted the Galilean reaction, but was a natural philosopher nonetheless. After this discussion on the aims of physical science died out, Newton, most famous for having applied the mathematical approach to produce the laws of motion and gravitation, actually followed a balanced approach; despite *"hypotheses non fingo"* there is really no reason to think he did not remain solidly behind aspect (1) questions in principle.

Advanced Causalist Physics

During the eighteenth and early nineteenth centuries the sciences of electricity and magnetism saw development in both aspect (1) and aspect (2) research. A variety of theories of electric and magnetic fluids, flowing through objects and through the space surrounding them, were put forth to account for the accumulating results of experiment, with some limited success. Leading intellectuals who were not experimental investigators or scientific specialists also contemplated the forces. As for aspect (2), inverse square laws corresponding to Newton's law of gravitation were established for electric and magnetic forces. Along with these successes came the practice of talking about action at a distance rather than about the adventures of the electric and magnetic fluids in the interspace of objects. Despite Newton's real views about this as expressed in the letter to Bentley, the era during which the idea of forces acting at a distance dominated physics is often referred to as the Newtonian era.

That in this "Newtonian" period there was a certain dominance of aspect (2) is evident in the suppression of the causalist conviction that mediating factors of some kind must exist. The extent to which aspect (2) dominated the thinking of these later researchers need not concern us here. But aspect (1) continued to have a role during this period, since the electric and magnetic fluids merely came to be thought of as confined within the objects, and even theories of fluids acting externally to the objects were not entirely abandoned—and in any case, the quasi-material medium of light transmission was alive and well.

Also noteworthy in this period were the experimental discoveries of interactions between electric and magnetic phenomena, which suggested that at least these, and perhaps all forces of attraction and repulsion, have a common basis in nature.

This brings us to Michael Faraday and James Clerk Maxwell. I discuss these scientists together not only because they are closely associated historically, but also because they shared important intuitions regarding the goals of science, though their individual contributions were very different.

Faraday (1791–1867) epitomizes the aspect (1) researcher whose life-project is summarized in the statement that "it is the cause of the forces that one wants to lay hold of."[8] Untrained in mathematics, he was a highly productive experimenter who developed rudimentary physical conceptions of the routes of transmission of electric and magnetic effects, involving structures in the space surrounding objects: the *lines of force* (manifested, for instance, in the patterns observed when iron filings are distributed on a sheet of paper and a magnet is placed underneath). Faraday thought of the transmission along the line of force (through empty space) as if it were something analogous to a state of tension in a piece of string, but he did not arrive at a more adequate physical specification. He came to regard the lines of force as more than a useful construct or model assisting experimental intuitions; he was devoted to the quest for a genuine understanding of how forces in general are brought about, and saw the lines of force as an answer or partial answer. That is, he clearly thought these structures were physical realities, whatever the limitations of his attempts to describe them. His basic scientific hope and expectation was that the common basis in nature for all the forces would someday be discovered and described.

What is especially significant about Faraday's work from the standpoint of the present inquiry is that he came to reject the other three modes of explanation that prevailed in his time: electric and magnetic fluids, action at a distance, and the mysterious ether filling all of space. In other words, he developed the idea that the transmission of forces must occur in and through *space devoid of matter*. This is the pioneering intuition of "field" physics, whatever one wants to make of the notion of "field"—and the usual account from our standpoint today would hasten to add that Faraday was nevertheless naive in thinking that the field could be given a "non-mathematical description." But from my own point of view, he leapt far beyond his time by envisioning potentially describable causal structure integral to space itself, and hence of the true direction for achieving fundamental narrative explanations. He *correctly* rejected material substrata, but at the same time he did not positively surpass the concept of localized entities or local "tensions" and "strains,"[9] though his thought did develop in the direction of such an overcoming (the significance of this particular limitation will become clearer in due course). Chapter 8 touches on Faraday's more speculative thinking.

Faraday's explanatory ideas might easily be dismissed today as those of an experimenter who was naive about theory; but no mathematical innocence can be attributed to Maxwell (1831–79), who was an avid promoter of Faraday's research and ideas. He claimed that Faraday's "lack" of mathematical training was not a hindrance to his having produced, with the lines of force and their variables of density per unit of space and individual intensity, a mathematical theory in its own original sense—a "new mathesis," as he put it. Maxwell himself followed a path decidedly balanced between aspects (1) and (2). The best evidence for this is not the fact that he constructed physical models of activity in apparently empty space as never before, for example, imagining a compact system of spinning vortices in the ether separated by smaller "idle wheels" rolling between them. He fully recognized the artificiality of these models, constructed as they were ad hoc in response to particular features or groups of features within a broad class of interrelated phenomena; he knew they could not finally be retained as the "true interpretation"[10] (the hope for which he did not at all view as chimerical), and claimed of them only that they helped suggest a formulation of the mathematical relations. By such means he was in fact able to produce the complete quantitative laws for the electromagnetic field, incorporating electricity, magnetism, and light in a single, integrated, aspect (2) theory (the only surviving vestige of aspect [1] or narrative explanation in Maxwell's electromagnetic theory is the simple schematic relation among the vectors of electric and magnetic forces and light propagation, though this schema originally had a physical dimension provided by the background idea of an ethereal medium). I will argue here that Maxwell's true commitment to aspect (1) was shown not so much by the use of physical models, but more by his defense and promotion of Faraday's explanatory ideas and his expectation that these rudimentary ideas would develop into a more adequate understanding in the future, one that he thought would amount to a science yet unnamed.

As in the case of Newton, twentieth-century commentators on Maxwell tend to emphasize the illustrious success of the unified field laws for electromagnetism, and to discount, overlook, or ignore what I would call Maxwell's unwaning commitment to aspect (1). Commentators often point to his sophistication in not believing his ether-based models to be finally correct explanations; but this is no argument that he thought that genuine physical explanations, for instance along the lines of those sought by Faraday, were not a proper business of physical science. There is an appearance of disagreement with Faraday in that Maxwell believed in some kind of an ether, but if this shows anything it is only that he retained a stronger element of materialism; more basically, the fundamental importance of both narrative and quantitative approaches in his views about physical theory can be seen in his writings, especially in his attitude toward Faraday. The quoted passages to follow do not lay out a *theory* of what "physical explanation" ultimately means, but show some intuitions and attitudes that identify strongly with the causalism of Newton and Faraday.

Maxwell explained his purposes in elaborating Faraday's lines of force into tubes filled with moving fluid as follows:

> By referring everything to the purely geometrical idea of the motion of an imaginary fluid, I hope to attain generality and precision, and to avoid the dangers arising from a premature theory professing to explain the cause of the phenomena. If the results of mere speculation which I have collected are found to be of any use to experimental philosophers, in arranging and interpreting their results, they will have served their purpose, and a mature theory, in which physical facts will be physically explained, will be formed by those who by interrogating Nature herself can obtain the only true solution of the questions which the mathematical theory suggests.[11]

Clearly he did not think the mature stage of theory in which real physical explanations first appear is attained with successful mathematical theory, which only "suggests" such physical explanations. What then is one to take as a model for understanding the *basic character* of the explanations that would eventually be produced by "interrogating Nature herself"? The penultimate paragraph of this same article lists some options for a physical explanation of electric and magnetic phenomena, and they are the three kinds of causal hypothesis existing at the time: motions in the ether, flowing "imponderable" fluids, and action at a distance. This confirms what is stated clearly enough in this passage, that he envisioned the future explanations at least as some kind of aspect (1) or causal narrative explanations. But Maxwell clearly does not here commit himself to any of these traditional options, and is in fact expressing doubts about them as a group. Therefore this article is of little help to us in understanding how Maxwell conceived the *positive* character of the future explanations.

But his writings do offer important guidelines about this. I think the soundest way to describe his views on the nature of physical explanation in the area of forces and radiation is as follows: He believed that genuine explanations of electromagnetic phenomena did not yet exist and that the future explanations would be the result of *an extension of Faraday's approach under some kind of transformation*. To show that this was the direction of Maxwell's thought, I bring together several separate pieces of text. First, let us note that he defended Faraday against the arrogance of the mathematicians who regarded their methods as superior to—more scientific than—Faraday's conceptions:

Up to the present time the mathematicians who have rejected Faraday's method of stating his law as unworthy of the precision of their science have never succeeded in devising any essentially different formula which shall fully express the phenomena without introducing hypotheses about the mutual action of things which have no physical existence, such as elements of currents which flow out of nothing, then along a wire, and finally sink into nothing again.[12]

Here Maxwell is supporting the idea of lines of force by pointing out that a limitation of physical reality to visible objects and *space as a void* will not yield explanations. He uses the word "law" in a sense that extends beyond quantitative analysis and overlaps with efforts at physical explanation.

In another paper he wrote about the special exertion of the mind that is really required of a successful explanatory endeavor, and also about the basic difference between the mathematical and experimental approaches to physical science and how each of these has its own built-in pitfall:

Each of these types of men of science is of service in the great work of subduing the earth to our use, but neither of them can fully accomplish the still greater work of strengthening their reason and developing new powers of thought. The pure mathematician endeavors to transfer the actual effort of thought from the natural phenomena to the symbols of his equations, and the pure experimentalist is apt to spend so much of his mental energy on matters of detail and calculation, that he has hardly any left for the higher forms of thought. Both of them are allowing themselves to acquire an unfruitful familiarity with the facts of nature, without taking advantage of the opportunity of awakening those powers of thought which each fresh revelation of nature is fitted to call forth.

There is, however, a third method of cultivating physical science, in which each department in turn is regarded, not merely as a collection of facts to be co-ordinated by means of the formulae laid up in store by the pure mathematicians but as itself a new mathesis by which new ideas may be developed.

Every science must have its fundamental ideas—modes of thought by which the process of our minds is brought into the most complete harmony with the process of nature—and these ideas have not attained their most perfect form as long as they are clothed with the imagery, not of the phenomena of the science itself, but of the machinery with which mathematicians have been accustomed to work problems about pure quantities.[13]

That he regarded Faraday's ideas as what he calls here a "new mathesis" is clear from another passage:

It is true that no one can essentially cultivate any exact science without understanding the mathematics of that science. But we are not to suppose that the calculations and equations which mathematicians find so useful constitute the whole of mathematics. The calculus is but a part of mathematics.

The geometry of position is an example of a mathematical science established without the aid of a single calculation. Now Faraday's lines of force occupy the same position in electromagnetic science that pencils of lines do in the geometry of position. They furnish a method of building up an exact mental image of the thing we are reasoning about. The way in which Faraday made use of his idea of lines of force in co-ordinating the phenomena of magneto-electric induction shews him to have been in reality a mathematician of a very high order—one from whom the mathematicians of the future may derive valuable and fertile methods.[14]

An interesting question is raised here about the extension of "mathematics" beyond its familiar associations, but this is a topic outside the present concern. Elsewhere he stated that the tendency of the purely calculative mathematician is to "entirely lose sight of the phenomena to be explained," and that the corresponding danger of the approach that would make use of a "physical hypothesis" is that "we see the phenomenon only through a medium [i.e., the hypothesis], and are liable to that blindness to facts and rashness in assumption which a partial explanation encourages."

We must therefore discover some method of investigation which allows the mind at every step to lay hold of a clear physical conception, without being committed to any theory founded on the physical science from which that conception is borrowed, so that it is neither *drawn aside from the subject* in pursuit of analytic subtleties, nor carried beyond the truth by a favourite hypothesis [emphasis added].[15]

And there are these intriguing remarks about the purely mathematical (in the calculative sense) approach:

> Mathematicians may flatter themselves that they possess new ideas which mere human language is as yet unable to express. Let them make the effort to express these ideas in appropriate words without the aid of symbols, and if they succeed they will not only lay us laymen [!] under a lasting obligation, but, we venture to say, they will find themselves very much enlightened during the process, and will even be doubtful whether the ideas as expressed in symbols had ever quite found their way out of the equations into their minds.[16]

To summarize Maxwell's views as traced in these quotations: A complete process of theoretical science that stays balanced and on track has three characteristics: (a) it keeps the phenomena in view, as the experimentalist does; (b) it develops quantitative-mathematical theory and puts it to the uses that are appropriate; and (c) it develops the elements of new physical conceptions suitable for these phenomena, elements out of which to build a language of description and explanation, as opposed to a "language" of mathematical symbols—and this project is mathematical in the sense of the origination of the elemental conceptions and construction out of these elements, not in the sense of the manipulation of "pure quantities." Faraday exemplified (a) but also the equally legitimate dimension (c). Maxwell thought that *both* what he called physical analogies—artificial models such as the system of spinning vortices—*and* quantitative formulas are properly employed for their *suggestive power* in furthering this program, though neither can securely orient inquiry. It is important to recognize the complete open-endedness with which Maxwell thought about the future of theory:

> We are probably ignorant even of the name of the science which will be developed out of the materials we are now collecting, when the great philosopher next after Faraday makes his appearance.[17]

(In the historical perspective taking shape here, this next philosopher would be A. N. Whitehead.)

Throughout his life Maxwell held Faraday and his theoretical methods in undying esteem, and evidently thought that the search for a genuine physical understanding was something that need always *complement* the methods which he himself had undertaken.

Another piece of evidence for my view of Maxwell as unequivocally causalist is his attitude toward Newton. He spoke of "the unimprovable completeness of that mind without a flaw."[18] The Newton he thus held up was not the Newton often associated with action-at-a-distance theories, but the Newton of the letter to Bentley. In his scientific papers Maxwell was given to quoting the famous passage from the letter quoted above together with a reference to a certain complementary passage from Newton's biographer Maclaurin. Maclaurin said that the reason Newton did not publish his attempts to explain gravity by pressures in a medium "proceeded from hence only, that he found he was not able, from experiment and observation, to give a satisfactory account of this medium, and the manner of its operation in producing the chief phenomena of nature."[19] Maxwell is at pains to uphold Newton as a natural philosopher concerned with physical explanation though aware of the inadequacy of his vision on the matter.

Though Maxwell is usually viewed as a precursor to the twentieth-century phase of physics, in which the broad-scale quest is for the complete field laws (pure aspect [2] theory), it seems clear that he would not have been at all comfortable with the complete absorption of physics within aspect (2) that was to come about. To him, "physical explanation" in the case of forces and radiation appears to have meant the fulfillment at some future time of the causalist intuitions of Newton and Faraday.

The New Era of Physics and the Reign of the Cult of Surfaces[20]

We have seen that the program of physical science that I have called "causalist" was resolutely upheld in Newton's and Maxwell's concepts of science (not to mention those of many other reknowned scientists), and is particularly well-exemplified in the life-work of Faraday. For most early researchers into electricity and magnetism, such as Benjamin Franklin, causalism was simply presumed at the philosophically naive level of experiment and physical conjecture, without being either explicitly stated or called into question philosophically as an approach to science. Faraday shared this naive causalism, but leapt beyond it scientifically by rejecting mechanistic models for the explanation of forces; but apart from cautious and occasional speculations, his positive ontology was limited to the abstractions of a "new mathesis." What I have traced so far is a solid intuitive commitment to the belief that the possibility of explanation/understanding in the case of forces and radiation lay with the description of some kind of complexes or configurations of entities or processes in (or of) what would otherwise be regarded as empty space, whether or not this would ultimately involve an "ether" of omnipresent substance.

Despite this powerful tradition, in the twentieth century aspect (1) evaporated from the official science of physics, as if never to be taken up seriously again (with the qualification that one can resort to talk of action at a distance to give the discussion a causal flavor while evading real explanation). Causal-narrative explanation appears deceased in spite of the fact that many physicists are speculatively quite venturesome in reaction to the new enigmas and mysteries. The most basic reasons for this are as follows: First, the notion of an all-pervading material serving as a medium for light transmission—the ether—was repudiated, for reasons discussed below. Second, the properties of light transmission as revealed by new experiments thwarted the effort toward a coherent and complete physical explanation using the idea of vibrating material or moving objects (for short, this is the "breakdown of mechanistic models"). Third, despite Faraday's intuition that space itself might transmit forces, the fundamental assumption has never ceased to hold sway that if a causal process is such that it cannot be described in terms of some kind of motion of matter, then it cannot be "narratively" described at all (and perhaps has no physical reality!). In Maxwell's words, with the new determinations and discoveries nature had presented a "fresh revelation" which required the "awakening" of "new powers of thought"; but with hindsight this proves to be an understatement. Richard Feynman described the new situation this way:

> . . . the more you see how strangely Nature behaves, the harder it is to make a model that explains how even the simplest phenomena actually work. So theoretical physics has given up on that.[21]

"Simplest phenomena" means, for instance, the fact that "when very weak monochromatic light (light of one color) hits a detector, the detector makes equally loud clicks less and less often as the light gets dimmer."[22] This seems simple enough, but under standard assumptions it appears entirely incompatible with the fact that light shows many features of wave phenomena. The abandoned question here—what sort of wavelike process could generate this pattern of effect?—is not a puzzle to be solved by a formula, nor does it call for a merely useful interpretive model, which is hardly unavailable to a physicist. This phenomenon with very weak light is the port of entry to the mystery of radiation in the first chapter of Part Two, where I argue that what "strange behavior" means initially is simply that light is *neither* motion of particles arriving at the detector *nor* vibrations in an omnipresent substance (though it is of course some causal process analogous to a wave). Considered in broader scope and context, the strangeness of radiation microphenomena means that physical science has encountered a level of nature that exceeds in its subtlety all possible applications of what had been assumed to be the irreducible and essential conceptual elements in any physical story: given local space and present material (or even immaterial localized structure such as lines of force). In view of this truly extraordinary encounter, physicists cannot be blamed for pursuing a technological (experimentally advancing) and mathematical science and turning away from the genuine "Why?" and "How?" questions, as a pragmatic alternative to an inauspicious and less technically productive effort to bring their discoveries into conformity with "common sense." The upshot is that a basic account of radiation is no more in hand than a basic explanation of forces, which makes it clear that a background or ontological lacuna of understanding exists.

A lone figure whose work in physical theory stands in contrast to the abandonment of causalist natural philosophy by theoretical physics is Alfred North Whitehead (1861–1947). The reader is advised that the ensuing discussion of Whitehead throughout this book is at some points a strongly extrapolative interpretation of this thinker, and I neither insist upon its definitiveness nor renounce all claim to its correctness. Whitehead's physical theory is best known as an idiosyncratic alternative to Albert Einstein's groundbreaking work on relativity; his own theory was inspired by Einstein's work but differed from it in fundamentals. He was not a theorist of radiation, and discussed the early "quantum" developments in physics (which I have referred to as the breakdown of mechanistic models) only in very general and somewhat metaphysical and metaphoric terms.[23] But he did in his own way respond concretely to the challenge to narrative explanation represented by the physical "field"; this response has to be wrested from the texts with some effort. Since it is known that the context of occurrence or "medium" of radiation is the electromagnetic field, it is clear that the question of the nature of light propagation would proceed inseparably from the question, what is the physical field? how is it to be identified in and for a causal-explanatory ontology?

Whitehead's approach to explanatory problems of physics was to adopt an ontology in which the category "natural events" has knowable extension beyond cases of given, present matter undergoing change of quality or change of place, so that the physical field would consist of events not conforming to any concept of a material event; but it is debatable whether and to what extent he managed to carry physical explanation beyond the representational imagery of materialistic explanations. The intuition, however, was to discuss events whose structure and interrelations are conceived in a way that dispenses with the requirement that systems of spatial position and material presences (the "extended substance" of Descartes) be pregiven. It is not that certain events are thought to comprise a realm apart from physical objects, but roughly that events are viewed as physical conditions of objects rather than the reverse. Whatever its original aim, such a procedure (if feasible) would open investigation both into events composing transmissions that propagate through or across *empty space* in a genuine sense, that is, such that this process does not involve present material, and into events forming a depth-composition of *atomic matter* which does not consist of a further subdimension of localized objects. He spoke of an "ether of events," conceived explicitly outside the conceptual framework of "grids" of spatial locality and occupying matter, that is, so conceived that its description cannot be constructed out of the elements of mechanistic causal stories.[24] Whitehead said that his conception of the field "is practically the familiar one of tubes of force, with one exception," this being the fact that a tube (or line) of force is "conceived statically as a simultaneous character stretching through space," whereas for Whitehead the entity in question is *essentially a structure of activity*, not involving *presence in* space, that is, side-by-side relations of parts, and as such is misconceived already if its concept entails a backdrop of extension in localized space (occupied or unoccupied) uniformly subsisting through time.[25]

In my understanding, Whitehead sustained an essential causalist intuition by treating the field as the trace or mark of generative antecedence in observed field phenomena such as the movement of iron toward a magnet. It is whatever generates the movement. This unobjectionable basic account of the field was elaborated not in mechanistic models, but in a general view about observation and about nature and the comprehension of nature: the truly elemental facts of physical observation are not motions and changes of and in material bodies, but the "passage" of events into other events. Thus an observed event (such as the acceleration of an object) discloses not only itself to observation, but discloses "by relatedness" other events, such as those of immediate causal antecedence. To give another kind of example, I observe that the beam from a flashlight makes a spot on the wall, or that it illuminates dust or smoke in between; I do not and cannot see the light transmission itself (which I am told is wavelike in structure); but *that there is an effective causal transmission* is disclosed to me just as surely as is the spot on the wall or the trace through intervening particles. As discussed in the ensuing chapters, the class of "things observed" for traditional theory of science is drastically restricted by comparison, so that this example would have to be viewed differently.

Concerning the field, on the Whiteheadian view one observes events marked by causal derivation from a concurrent background of events. *Transition* in what Whitehead called the "passage of nature" is the irreducible structural element by which nature is ultimately known, and is the essential basis for a new understanding in physics. According to this approach the field is causally enigmatic for traditional science because of its status as a physical ultimate in the sense that this unique transition, *which alone identifies the form of activity of the field*, is a natural structure lying physically and genetically prior to the merely manifest or "derivative" linear stretches of space and intervals of time, for example, those entering into a physicist's measurements; thus the field consists of generative physical activity in a different category from local motions, since the concept of the latter presupposes such linear spans of time and space. The intuition that a motion attributed to a force is the terminus of some antecedent physical occurrence is to be steered entirely away from materialistic models and toward the suggestion that physical actuality is fundamentally "passage," intrinsically transitional in the sense of a process, and thus has a dimension extending beyond simple material givenness that is not itself reducible to any postulated material givens. This fundamental "process" dimension of the physical is the context and constitution of fields.

Presumably the concept of this "causal past" observed as a generative trace in the phenomena must somehow support a variety of specific structural possibilities, if the various species of field and modes of effect are to be accounted for; but Whitehead discussed the composition and interrelation of his "events" only in highly abstract terms, being oriented instead toward developing his own theory of space and time in response to the new relativity physics and deriving therefrom an alternative mathematical model. My own neo-Whiteheadian physical inquiry, elaborated in Part Two, seeks to counterbalance this emphasis by focusing on the understanding of radiation. It suggests a way in which the background events of the field and radiation can be conceived concretely as physical transitions in a sense that escapes the conceptual precondition of a purely spatial extendedness that comes with lines or tubes of force (and with "matter in motion"). But I do not employ Whitehead's abstract methods, nor do I promote his metaphysical ideas.

Thus Whitehead's physical ideas are well situated candidates for those new fundamental conceptions that Maxwell predicted would follow upon Faraday's scientific thought. He continued the causalist program—in spite of the fact that its abandonment by physics was in progress—in an original way by suggesting that a concept of physical "events" needed to be developed explicitly outside the confines of mechanism. Even so, these ideas—found mainly in the three works *An Enquiry into the Principles of Natural Knowledge, The Concept of Nature,* and *The Principle of Relativity with Applications to Physical Science*—are another kind of "new mathesis," limited to the abstraction "events" and the geometry of specially defined "event" relations. It needs reformulating in a fleshed-out interpretation before it amounts to narrative causal explanation, that is, to something fully comparable to old-fashioned aspect (1) science. (For this reason, physical ontology cannot conceive its task simply in terms of the explication and rehabilitation of Whitehead's physical theory.) Here I wish only to outline those aspects of Whitehead's work that stand out against the rise to dominance of scientific nihilism; there are other aspects of his work that are more or less compatible with this development and even suffer from closely allied assumptions. Whitehead's theoretical ideas were never taken up by official physics, though a handful of physicists have made a hobby of working with the formulas of Whitehead's alternative theory of relativity, apparently with the result that they proved generally on a par in their predictions with their rivals, the universally acclaimed formulas of Einstein.[26] And some physicists have drawn suggestions from Whitehead in their work on speculative problems.[27] But though a few experts have promoted the *approach* known as "event ontology," Whitehead's broad and philosophically grounded theory as he constructed it lies in obscurity, without official sanction or genuine advocates in science.

Einstein's theory of relativity was itself one of the major blows to the status of natural philosophy in official physics. This stroke against causalism did not come out of nowhere; Einstein was drawing on the ideas of Ernst Mach, who was an outstanding expositor of a certain tradition of general methodology in physics. Mach thought that claims to the physical reality of all kinds of conjectured entities whose function was purely explanatory, including the ether along with the Newtonian system of "absolute" reference for motions that a material ether implied, could not be allowed as scientifically valid. Along with this he developed as an explicit methodological principle of scientific knowledge the idea that understanding in physical science might ultimately concern itself solely with perceptible material objects and their spatial relations, exactly contrary to the Whiteheadian approach as I understand it. Einstein followed Mach by effectively adopting what came to be called the *operationalist* method toward his special theory of relativity. Operationalism follows two major principles: that all motion of bodies be regarded as purely relative to other bodies, and that all places and times of events coming under the treatment of theory be identified by reference to marks on measuring rods and clock readings (this is the "coincidence assumption"). That these stipulations seemed natural and unobjectionable reflects the absorption of physics within the practical sphere of measurement and quantitative laws. By proceeding thus Einstein was able to construct a mathematical model for the prediction of measurements under conditions of motion or propagation at all possible velocities, up to and including that of light. This model nicely simplified and integrated some kinematic phenomena of electromagnetism that were anomalous within the operating ideas of the time. And in it there was no longer a place for the ether *as a spatial system of reference for motions.*

But the function of the ether *as a medium for light propagation* was not replaced with any other conjecture that could fill the need for a medium or other physical context for interspatial processes that occur independently of matter; indeed, the whole question of the physical nature of radiation is untouched by the success of Einstein's mathematical models (though the models do contain clues regarding the relation of the velocity of light to space and time). The superfluous status of the ether after Einstein only affects the status of aspect (1) physical theory insofar as the latter employs the idea of an ether in the traditional sense of all-pervading substance. It does not affect consideration of empty space processes conceived, as in Whiteheadian theory, outside of the framework of description tied to locality and local motion of matter.

Einstein's own basic view of field theory is conveyed in these remarks:

> Could we not reject the concept of matter and build a pure field physics? . . . We could regard matter as the regions in space where the field is extremely strong. In this way a new philosophical background could be created. Its final aim would be the explanation of all events in nature by structure laws valid always and everywhere. . . . Our ultimate problem would be to modify our field laws in such a way that they would not break down for regions in which the energy is enormously concentrated.[28]

I wish to point out three things about this. First, Einstein's philosophical view is that the field is not only real (if this need be said), but is the fundamental reality of which matter is only a particularly "concentrated" condition. Second, physical explanation (of the field) is viewed as entirely a matter of finding a satisfactory set of laws. This same project of the unified field theory is elsewhere characterized by Einstein as a continuation of the "Maxwellian" program;[29] according to the discussion above, this seems to be an attenuation of Maxwell's actual views. Third, this basic intuition about the need for a "philosophical background" is shared by the

Whiteheadian approach, with the difference that in the latter the aim is to discuss the events comprising the field itself and not merely to construct the formulas by which all of its manifestations can be predicted. As Whitehead stated it (and this is reminiscent of Maxwell), "There can be no true physical science which looks first to mathematics for the provision of a conceptual model. Such a procedure is to repeat the errors of the logicians of the middle ages."[30] This very procedure has been routinely adopted to the point that it has become a distinctive mark of the present century. The vague, scarcely examined notion prevails that mathematical models are not only technical/mathematical marvels in their own right and useful tools of research, but also somehow describe the field activity itself and achieve explanation. This notion assumes a variety of forms. So strong is the inclination to view the theoretical achievements of recent physics as achievements in explanatory description of nature that it seems, in the words of one writer, that "one can no longer think physically without thinking mathematically."[31]

The other major blow to causalist physics (though it should really be called a gradual transition) took place with the development of quantum theory from Max Planck through quantum mechanics. Now, Einstein's special and general theories of relativity have surface features that can obscure its aspect (2) character, to the confusion of those who reflect on it; after all, Einstein talks about ordinary motion (though under the constraint of a pure relativity), he defines routes of light transmission as abstract trajectories, and he retains continuity in "space-time" (it is typically supposed that the gravitational field is finally "explained" as a mathematical property of this mathematical space-time). And the general theory of relativity can easily *seem* to yield a cosmologically satisfying "global" theory, that is, a description or model of the universe as one spatial and temporal whole. But it is almost too well-known to need stating that quantum theory apart from its various unofficial interpretations wears no such disguise of narrative-explanatory significance, and that its orthodox "Copenhagen interpretation" amounts to an explicit renunciation of aspect (1) for detailed phenomena of radiation, allowing only perceptible interactions with the material apparatus as the scientifically acknowledged "reality" of the physicist's experiment, and predictive formalism as its sole positive theoretical treatment. In this case the founding physicists themselves derived fully nihilist conclusions from the theory, whereas for Einstein the corresponding philosophical extrapolation was carried out by a century of scientist-philosophers, of whom Bridgman, Popper, and Kuhn are convenient and prominent examples.

Niels Bohr is the physicist most closely associated with the "Copenhagen interpretation," and he is famous for his rejection of any "quantum world" of entities knowable by science. This position limits theory to the measureable aspects of events in "quantum" experiments, conditioned by the "classical" (material) character of the apparatus, and the formulas derived therefrom. An associated approach is that of Max Born, who originated the idea that the physical processes of the field be interpreted as "probability waves." This notion is sustained by the spurious procedure Whitehead mentioned: the formulas predict certain outcomes of experiments as probabilities rather than deterministically, and there is a suggestiveness in the fact that these probabilistic predictive models can take the form of wave equations. Born thus managed to form the idea that the mathematical procedure yielding a distribution of probabilities for a set of outcomes of an experiment itself displays in its form the nature of the underlying physical process in the experiment, that is, that the predictive formula itself can be interpreted as affording a correct understanding of this process, which is then a "probability wave." Regarding the question of the "reality" of transmissions so conceived, he said this:

> The question of whether the [probability] waves are something "real" or a fiction to describe and predict phenomena in a convenient way is a matter of taste. I personally like to regard a probability wave as a real wave, even in a 3N-dimensional space, as a real thing, certainly as more than a tool for mathematical calculations.[32]

Notice that for him there is no question whether the mathematical equipment embodies a physical description; what is a "matter of taste" is whether this is the true "description" or purely a useful fabrication. Based on this quotation one might object to my reading of Born; he seems by his own words to be a kind of "realist," so how can he be considered a nihilist? This realism, however, depends on the basic tenability of thinking that some distinctive physical process can be described as a wave of probability, which seems to me conceptually flawed (more on this in Chapters 4 and 5). A later passage in the same book displays a lucid acausalism:

> ... if reality is understood to mean the sum of observational invariants—and I cannot see any other reasonable interpretation of the word in physics. ... [33]

This is operationalism in another context. Elsewhere he says, "particles are real, as they represent invariants of observation."[34]

The notion that the field is a physical probability or "propensity" was taken up and developed by Karl Popper and many others. It is one of the chief anodynes that ameliorate scientific nihilist conclusions and help solidify their acceptance.

A later development of quantum theory is associated with a thought experiment of Einstein, Boris Podolsky, and Nathan Rosen, and with Bell's Theorem and the term "nonlocality." This development appears—but only appears—to lend a final, decisive blow to all causalist hopes, because it does in fact rule out explanations of certain field effects in terms of local causal transmission through space. Nonlocality tends to lead investigators to the conclusion that nature at this level is simply bizarre. This matter is taken up in Part Two, along with other famous "quantum" problems and some semi-dormant scientific/philosophical questions.

A cultural phenomenon with its sources in the triumph of acausalist physics (particularly quantum physics) will be recognized by all who have browsed in the science sections of bookstores in recent years. There is a general fascination with a certain mysticism of inscrutability that arises from physics in its contemporary results and from the things certain physicists are led to say about these results. This mist shapes itself into a variety of intriguing suggestions: that the meeting ground between Western science and Eastern thought has been discovered; that ancient religious intuitions about a world or realm of the spirit have been confirmed; that an opening has been secured for the revitalization or "reenchantment" of nature after centuries of mechanistic science; that perhaps science has revealed the active role of mind, consciousness, or God at the deepest physical level; that this inscrutable realm might account for psychic phenomena. Without passing judgment on any of these suggestions, I only wish here to point out that underlying this speculative free-for-all has been an authoritative conclusion that the explorations of physics have permanently out-stripped all narrative physical understanding by discovering a realm inaccessible to "ordinary [or commonsense] concepts"; refer-ence to an *ineffable reality* is thought to account for or explain what science cannot account for or explain. Now, it is accurate enough to say that physics encountered a realm escaping "ordinary concepts" insofar as this is an alternative formulation of "the breakdown of mechanistic models," but then it does not occur to anyone that this might not entail "the permanent collapse of narrative explanation." The popularization of the "quantum" mysteries of a seemingly postcausalist physics came about as follows.

Copenhagenism is a compelling position from a pragmatic standpoint, but is rather strained and problematic from a general philosophical standpoint. It involves a bifurcation of the experiment into a "classical" physical world and a "quantum system," which is ultimately untenable. A more basic critique occurs when outstand-ing physicists maintain the intuition that what is needed is a fresh understanding of physical reality rather than a renunciation of the possibility of a fresh understanding.[35] Thus, by crossing a fuzzy boundary within the community of quantum physicists, one finds theorists who are motivated by strong dissatisfaction with Copenhagenism for such reasons and who share my view that a dimension of explanation is needed in addition to the predictive formalisms of quantum theory. But with a regression to the naive mechanistic models apparently ruled out, many in this group see no alternative but to resort to mystical or metaphysical explana-tions, even though, on the side of Copenhagenism, these would seem to drastically stretch or exceed the normal scope of physical science.[36] Of course these various speculations all in some way man-age to account for something, to resolve some particular problem or other, but at the same time they tend to have serious philosophical problems of their own. They may be ad hoc and regressive, or unhelpful mind twisters. Some would clearly exit the normal sphere of physical science, for example into mentalism (talking about "consciousness" and claiming a counterintuitive result re-garding the observer-observed relation). In contrast to anything I will be doing, they typically fix on one or another formulation of quantum mechanics and seek to *interpret* it. Far from being isolated and unbalanced eccentrics, the quantum ontologists currently dominate philosophical discussion in the field, and constitute a genuine speculative branch of quantum theory. They are discussed further in Chapters 5 and 8.

Naturally, it has not been the pragmatic, sober nihilism of the Copenhagenist orthodoxy that has caught the popular imagination. Given the high cultural status of physicists, and given a widespread feeling of the spiritual importance of cosmological understanding, it was inevitable that mystical and metaphysical ideas articulated by physicists as appearing to flow from physics at the margin of its explorations would enjoy a degree of popularity. Hence the marketability of quantum mysticisms, from Fritjov Capra's *The Tao of Physics* to the current flurry of titles. These are counternihilist in spirit, but inescapably nihilist in assumptions and conclusions.

Whatever one may think of the taste for quantum mysticism, it should be appreciated in light of historical conditions under which such ideas can feel enlivening and refreshing. Ideas and opinions today, disseminated and homogenized by commercial mass media, are increasingly marked by a certain sophisticated unthinkingness subliminally enforcing uniformity of thought. This derives a special solidity from the epochal myth of a modern fulfillment of scientific rationality. These are forces directly inimical to science in the broad sense of an ever-nascent advance in penetrating to the truth of things. Behind the general facade of wide-eyed sceptical rationality, the various conpartments of the monolith of the sciences develop dogmas of specialization insulated from criticism; for example, "the human is an acquisitive animal" (just as bees are nectar-gathering animals and bighorn sheep are mountain-dwelling animals) is the foundational doctrine of economics, yet it has no such status in anthropology and is arguably based on quite narrow evidence, apart from the question whether an approach to evidence and conclusion borrowed from natural science is at all appropriate. The unspoken agreement is to leave each discipline free to settle into its own methodologies and working assumptions in a system of absolute mutual deference. In the social sciences the sum of specialized outlooks is a concept of society as a blind process of contractual struggle. Given this, the enlightened society is the one that harnesses this "natural" antagonism by letting it rampage to some extent (often to a major extent) in the economic sphere, while in

politics the unprincipled scramble of multiple interests and spheres
of power for the sanction of majority opinion is widely accepted as
"the way democracy works;"[37] the logical conclusion is that our
"nature" is attaining its final expression in the present time! A re-
current expression of the half-conscious process of securing uniformity of thought is the all-fronts attack on social idealism and ideals
in general. One finds at the heart of such campaigns the
pseudoscientific notion that the values manifested in a particular
society are immutable "cultural" givens, determined by religious
traditions taming or constraining "biological drives"; whether or
not it is explicit, a biocultural determinism becomes the new realism
and the new sophistication. The impulse to lay hold of a scientific
procedure inspires a misguided quest for "human nature" that
leads us to our reflection in a stagnant pond of historical contingency, or in the case of a counterproposal, to some new rationalized
prejudice. The general historical wisdom invokes a "scientific
worldview" of modern times; but what is truly characteristic of the
present time is an encompassing philosophical prejudice about
what constitutes science: the quest for laws and prediction. All
thinking and questioning is directed along lines established by the
system of official sciences with its hierarchy of authority in establishing a concept of science (and there is a suppressed backdrop of
traditional metaphysics, discussed in Chapter 7). A core tendency of
this contemporary scientific medievalism—and it is bound to surface at some point—is to attack the whole purpose and value
of philosophy, as discussed below. These are some of the ways
in which a distorted relationship with science helps to undermine
human aspirations.

My point is that although quantum mysticisms and other sci-
entific/metaphysical ideas associated with "New Age" thinking
may be fundamentally misdirected, their impulses are those of
spiritual life, with the freshness of revolt against a monolithic system
of official research in which humanity's gift of primitive thought
and wonder, however much it may motivate individual scientists, is
buried in a snowstorm of specialized knowledge. Associated impulses are asserted against the near-total hegemony of consumption
and wealth in our effective values with its generally destructive
consequences. East-West synthesis, a global spiritual center, a
"planetwide mythology"—these are optimistic thoughts, enlivening
an idealism that is literally vital to the social development of values.

But such impulses of spiritual life must aspire to the highest standards of critical reflection, even at the risk of seeming severe.

One form of mysticism in the bad sense is the belief that the basic relation of observer and observed (or subject and object) is shown by quantum physics to be not at all what humanity had previously thought. Principally from speculative ideas on the part of certain physicists themselves one obtains the idea that the consciousness of the observer intercedes in experiments and affects their outcomes. "The observer and the observed are no longer distinct. This shows that the entire world of separateness and division, the root of political and psychological evils, is only a limited aspect of things, perhaps even an illusion produced by the innate prejudice of an ordinary human perspective." It may be philosophically exciting to think that news about subject and object (in this context) could come out of "hard" science, and I do not wish to be unconvivial, but in this case the idea rests on quicksand. It has a combination of sources, one of which is the assumption that the term "particle" as used by physicists must refer to a localized, present object if it refers to anything physically actual (or more commonly, to something "having definite properties") prior to interaction with the detector: the fact that the "particle" propagation demonstrably does not have this basic ordinary object character seems to speak against its "objective reality." And the "probability wave" notion has already pulled thinking away from actual physical reality into "potential" reality, which is then thought to "actualize" as determinate and definite events in and with observations; thus the "act" of observation is seen as in this weird way physically efficacious. Thinking is misdirected when certain terms of discourse—"particles," "probability waves"—are accorded descriptive functions they cannot perform, and this widespread idea about a drastic alteration of the subject-object relation is only one of many consequences.

Sources of quantum mysticisms in general also include (a) the core assumption of scientific nihilism, that narrative causal explanation has permanently broken off in the domain of physics, and (b) the vague, subliminally effective idea that modern physical theory is a new, mathematical way in which things are *figured out* in this science, an *understanding* forged in symbols and accessible to initiates; once (b) is established the encoded explanation is viewed as a framework for, say, a metaphysical "interpretation"; all of which covers up the essentially instrumental character of the theories. Source (b) corresponds to a universal turn toward the successes of mathematical theory by which major blank spots in understanding are turned away from and left to history along with the traditional answers; the general vortex of thinking in which this occurs is one of the primary topics of this book. From the point of view of this work insofar as it presumes to propose and follow out a fully naturalistic (though postmechanistic) approach to the physical problems, the more extreme nihilist/mystical and metaphysical approaches hold no appeal. My long-run proposal is to replace all mysticisms of inscrutability with renewed project toward narrative comprehension.

Philosophers of science have been no antidote to popular notions derived from physics; on the contrary, they have by and large shared the same root assumptions. The overall reflective determinations that have taken place in recent philosophy of science can be summarized by saying that a concept of science and of what is truly scientific has been forged in the image of relativity theory and quantum theory. Where this determination reigns the importance of narrative physical explanation to science is submerged. The bulk of the literature in philosophy of science is characterized by a relentless fixation on mathematical physics. The discussion of philosophers of science presently and throughout Part One should be understood not so much as criticism of the work of individual philosophers as criticism of an authoritative twentieth-century outlook through some of its monumental expressions. The intent is not to suggest that "philosophy of science" names a fully monolithic view, but only that the thrust of this discipline has been theory of science modeled after physics.

P. W. Bridgman, an avid student of Einstein's theory and an influential philosopher of science, shaped Einstein's operationalist method into a full-fledged scientific nihilism.[38] Specifically, by appropriating operationalism in the form of a criterion for the meaningfulness of physical concepts and for the scientific status of a physical theory, he came to the conclusion that only matter had physical reality, whereas the other ontic region for the physicist, the field, did not (despite what Einstein actually said about fields in the quotation above, page 36). The strength of this antirealism is its clarity in recognizing the permanent demise of mechanistic models in this area; but then the possibility of actual empty space events is never considered. Given this limitation, which is most understandable, antirealism can seem compellingly antidogmatic. Quite consistently with his philosophical operationalism, and in company with the philosophical empiricists among the physicists, Bridgman thought Einstein somewhat naive in continuing to believe that light is "a thing that travels."[39] These comments are instructive. Einstein obviously did not hold a crude mechanistic doctrine about light, but spoke only about transmissions of influence or "signals." Therefore Bridgman's remark has to be construed to mean not only that light is not *thinglike*, which is entirely cogent, but that light is no physically actual transmission at all, suggesting that a light "wave," which certainly has the status of an enigma in today's physics, is in fact *no real physical occurrence*. Put this way, the view seems extreme and unlikely; but this very claim is entailed in any antirealism directed, as they typically are, toward the technical term "particle" and its seemingly obscure referents. An antirealism about light transmission would be a vivid form of scientific nihilism, and here the pressure toward it stems from viewing Einstein's operationalist methodology as a desideratum of scientific understanding, rather than as a practical stipulation with an eye toward useful predictive models. Bridgman is representative of operationalisms in philosophy of science that have prevailed in the wake of the triumph of Einstein's work.

The Vienna Circle of positivism, under the influence of Mach, Einstein, and emerging Copenhagenism, concluded that in order for a theory or a proposition to be scientific—which for the positivists meant that it avoided being "metaphysics" and therefore "meaningless"—it had to consist of or entail a prediction about observations that can be tested and confirmed ("verified") or disconfirmed. Karl Popper modified this with more lasting influence: a *scientific* theory is a theory that invites specifically disconfirmation or "falsification" by yielding predictions in a domain currently being explored, thus providing immediate prospect of further advance in technical and theoretical knowledge.[40] Without intending to minimize their differences, these influential doctrines can for my purposes be placed together under the heading of "confirmationist." They share the conclusion that whatever deserves to be called a scientific theory (as opposed, say, to "metaphysics") is something that presents scientists with a set of predictions that advance science by being investigated. This is because both take their model of science from theories that *are* essentially tools of prediction applied in furthering experimental knowledge, theories that belong to aspect (2) of physical science and are confined, despite "realistic" guises, to formulations of general laws of measurement. Popper, for example, relates that Einstein and the success of his relativity theory was an important stimulus for his ideas.[41]

One should not, because of familiarity with the names and basic contents of confirmationist doctrines, be unmindful of their peculiar consequences. Suppose a scientific theory is proposed that explains a variety of *known* phenomena at a stroke, but predicts none that are yet to be encountered by science. The theory, in other words, explains a set of observations that have already been made and have not previously had a satisfactory explanation, but it predicts no new findings. Such a theory *would not qualify as science* by confirmationist criteria, merely because all of its various contributing evidence is explained in one stroke, that is, merely because without testable predictions, it would not be subject to any experimental/technical confirmation or disconfirmation subsequently to its formulation. This incredible a priori conclusion about the theory bears the clear stamp of the standard fixation on mathematical physics, where the actual function of theories (apart from their intellectual/aesthetic appeal as ingenious, broadly unifying, and so on) is to guide and further research in experimental observation. The hypothetical theory would not fit the model because it would only purport to explain and therefore would have, as Popper said, a questionable "status" as science—no matter how completely appropriate, coherent, wide-ranging in its explanatory power, and highly probable it may seem.

As a result of this disturbing consequence Popper had serious trouble with Darwin's theory of natural selection, which did not seem to fit the criterion of falsifiability because it is exactly such a theory as I have just described. Going by Popper's theory of science the very basis of the modern theory of evolution, an accepted, basic explanation in biology, falls somewhere outside science, in the traditional sphere of "metaphysics," for good or ill. In his attempts to deal with this problem he was attracted to a tradition that viewed the theory as a tautological law—"whatever survives and reproduces survives and reproduces" (as if this rendered the explanatory narrative of evolution)—and he ended up concluding that the theory of natural selection is merely a "research program" that is presently useful![42] In my view this is a remarkable inversion of the truth. Rather than recognize today's *physics* as advancing in a one-sided condition, as *lacking an explanatory dimension, that is, in order* to preserve as a starting point the paradigmatic status of the current theories of physics as examples of achieving science, Popper was forced to adopt the view that the general theory advanced by Darwin—which is certainly science, certainly explanation, and moreover indubitably *correct*—actually has a status and role quite specialized, instrumental, and transitory. A late effort to extricate himself from this problem still drifted into a quasi-teleological interpretation of the theory.[43]

The fact that a specially concocted explanation without independent confirmation is weakly supported does not disallow the possibility that a theory as it stands may be *broadly* explanatory enough that it does not have to be *tested* by any newly devised means to compel acceptance.

Another highlight of the effects of acausalist physics on philosophy of science is Thomas Kuhn. In an extremely influential work, Kuhn appears to have concluded from the history of physics that explanations in science (in general!) can never be considered settled truths, in opposition to the conventional "image of science"[44] according to which new explanations ongoingly expand upon and/or add to those previously determined. He claims that instead of this "process of accretion," or "cumulative process,"[45] there is general "progressive" development in science only in the sense of a succession of radically different "paradigms," with no fundamental continuity of understanding (the term "paradigm" is contrived, quite outside its ordinary meaning, to cover a variety of models, theories, and frameworks of research, and this indiscriminate utility has made it popular). The pressure toward this rebellious conclusion stems directly from confining the discussion to examples from physics by explicit intention;[46] if one focuses instead on narrative explanation in natural science in general, it seems clear that as a rule natural sciences fit the "image" Kuhn attacks. Biology, geology, and general astronomy have over the centuries established the basic accounts of nature in their respective domains, and have continued to fill in the detailed understandings (disregarding special arcane problems). These understandings emerge as integrated into an overall cosmological history. This is precisely a "cumulative" progress in the sense that more and more of the natural world, how it has evolved and how it transpires, is scientifically disclosed, as these sciences leave behind the errant general theories to which sciences are naturally subject in their early stages and settle upon fundamentally correct basic explanations. Kuhnian nihilism appears, again, to derive particular inspiration and support from the Einsteinian transformation. In short, Kuhn examines a particular history of science, one which is indeed characterized by a succession of theories utilized and then cast aside and even by major shifts of approach, and concludes (apparently) that a genuinely scientific theory is a radically historical and transitory piece of knowledge.

Whatever Kuhn's basic conclusion is exactly, its "revolutionary" import comes from mistaking the instrumental value of theories for explanation/ontology. I say this because if certain passages are ignored, there are grounds for understanding the argument as an instrumentalism, as if the claim were that the theories in question are not explanations, hence, trivially, not definitive and secure explanations that would add to a cumulation of established truths; however, a thesis that continuity or cumulation is absent from the history of the cognitive *instruments* of science would be an academic curiosity at best, certainly unsuitable to the dramatic implication that a revolution in the "image of science" is being forged. Considered overall he must mean that science in the proper sense of the word produces no explanations that are simply correct ahistorically and hence accrue with scientific advance. But it is after all only aspect (2) or *technical-instrumental* theory that is always a choice among empirically equivalent constructions, subject to substitution at every stage in technical advancement.[47] Kuhn's argument relies on a vacillation between the two basic senses of "theory," assisted by the collapsing of differences under a made-up general term. Though Popper and Kuhn are normally viewed as polar opposites, there was a strong foreshadowing of Kuhn's ideas (and also of other later nihilist stances) in Popper's *The Logic of Scientific Discovery*, which concluded that science is not a "body of knowledge," but a "system of guesses," which "must remain *tentative for ever*."[48]

Whether or not Kuhn's ideas have been distorted as they have diffused into the public consciousness, the post-Kuhnian climate of thought has been favorable to relativism about science all the way to its nonsensical extremes. This tends to play into the hands of various forms of antiscience, such as that stemming from religious fundamentalism. Drawing support from Kuhn's ideas (whether legitimately or not) by analogy has been the ascendance of the view that *philosophy*, having deluded itself for centuries as to the possibility of a general progress, is now in the situation of a Sisyphus in his doom, investigating forever the same traditional problems, ever compelling and without prospect of solution; the conclusion, of course, is that this grand icon of European culture—philosophy—should be given up and left to history. Allied with popular suspicions, this antiphilosophy has established a niche in academia, one associated with a fashionable mishmash of literary criticism and philosophy. But this view is drastically out of touch with the course of philosophy itself. Where "postmodern" thinking leads to antiphilosophy this is really only a new way of caving in to the traditional impasses of epistemology, whose overcoming is already well underway as part of the underrecognized positive achievements of various streams of twentieth-century philosophy (more on this in Part Two); thus it amounts to a retrograde modernism. (I would not deny the unfortunate persistence of certain quandaries of philosophy beyond their historical life.) This also fits the Nietzschean model of nihilism: discerning just enough to recognize that an important transitional stage has been entered, and wildly misreading this transition as philosophy coming to an end by losing all semblance of purpose.

Kuhn and "postmodern" antiphilosophy are particular expressions of a broad contemporary thrust of radical pragmatism-historicism, in which results of scientific, philosophical, and scholarly enterprise in general are treated as essentially *artifacts*. For example, the constructive physical ontology comprising Part Two might be successfully corralled within special scientific frameworks by automatically viewing it as an imaginative story spun out of a psychological need or proclivity, weaving together a set of literary influences. The doctrine now on the march is that whatever philosophy does is already a textual object of history and anthropology; to think it might come to an unhistorical truth, or advance clarity or understanding purely and simply, is seen as unsophisticated (much of this is the curious phenomenon of an educated ignorance running amok). Consider contemporary treatments of *understanding*: where the reading of written texts supplies the model there is a tendency to reduce all instances to *interpretations*, as though it were not possible to simply read and comprehend a text, let alone express one's understanding in an exposition or paraphrase, without thereby producing an *interpretive* text, that is, a *particular reading*; this of course reduces all discussion of texts to a plurality of interpretations. It is probably to little or no avail that philosophers point out that if there is not an independent role for understanding, interpretation cannot take place.[49] A related claim enjoying a rather wide consensus is that we must finally put behind us the traditional notion that science leads to "objective" knowledge, which can only mean giving up the idea that a scientific explanation might actually be a *disclosure of what is, as it is*, independently of "interpretive" constructions of scientists at particular historical stages of science, which necessarily reflect particular values and prejudices. (This view will wax unconstrained as long as the whole notion of a scientific theory or description is universally narrowed to the rubric "laws of nature." As with Kuhn, to put the view to the test is to consider examples such as plate tectonics in geology, or the description of certain celestial nebulae as aggregations of stars beyond the visible stars.) Clearly discernible in these intellectual trends is a late-developing, dispersed impact of the transformation of physics into its twentieth-century phase.

   The general reaction ostensibly against "metaphysics" in the twentieth century, a reaction linked to the ascendant doctrines of procedure in physics, eventually gave rise to a fruitful tradition of philosophy confined to a practice of critical analysis that consciously avoids all scientific or metascientific aspirations to describe, explain, or theorize about the world. For this school (which is conservative from a "postmodern" viewpoint), philosophy ultimately does nothing more than demonstrate that "the world makes sense" in spite of the conceptual confusions and misuses that give rise to philosophical problems; philosophy as a therapy of thinking. This is powerful from a critical standpoint; and to follow the motto "the world makes sense" is to stay well away from certain questions arising from advanced physical explorations, questions which to all appearances are matters for advanced specialized science. A trained confinement of the general scope of questioning in philosophy comes about along with the ascendance of a certain critical-technical mastery peculiar to the field.

A factor in the hegemony of scientific nihilism has been a certain fragmentation of disciplines, which can be illustrated by the author's own experience. A visiting speaker, a philosopher of science, has the originality to propose that some philosophical problems in the margins of physics be addressed constructively through a "descriptive metaphysics," which, he asserts, is not a scientific pursuit, and he hopes in the course of this to build a bridge between the fields of philosophy and physics. In discussion I raise the question as to the reasons for calling this theorizing pursuit "metaphysics" as opposed to "physical science." A third philosopher whom I will loosely class within the Wittgensteinian tradition protests against the speaker's notion of a bridge that would bring the general philosopher any closer to recondite physics, and chides us both for our "hubris" in seeking to perform some complementary function as philosophers for this science. This partially illustrates how strikingly disparate, in my experience, are the broader professional attitudes toward, and understandings of, the general matter I call physical ontology. Bringing another group into the discussion, working physicists generally agree that major *philosophical* problems in fundamental physics exist, though they differ sharply as to whether there is any point in pursuing them; across this difference there tends to be an intolerance for the suggestion that constructive solutions might be sought outside of their trained methods (as I see it this shows the need for basic clarification of the difference between a philosophical problem or problem of understanding and a technical-formal problem). Nonphysicist philosophers for the most part assume that whatever problems remain unresolved in physics are the business of that science, and it is likely that they have not looked into the matter enough or in the right way to have the insight that the crucial unresolved physical problems are both matters for philosophers and nontechnical in their basic character; because of this particular rift some valuable skills fail to be utilized in consideration of the background questions. Philosophers of science are typically trained in physics, and therefore know something about the problems, whatever they may think of them, and this knowledge in the present case is enough to inspire an adventurous transformation of the orthodox nihilism into the proposal that these problems might warrant a specially designed rehabilitation of "metaphysical" inquiry.

I said that Kuhn proceeds as if there were not two distinct modes or functions of theory, empirical/instrumental and explanatory (or ontic-ontological); but Kuhn is certainly not alone in this particular loss of bearings. Philosophers of science generally have been troubled by the specter of instrumentalism. The overall thrust of the twentieth-century appropriation of that epistemological tradition passing from Hume and Kant through Mach and other Galilean physicists has been as follows: Given the revolutionary and seemingly permanent status of the new theories in physics, must not these great theoretical achievements constitute explanation of phenomena, and not merely tools of technical progress? If one would preserve the assumption that science (conceived on the paradigm of physics) explains, must not "scientific explanation" be understood differently than it was—for instance, by Maxwell—in the previous era of physics in which the quest for narrative causal explanations had an official legitimacy?

This train of thought led to what I call the neo-Humean outlook that pervades philosophy of science in different guises. Its core is a twofold philosophical strategy in response to the new physics: (a) a Humean analysis of the causal relation as consistent succession in discrete (possibly contiguous) events, with or without a contrived "necessary connection"; and (b) a special "model of explanation" in which scientific explanation is defined as deduction of phenomena from laws plus given conditions. The view of the causal relation reinforces the law-based model of explanation by entailing that knowledge of causes in any case effectively reduces to recognition of recurring patterns of succession in observations (this view is actually reflected in the technical meaning of "causality" for a physicist). The anticipated conclusion is that in light of these special analyses of causality and explanation, today's physics can be considered to achieve "causal explanation" in addition to the dramatic instrumental/technical advances; but this only emerged in full clarity in the maturity of this tradition. In my rough and perhaps somewhat idiosyncratic categorization there are two types of neo-Humean philosopher of science, and both are inspired by the new methodological and theoretical decisions in physics: the *logico-empiricist* believes that extant theories of physics are paradigm cases of scientific explanation (in accordance with the "clarified" concepts of causality and explanation), and the *positivist*, while agreeing that "causality" refers only to predictability, takes the position that science ultimately does not explain, but at best achieves "exact" and "unifying" empirical theories. In truth, however, most philosophers move between these positions in one way or another. The main pillars of the neo-Humean outlook are the objects of the critiques of the next two chapters, where it is spelled out how the whole rationale for this outlook would be undermined if a compelling case were made for revival and success of narrative causal explanation in the area of fundamental physics.

I have so far given no substantive argument for my incredible claim that up-to-date causal-narrative explanations for phenomena of radiation/fields can actually be given; this falls almost entirely to Part Two. But I hope to have stimulated some optimism about prospects through the discussion of the contrasting outlook of Whitehead, and more generally helped recover a sense of the importance of the causalist quest in physical science. In particular, the basic historical survey of the causalist question should have rendered it plausible that the means and methods for a revitalization of aspect (1), assuming this is possible, are not to be taken from or modeled after contemporary technical methods and language of theoretical physics. If this is the case, then training in technical and mathematical procedures may not be a prerequisite either for carrying out or for comprehending such a project. Powerful prejudices speak against this possibility, but their remedy is a careful consideration of what took place with the transformation of physics: the basic aspect (1) questions—How does light propagate? What is the underlying and mediating cause of a force of attraction?—simply *passed out of view* with technicalization and mathematicization. Pursuing narrative causal explanation for phenomena of forces and radiation is generally not a part of what being a physicist means today; as Feynman stated with great clarity, physics has actually *left this particular turf entirely* in its aims and in its methods. So utterly disparate is the causalist approach from the accepted concept of physical theory that a shift to an acausalist conception of advanced science (adopting physics as paradigmatic) has prevailed under the implicit assumption that the demise of mechanistic explanation is the demise of any narrative causal explanation. It is clearly a mistake to assume that revival of aspect (1) would involve advanced specialized methods that have evolved in the context of *these* epochal decisions.

It is a settled doctrine that "natural philosophy" as I use this expression has lost all scientific legitimacy, but I urge a full exploration of the intuitive possibility that the whole source of this doctrine lies in purely pragmatic decisions of technicalized research. My claim, contrary to this standard doctrine, is that the breakdown of traditional *physical concepts* in a certain area really means, quite naturally, that problems of explanation in this area have passed into the hands of philosophy with its long history of experience in the scrutiny of meanings and in the crafting and critical evaluation of highly general understandings. Needless to say, philosophers' current conceptions of their field have not favored recognition of this bequest, let alone any venture into its unique challenges. As for physicists, it could be said that as far as *official* physics today is concerned, achieving a fundamental and general understanding of the universe is essentially a matter of thinking up a formula and writing it down, which is not to deny the widespread interest among physicists in speculative and interpretive theorizing. But physical ontology aims at background explanation and has no immediate interest in mathematical formulas or experimental applications. Far from meddling in physics, it formulates and investigates scientific questions that have been thoroughly abandoned as a business of physics; accordingly the inquiry will differ in both character and aims from what is called physical theory today. At least in its initial phase of philosophical and conceptual adjustment that is eventually thought through in this work, the constructive project finds its own way with the help of the isolated and obscure strategy of Whitehead, and also finds resources in the clarities achieved by the tradition of phenomenology concerning questions of ontology in general.

In the meantime some objections and challenges from my professional opposition—neo-Humean philosophy of science with its basic doctrines and precepts—will be placed under historical and critical magnification. To summarize what has been said about this tradition so far, it has arisen at least partly in reaction to the fact that "theory" in physics came to mean quantitative laws constructed for prediction; reaction to this fact has usually been coupled with the unspoken assumption that successful physical theory must always be explanation (except where a clear instrumentalism takes hold). The contemporary neo-Humean tradition, dominated by logico-empiricism, will object to my project of physical ontology or philosophy of nature from the very start by saying that the default of what I have called aspect (1) was actually not a breakdown of causal explanation. Its claim will be that what actually happened when mechanistic models of light permanently collapsed was that the true meanings of "causality" and "explanation" forced themselves upon science, meanings that were anticipated by David Hume. I therefore face the task of evaluating these neo-Humean conceptions of causality, explanation, and science in general, in order to show that the intention to resuscitate narrative physical explanation in physics is not a misguided nostalgia for scientifically unsophisticated meanings of words. Guided by this rather academic critical purpose, the next four chapters find some adventure in tracing the course of scientific nihilism in philosophy of science, with special attention to counternihilist efforts within theory of science. The point is to think through and dismantle some pervasive doctrines that obstruct and occlude the questions of fundamental physical explanation.

2

"Law Explanation"

> At bottom the whole *Weltanschauung* of the moderns
> involves the illusion that the so-called laws of nature
> are explanations of natural phenomena.
> —Wittgenstein, *Notebooks 1914–1916*

> The calculability of the world, the expressibility of
> all events in formulas—is this really "comprehension"?
> How much of a piece of music has been understood when
> that in it which is calculable and can be reduced to
> formulas has been reckoned up?
> —Nietzsche, *The Will to Power*

DESPITE EAGERNESS TO GET TO the crucial scientific questions, a resurgent natural philosophy cannot avoid the question raised by the well-developed and widely established view that scientific explanation can be said and thought not only to take place within, but to be encompassed wholly by, the mode of theory designated aspect (2) (see Introduction). Where this is not an explicit claim it is an unexamined assumption stemming from the impressive successes of physics.

Foundations of Logico-Empiricism

Extraordinary though it seems to a naive expectation, the basic tenets of the recent tradition of philosophy of science have formed around the conclusion that advanced knowledge in natural science has a distinctly different character from that of a body of understandings of the processes that generate things and phenomena. An important source of this shift away from a causalist understanding of science is the fact that what is called "theory" in the present era of physics is made up solely of conceptual models and laws falling under aspect (2), this science having actually left behind its whole dimension of causal stories. Such a narrative dimension, however, has in the broader historical tradition been part of physics, and it remains essential to sciences other than physics. The impressive powers of organization, prediction, and technical application embodied in the inventions of mathematical theory have led philosophers of science to regard the quantitative/empirical method of mathematical-technical physics as paradigmatic science. The result has been devastating to the status of causalism, which gave way to the standard view that the invention of formulas for quantitative relations in the data of observation, *in contrast to narrative causal explanation,* identifies the *essence of science.*

Philipp Frank, a physicist and philosopher of science associated with the Vienna Circle, says the following about the transformation in physics which actually began in the late nineteenth century:

> [The theoretical models of Maxwell and Lorentz for electromagnetic phenomena] were accepted only because the observed facts about the motion of bodies and the propagation of light could be derived. . . . Principles of physics were accepted if they could stand the test of logical consistency and empirical confirmation. The era of mechanistic physics was reaching its end, and the era of logico-empiricist physics was beginning. Roughly speaking, we may say that the mechanistic era had extended from 1600 to 1900, and that the twentieth century opened with the logico-empiricist conception of science in the making.[1]

We can understand "logico-empiricist" here to mean the neo-Humean outlook generally, positivism included.

What Frank means by "mechanistic" science here, a mode of science that he says came to an end in the twentieth century, would at a minimum include causal theories of light and electricity involving some kind of motion of postulated matter, such as vibrational disturbances or spinning vortices in the ether. But "the logico-empiricist conception of science" suggests a generalization not only beyond this particular area of physics but beyond physics to *science*. It is clear, however, that his claim about the end of mechanistic science cannot be generalized over all of science. Suppose he means "mechanistic" in the strict sense, in which explanations rely on the idea of matter undergoing motion, stress, collision, and deformation; then it would not be correct to include, for example, geology under the postmechanistic conception of science, since in geology explanations of this kind have in no way met their demise. It is not uncontroversial to claim that explanation in biology, such as the theory of natural selection, is mechanistic or reduces to mechanism in the strict sense. But if Frank is using "mechanistic" in a broader sense, to mean any physical story describing the genesis of facts or phenomena, then the logico-empiricist conception of science must exclude biology as well. Integral to biology and geology are narrative accounts of natural processes having brought about some existing conditions, and this fact is certainly not affected by any changes in the meaning of "theory" in physics.

From now on I will use the terms "mechanism" and "mechanistic" in the narrow sense, as causation which is described using the elements matter, motion, stress, and collision. To designate a broader concept of narrative causal explanation, a concept to be gradually developed, which may—and, I will be arguing, does—extend beyond mechanism, I will reserve the term "productionism." A "productionistic" explanation is a causal explanation of which the only essential components are events arising out of one another in succession and/or giving rise to (in a perfectly innocent and literal sense) the fact, entity, or phenomenon that the particular story explains. By this definition the causal process involved may or may not be mechanistic. "Productionism" asserts the unreduced meaning of narratively comprehended causality, encompassing its broadest possibilities, that is of the essence of aspect (1) physical science and is the basis of explanation in natural science generally. It is counterposed to the Humean and neo-Humean conceptions of causality and explanation. The productionistic–mechanistic distinction anticipates the need for causal explanations to be formulated outside of mechanism for certain physical domains.

There should be no disputing that Frank puts his finger on something important about the history of *physics*. About the two distinct eras of this science, divided roughly at the turn of the century, this much can be said: There was in the earlier era a living expectation among physicists that the regularities which their equations quantified, notably those occurring in the context of experimentation with light, electricity, and magnetism, had their explanations in underlying physical processes whose description "in ordinary language" might someday be carried out. But in the new era this science came to have the whole of its progress and success in predictive equations developing in synchrony with experimental technology, while the prospect of a narrative description of the causal processes exploited in this technology floundered utterly amid philosophical turmoil. This was not so much a smooth and natural maturing of knowledge, as Frank suggests, but more like a difficult birth, namely, of the pragmatic renunciation of a certain traditional quest for narrative and naturalistic explanation, down to its last remnants.

But in light of the actual philosophical views represented in the term "logico-empiricism," Frank intends just the generalization over all of science which, I have noted, appears untenable. This view is one result of the positivist striving to "reconstruct" the whole of science as a logically interconnected system of lawlike propositions—which is to say, of conceiving the content of the whole scientific account of the world after the model of modern physics. Positivists thought that such a single "unified" system of science, with its well-defined allowable content of laws, would establish the clear partitioning of scientific knowledge as against metaphysical speculation or otherwise unfounded popular beliefs, which they regarded as intellectual and social evils.[2]

Frank's "logico-empiricist" conception of science sees as genuine and permanent results of *science in general* only systematically organized, lawlike propositions about observed regularities. From this point of view a pervasively causal-narrative approach to explanation, which I am arguing for, will be seen as an errant nostalgia for "mechanistic" science, even though my aim is explicitly to conceive of causality as having meanings that extend beyond the constraints of mechanism (in the narrow sense). Whether explanations in physics can be pursued outside the mechanistic mode through a transformed conception of natural productive activity remains to be seen. This aside, the logico-empiricist point of view may pose a challenge to my assumption that to produce explanations in science is to produce narrative causal explanations. A consideration of this possible challenge is the primary concern of this chapter.

I will formulate the logico-empiricist's objection to my project in this way: Suppose we relieve logico-empiricism of the problematic generalization over all of science that I have made note of, by restricting our attention to physics and setting aside the hope for "unified science." Suppose further that the tradition of positivism and logico-empiricism can produce a sense of "explanation" applicable *to the domain of physics* in accordance with the position that not causal stories, but applicable laws, will be its essence. If this cooked-up meaning of "explanation" is clear and convincing enough, they may have a case against my view that where narrative accounts of physical processes are not produced, the explanatory dimension of science is lacking. They could claim that it is open to question whether such accounts are necessary for science to have explained known phenomena. Perhaps physics *just as it is* displays the real, final, and complete character of scientific explanation, through its status as a fundamental science, as the one, true "unified" explanation of nature!

The tradition of logico-empiricism has in fact embodied this objection to my view: it has been deeply occupied with the problem of explanation, that is, with the construction of models of explanation based on laws and the internal criticism of those models. I will examine a classic text of logico-empiricist explanation theory after first providing some background discussion.

Instrumentalism and Antirealism: Breaking the Link

The distinction on which this study so far has been based, that between the causal-descriptive and mathematical approaches to physics, can, if adequately clarified, illuminate a great deal about traditional problems of philosophy of science. The distinction underlies the fact that precise, predictive knowledge of reliable correlations among observed and measured events in experiments can develop into advanced and highly efficient mathematical models, as part of a dramatic surge forward in the technical dimension of a

particular science, *independently* of the quest for causal comprehension whether it succeeds or fails, whether it is or is not pursued. In
other words, inquiring about the connecting process, if any, that
accounts for a certain regularity or "invariant" in observation can
be a concern wholly separate from the formulation and application
of its law. That laws alone without corresponding causal explanations not only can play a central role in the development of science,
but can seem to embody a complete and independent knowledge of
the phenomena, has been basic to the predominant operationalist
physics and neo-Humean philosophy of science.

This independence of successful laws from knowledge of
causes has traditionally harbored two forms of epistemological
skepticism; I call them forms of acausalism because they both in
different ways stand in opposition to my "productionist" point of
view.

There is first of all the general acausalism, associated with
Hume, which says that apart from regularity of succession in separate events there is nothing in nature answering to the notion of
causal connection.[3] On this view the inscription of a regularity in
the mind (or in an equation) is really the whole factual content
involved in saying that one event causes another. For contrast, it
might be supposed (innocently of this determination) that a causal
process is not merely a set of spatially and temporally separate
events (in physics experiments and elsewhere), but that there are
also causal connections in the form of continuous physical transitions from one state, stage, or event to another. To claim, as the
present work does, that unreduced causal narrative is essential to
scientific understanding of the natural world, is to maintain the
supposition that there is more to causality than regular succession
in events, namely, there are connections which can be spoken of in
general as the "production" of one event by another and which
have a role in any causal account of how some natural feature or
phenomenon—say, a mountain range—is brought about. This is
one essential difference between productionism and the neo-
Humean view.

The second and more specific form of acausalism is the view
that there is no justification for believing, as a matter of scientific
truth, in the physical reality of certain causal occurrences in a special class investigated by modern physics—those that are called

"particles," "radiation," and some other names in different contexts—*because they are outside all possibility of direct perception.* Behind this conclusion lies the view that a body of scientific knowledge is only a systematization of facts of observation, so that all that is available to comprise scientific explanation of phenomena are the measured quantitative interdependencies found in them. As is the case with the first form of acausalism, one can take a position contrary to this fictionalist view of "unobservables," and furthermore this position will have considerable weight of commonsense expectation behind it. Innocently of the skeptical antirealist arguments, there is a forceful presumption that there are some occurrences in nature giving rise, for example, to forces of attraction—Newton stated this presumption in the third letter to Bentley—and with even more surety that there are physical propagations intervening between what are called emitters and detectors of radiation; even apart from specially scientific contexts, there is a strong presumption that *light is some physical propagation.* Causalism/productionism goes a step beyond this to the conviction that knowledge not only of the existence, but also of the nature of these physical processes, would explain the specific features of whatever is observed and determined by measurement.

The Humean view about causality in general and the antirealist view about causal entities of physics can be combined into the claim that mastery of regularities in observation is the substance of knowledge under the rubrics of causality and scientific explanation. The productionist view, by contrast, will not cease to expect that real causal connections and connecting processes pervade nature and that to understand and explain them is to describe them in detail and in their interconnections as physical events. My use of the term "acausalism" takes the standpoint of productionism; if the regularist does not think that "causality" is an entirely empty concept, he is free to claim that it just *means* regularity in observed and discrete events, rejecting the label "acausalism." The productionist, though, claims that to throw out the reality of connection in the relationship of production is to throw out causality.

The assumption that formalization of regularities comprises the whole essential content of successful physical theories raises the disturbing question of *instrumentalism* for philosophers of science. Might this not mean that science, or at least physics, does

important

not actually explain anything, but only provides us with the power
to predict phenomena and manipulate nature? But the logico-
empiricist tradition has shielded itself from this conclusion by continuing to think that the impressive intellectual and technological
successes of physics *must* mean that this science explains phenomena and achieves understanding of its domain of nature, and that
therefore it must be possible to conceive how explanation can take
place through the formulation of laws and their integration into
broadly applicable mathematical theories.

But this "must" has remained unargued. Productionism would
recall *"hypotheses non fingo"* and would understand it as meaning
simply that despite having determined the law of gravity, Newton
could conceive no *way to explain gravity* that had enough plausibility to put before the public. Apart from the usual assumptions and
procedures, there is no particular reason to think that any degree
of success in the determination of laws, such as their integration
into encompassing mathematical theories, is equally a success at
explanation. From the productionist point of view, the very fact of
the acausal character of contemporary physics indicates that
mathematical theories do not themselves explain, however powerful their utility. The claim that such theories are artifices and not
explanations—instrumentalism—is quite defensible, just because
of the independence of knowledge of laws from knowledge of
causes.

My assertion that instrumentalism about theories fits well with
productionism may generate confusion, especially to those initiated
into these issues. This is because instrumentalism, associated with
positivism, has traditionally been assumed to be tightly bound to
antirealism, the complete contrary to productionism. The thinking
behind this linkage has been as follows. Given an unexamined assumption that it is always theory that embodies the first and only
knowledge of the "unobservable"—that theory is, in other words,
an exclusive transcendence of mere description of phenomena—it
seems that if such a theory were only an instrument of prediction,
this would render the "unobservable" entity it supposedly postu-
lates a *pure construct*. Instrumentalism has become unpopular because, due to the unexamined link with antirealism, it seemed to
have "consequences too irrealist for many to swallow," as Lawrence

Sklar says.[4] Philosophers wanted to think that what on the face of the matter appeared to be real entities (where this refers, safely, to forms of causal propagation) studied and named by physicists, were in fact such real entities. "Realism" asserted itself because the claim that the electron and the photon are not known by science to exist but are instead useful fictions seemed extreme, though the neo-Humean outlook remained unchallenged.

As for positivists, it was not that they necessarily believed electrons were fictitious, if they were to state their view of the matter "off the cuff." Nor were they necessarily explicitly instrumentalist about theories. But their neo-Humean empiricism, in particular the focus on mathematical laws, irresistably suggested both. Rudolf Carnap's arguments provide an illustration. He started out a passage intending to show the *falsity* of the view that mathematical theories are " 'mere formalistic constructions,' 'mere calculi.' " He claimed a theory must have in addition an interpretation which is an understanding of nature. But his discussion ended up with an "understanding" for theories in the domain of the electron which is actually no understanding, but only utility. This is the "kind of understanding which alone is essential in the field of knowledge and science."[5] This quagmire shows the confusion stemming from the modern use of the term "theory."

The resurgence of realism helped to discredit certain aspects of positivism and displace instrumentalism. But I want to suggest that instrumentalism embodies a clarity that has been suppressed in the process. Productionism says that explanations dropped out of physics, instead of saying that the meaning of "scientific explanation" must be reconstructed along Humean lines. With this, mathematical theories can be allowed to have a purely instrumental function, while explanation and its prospects can be a different matter entirely, one that is far from resolved, namely, the question whether genuine causalist physics has any further possibilities. I am thoroughly convinced that it does; but even if productionism were forced to admit that causal explanation in this area is indeed no longer possible, it could continue to insist upon a narrative concept of explanation, since it could still be cogently maintained that physics at this point has permanently *ceased to be an explanatory science* and has become a technical and technological science.

Two contemporary philosophers have already been effective at loosening the bond between instrumentalism and antirealism. Nancy Cartwright makes the simple point that in the identification of causes, while there may be several competing hypotheses, there is only one correct one, whereas it is often the case in physics that there are several existing formulas for a particular measurement phenomenon, without any one of them being in any legitimate sense "correct" as against the others.[6] This suggests that theory in the modern sense may not be something that in any way identifies causes. (For productionism, this just means that it does not have the character of explanation.) The disclosure and naming of entities efficacious in experiments (short of their understanding) might then be a process distinct from theorizing in any sense of the word (though this result is not clearly brought out in Cartwright). To take up a theme of Ian Hacking's, why should the electron, for instance, be permanently a "theoretical entity," as is traditionally presupposed? The atom of electric charge was at one time "theoretical"—which is to say, a hypothesis. But the subsequent history is that this hypothesis was experimentally verified by the Millikan oil-drop experiment. Should one view this standard, uncontroversial textbook account as philosophically naive? Did not the experiment detect the electron by measuring its effects directly?

There is a perfectly good candidate other than what is called "theory" for what identifies and differentiates "particles" in addition to disclosing their actuality, namely, the fully specified experimental conditions in which they occur, including their characteristic effects; this is how I understand Hacking's critique of antirealism. He presents the matter from the standpoint of the experimental physicist who is making use of specific "unobservable" transmission, a species of the entities in question, in his investigations.[7] He knows the transmission consists of positrons because he has designed just the emitter that will emit positrons, and what is useful to him is that this kind of transmission has a highly specific set of causal properties or "powers." "Positrons" identifies a species of physical process that is effective in just these ways. In light of this practical, prereflective knowledge, antirealist skepticism appears just as untenable as it would if it claimed that light is no real causal transmission across space.

In these ways Cartwright and Hacking make a compelling case at least that the question of the reality of "unobservables" is quite separable from the question about the status of theories, so that one need not reject instrumentalism in order to avoid antirealism about causal entities. Productionism would interpret their results by pointing out that if theories in physics are not explanatory descriptions of nature, but only instruments, this may mean (and I think it is quite plausibly the case) that the "particles" of physics simply remain undescribed in spite of thorough quantitative knowledge of the observable conditions of their occurrence and of their effects. This removes the pressure to think that, if the entities are real, then the theories in question must be some kind of explanations. Hacking's and Cartwright's ideas are discussed in more detail in Chapter 4.

The "Covering Law" Model

The orthodox view in philosophy of science, however, has been that physics establishes a standard for all scientific knowledge by *explaining acausally*. This leads to the conclusion that an acausal theory of scientific explanation in general needs to be formulated. As I have already remarked, this is strange in light of the fact that other branches of science explain causally. Logico-empiricism shows intellectual resistance to the possibility that physics may be unique among the sciences in that, though it masters through mathematical models the regularities in measurement that it encounters, it does not, in a sense basic to natural science, explain them. But there is nothing incongruous in this possibility. The thrust of my argument at present is to set alongside standard acausalism the possibility that physics has not achieved explanations concomitant with technical advance in exploring phenomena, and that its "theories" are simply artifices of mathematical ingenuity with which research equips itself.

The law-based model of explanation which I will presently consider is one outcome of the acausalist tradition, transmitted through positivism. It is the culmination of the view that a genuinely scientific explanation is comprised of laws. Philosophers who have committed themselves, for the reasons I have discussed, to an acausal theory of scientific explanation, have endeavored to develop a meaning for "explanation" by which it contrasts sharply with "description," so that explanation of phenomena can be a matter of the laws covering them, superceding the idea that descriptions of causal events can themselves constitute fully scientific explanations. Through a formal definition, they thought, "explanation" could be given a meaning confined to the citing of a law or laws specific to the circumstances of the thing to be explained.

The standard account of explanation according to this prescription is found in a famous paper by Carl G. Hempel and Paul Oppenheim on "The Logic of Explanation."[8] These authors describe explanation as involving a kind of syllogism. Suppose we have an occurrence or fact we wish to explain. We take as one premise a proposition describing in adequate detail the circumstances surrounding the occurrence or fact; we take as another premise the law stating what will come about in those exact circumstances—namely the thing we are trying to explain; and there follows deductively a proposition stating that what we are explaining is indeed the case or does occur. A law of science thus becomes a conditional statement logically implying (given the required conditions) the phenomenon to be explained. This is the "covering law" model of explanation. In short, to explain is to deduce the thing-to-be-explained from a general law.

An initial response one might have to this claim is to ask, "Why does this deduction of something from a covering law (plus initial conditions) *explain* the thing? Does this not simply amount to saying, 'It always occurs in those circumstances?' What is explanatory about that?" A background view, already alluded to, is what seems to alleviate this problem in the minds of the proponents, namely, a view of science as a *total interconnected system of lawlike propositions*. With this picture, what explains a phenomenon is not just that one law. Its covering law explains the phenomenon with the backing of the whole system of laws, once disclosed (I will return to this).

Hempel and Oppenheim offer examples of explanation which they claim illustrate this "covering law" or "deductive-nomological" model. First there is an introductory passage in their paper intended to drive home the distinction between description and explanation, which had been formulated by earlier positivism:

To explain the phenomena in the world of our experience, to answer the question "why?" rather than only the question "what?", is one of the foremost objectives of all rational inquiry, and especially, scientific research in its various branches strives to go beyond a mere description of its subject matter by providing an explanation of the phenomena it investigates.[9]

The following example is supplied to illustrate the deductive model:

> A mercury thermometer is rapidly immersed in hot water; there occurs a temporary drop of the mercury column, which is then followed by a swift rise. How is this phenomenon to be explained? The increase in temperature affects at first only the glass tube of the thermometer; it expands and thus provides a larger space for the mercury inside, whose surface therefore drops. As soon as by heat conduction the rise in temperature reaches the mercury, however, the latter expands, and as its coefficient of expansion is considerably larger than that of glass, a rise of the mercury level results.[10]

The authors then attempt to identify the features of the deductive model which occur in this explanation. It consists, they tell us, of two kinds of statements: a description of some antecedent or co-occurring conditions—a mercury thermometer is dipped in hot water—and some laws covering the case. Specifically, these laws "include the laws of the thermic expansion of mercury and of glass, and the statement about the small thermic conductivity of glass."

It is not clear that these laws—if they are laws—are central to the explanations presented. The "small thermic conductivity of glass" is simply not mentioned in the text as part of the explanation. The coefficients of expansion of mercury and glass are mentioned. But, to start with one of the problems, why should these coefficients be called "laws"? They are quantitative properties of substances; at least in this sense they represent "regularities"—always the same quantity for a given substance. But this regularity is not a variation in quantity A as a function of some other quantity B. It seems unlikely that scientists would have a reason to refer to these quantitative properties as "laws."

But let us try to view them as laws in the sense which Hempel and Oppenheim would need for their model. The model calls for the conditional, in this case, which says, "If this substance is mercury, then it will expand a specific amount with a given rise in temperature," and the corresponding conditional for glass. Now, the explanation they give does not say what these coefficients *are*. It only says that one is "considerably larger" than the other. What is used in the explanation, the authors claim, is the "considerable" difference of the two. These coefficients are brought up in the explanation for why the mercury level rises *after* the brief dip in the level—the original thing to be explained—occurs: " . . . as its coefficient of expansion is considerably larger than that of glass, a rise of the mercury level results." The authors can be taken to be explaining by this difference of coefficients how the thermometer works, telling why the mercury level changes with the temperature.

But it seems off the mark to cite this difference as the principle of the thermometer, which it would be more accurate to describe as follows: The space in which the mercury can expand is flattened in one of its dimensions as far as is practical. An increase in volume cannot spread in this dimension, so that the expansion along the length of the space will be observable for a usefully small variation in temperature. The truth in what Hempel and Oppenheim say about the coefficients of expansion is that it makes sense to use a liquid that expands a relatively large amount. However, for all we know given their explanation, if enough flattening of the expansion of space were achieved, a moderately useful thermometer might be made out of materials whose coefficients of expansion were not so "considerably" different. It is more a matter of geometry than of exact properties of materials. If one thinks carefully about it, one does not find that these numbers and their comparison explain; rather, to focus on them as a centerpiece of the explanation only draws one into questions. Why should the basic functioning of the thermometer depend on just *these* quantities?

And we have been led astray here anyway, because the original thing-to-be-explained was the brief, initial dip in the mercury, not its expected rise. Or *was* it?

Hempel and Oppenheim aim to present a scientific explanation, and to describe its structure. For them what will make the explanation both explanatory and scientific is that it will go beyond merely describing the observations as any ordinary observer could. As a result they take as the heart of the explanation some scientific-sounding things, coefficients of expansion, thought of as laws in order to fit the covering law model. But our examination has arrived at the question: Why should it be necessary at all for the explanation to talk about these coefficients? This talk sounds slick and scientific, but to try to decide whether knowing these quantities (which we do not) is actually important for being able to explain either the dip phenomenon, or how a thermometer works, is deeply confusing. This situation does not bode well for their account of the explanation.

Going back to the starting point in this text: After describing the mercury dip, they ask, "How is this phenomenon to be explained?" Note that they do not specify that "explain" has any specially scientific or specially philosophical sense. They just ask how it is to be explained. Now, despite the problems with the explanation and commentary together, we were given to understand why the dip happens. So let's try to put it in our own words. What *is* the explanation? The heat from the water, of course, reaches the glass before it reaches the mercury. The expansion of the glass bulb slightly increases the volume available for the mercury filling it before the mercury can expand from the heat, drawing the column down slightly. There is the explanation. With it, I understand how and why this happens. It does not cite any scientific laws. But then, it does not fill the bill for an acausal sense of explanation. It is simply a narrative account of the mechanism involved.

To be fair to the Hempel and Oppenheim explanation and commentary, the unmentioned "law" about the "small thermic conductivity of glass" should be considered. This seems to get something right. The heat takes a certain interval of time to get through the glass, and this is crucial to the explanation. But the exact heat conducting properties of glass as compared with other materials is no centrally important fact—or at least, it is not clear why it should be. Would the dip phenomenon not occur *at all* with a bulb made, say, of metal? If what we were explaining to begin with was an observation that a dip occurred with glass but not with metal, then a comparison of conductivities would be just what is called for. But in the case at hand, the only relevant fact along these lines, and it is understood in the explanation just formulated, is that conduction through the glass takes enough of an interval of time to produce an observable effect. The specific measure of the conductivity of glass is not really to the point in explaining the phenomenon.

We are left with the simple causal story above as a solid version of the explanation. (Along the same lines as my promotion of narrative causal explanation in physical science, William Dray rejects the covering law model as an answer to the problem of explanation in history, and proposes instead the "model of continuous series," according to which explanation is a tracing of a course of events.[11]) The objection might be raised that as a commonsense explanation, ours is not *scientific*, and therefore not what Hempel and Oppenheim are after. This could not be taken to mean that the explanation is unscientific in the sense of *unsound*, as if it were superstition or futile guesswork. It is not subject to rejection, scoffing, or skepticism from science. So what *does* this handy objection mean? When philosophers of science seek to understand a specially scientific sense of "explanation," what *are* they after? For one thing, they want this sense to sharply contrast with description, and not only with description of what is observed, but also with description in the sense of the narrative causal explanation of what is observed. The objector is thinking that an acausal theory of science is needed. If it is not needed, then causal explanation bridges between scientific understanding and ordinary understanding, and does not allow this sharp distinction. I am contending throughout this book that the whole search for an acausal theory of science (in the productionist's sense of "acausal") is misguided. At present we are examining the constructed theories of explanation that were put forward once acausalist assumptions were well ingrained in tradition.

The upshot of the discussion so far is that a classic text has (a) provided a correct, if cumbersomely expressed, explanation of a phenomenon, and (b) adopted a quite implausible strategy for understanding that explanation.

Carl Hempel gave a revised presentation of the deductive model in 1965. Here he provided another example of an explanation, citing a genuine scientific law as its essence, and this time the law had a closer actual relevance to the explanation than in the above example. The text reads as follows (emphasis added):

> In a beaker filled with water at room temperature, there floats a chunk of ice which partly extends above the surface. As the ice gradually melts, one might expect the water in the beaker to overflow. Actually the water level remains unchanged. How is this to be explained? *The key to an answer is provided by*

Archimedes' principle, according to which a solid body floating in a liquid displaces a volume of liquid which has the same weight as itself. Hence the chunk of ice has the same weight as the volume of water its submerged portion displaces. Since melting does not affect the weights involved, the water into which the ice turns has the same weight as the ice itself, and hence, it also has the same volume as the displaced water; hence the melting of the ice yields a volume of water that suffices exactly to fill the space initially occupied by the ice. Therefore the water level remains unchanged.[12]

This is an example of making an observation, being puzzled about it, then producing an explanation in almost obsessive logical detail. Despite this diligence, not much life-context is given for this case of being puzzled and having the puzzle resolved. Suppose we give it the context of a high school science classroom. The teacher places a beaker full of water on top of a costly and undamaged textbook. There is a chunk of ice floating in the water, one of those oblong, irregular pieces that come in ice commercially bagged. The teacher carefully adds more water with an eyedropper until there is a rounded bulge at the edge of the water from surface cohesion. It appears that another drop of water, no matter how delicately added, would cause it to spill over and damage the textbook. The chunk of ice projects a point above the water like a tiny iceberg. With the students on the edge of their seats, the teacher forgets about the beaker and delivers a lecture.

At the end of the lecture, the attention of the class turns again to the beaker. The ice has melted, the edge of the water is still bulging, and the textbook remains dry. In keeping with Hempel's focus, suppose that the teacher's intent is to use this demonstration as a way of introducing Archimedes' principle of displacement. Why were the students on the edge of their seats? *Why* "might one expect" the water to spill over, so that when it does not, there is something to be explained? If I expect it to spill over, I have to be thinking that when ice melts it turns into a volume of water equal to its volume as ice, so that when melted, just as in the frozen state, it will take up a little more space than a beakerful of water up the level that it has. I am not cognizant of the fact that water expands slightly when it freezes, making it less dense than the water, which is why it floats. Hempel does not talk about this when explaining the "phenomenon" or when giving his account of the explanation in terms of deduction from laws. But it seems that, far from introducing laws either explicitly or implicitly, the explanation is nothing more than the correction of an erroneous and poorly thought out assumption. If there is puzzlement it probably results from assuming that whatever increases in temperature necessarily expands. This case of an explanation recalls Hilary Putnam's remarks:

> Explanation is interest-sensitive and context-relative. We expect an explanation of a fact to cite the factors that are *important*, (where our notion of importance depends on the reason for asking the Why-question).[13]

In the above example the reason for the why-question is an expectation based on an error.

Before she talks about Archimedes' law, the teacher asks a bright student, "Why did you expect it to spill over?"

"I assumed that the part of the ice that was sticking out above the water would be added to the volume of water in the beaker."

"And what caused you to assume that?"

"Not realizing that water expands when it freezes." Even with this much understanding of the experiment, to the point that he understands why he was wrong in his expectation, the student has not yet learned about Archimedes' law; he is not yet even thinking about the relation between displacement and weight.

Archimedes' principle says that a floating object displaces an amount of liquid with the same weight as itself. When I said that this law has a closer relevance to the explanation than those in Hempel's earlier examples, I meant that one could go about explaining why the beaker does not spill over by starting with a reflection on this principle. The science teacher might in her next words say what Hempel said, that the principle implies that the floating ice turns into the same volume of water that, as floating ice, it displaces, necessarily leaving the water level unchanged. In taking this tack of explanation, she would at the same time be saying something about science. Hempel did so to try to establish an *uniquely scientific* sense of explanation. This made a better example of the importance of a law to an explanation than the thermometer example, where the "laws" of the exact thermal properties of certain materials seemed distinctly peripheral if not irrelevant.

It is problematic, however, to say that in the ice example Archimedes' principle is the "key" to an explanation, when the reason an explanation had to be given at all arose from a natural, casual, mistaken assumption. It seems that Hempel's statement comes only from the pressure to see scientific explanation in terms of laws and their logical entailments.

Another example Hempel uses in this same work[14] is taken from John Dewey's book, *How We Think*. In his book Dewey relates an explanation he produced for himself after making an observation one day while washing dishes. He noticed that when he placed a tumbler, fresh from hot soapy water, upside down on a plate, soap bubbles would emerge from under the rim and expand for a short time, then come to a stop and recede. Dewey came up with the following explanation of this occurrence: When a hot glass is first set down, it traps air which is cooler than the glass. The air is warmed by the glass, so that it expands and produces the bubbles. As the air recools, the bubbles retract.

Naturally, Hempel wants to construe this explanation as consisting of two parts: first, a detailed scientific description of the case, including temperatures, materials, and so on; and second, laws of various kinds—gas laws (relating pressure, volume, and temperature), laws about the behavior of soap film, and so on. Now, he does suggest in this text that he is offering a special, scientific kind of explanation. But he is also talking about Dewey's explanation. How does he manage to translate Dewey's explanation into such very different terms?

The answer is that he thinks all these laws which are not stated in the explanation really belong to it as its suppressed detail. When like a blossoming thistle this detail is brought forth, the case fits a deductive model. He writes (emphasis added),

> While some of these [laws] are only hinted at by such phrasings as "the warming of the trapped air led to an increase in its pressure," and others are not referred to even in this oblique fashion, they are *clearly presupposed in the claim that certain stages of the process yielded others as their results.* If we imagine the various explicit or tacit explanatory assumptions to be fully stated, then the explanation may be conceived as a deductive argument. . . . [15]

Later in this discussion he says more specifically what it is he refers to as a "claim that certain stages of the process yielded others as their results" (emphasis added):

... reliance on general laws is essential to a D-N [covering law] explanation; it is in virtue of such laws that the particular facts cited in the explanans [the description of initial conditions] possess explanatory relevance to the explanandum phenomenon [the thing-to-be-explained]. Thus in the case of Dewey's soap bubbles, *the gradual warming of the cool air trapped under the hot tumblers* would constitute a mere accidental antecedent rather than an explanatory factor for the growth of the bubbles, if it were not for the gas laws, which connect the two events.[16]

This is quite a startling thing to say. Consider Dewey's reasoning, which was as follows:

Why should air leave the tumbler? There was no substance entering to force it out. It must have expanded. It expands by increase of heat, or by increase of pressure, or both.[17]

He tried to think why it happens. And he came to see why, to realize what makes bubbles arise and recede. And that of which he became cognizant was precisely "the gradual warming of the cool air trapped under the hot tumblers." For Hempel, this process that Dewey found explanatory, the very thing he identified as a *cause*, after which he felt satisfied that he understood what a moment ago he did not understand, would fail to explain, would be explanatorily impotent and empty, were it not for the fact that the gas laws are "hinted at," "presupposed," and/or "assumed."

Hempel thinks that to realize that the air inside the tumblers is gradually warmed would be *not relevant to understanding the occurrence* "if it were not for the gas laws." Now, scientists have been able to formulate the gas laws because the pressure, volume, and temperature of a gas are functionally connected. These variables depend mechanically on each other according, it is found, to more or less precise quantitative rules—the gas laws. These laws might prove indispensable for designing equipment such as steam boilers and pneumatic chambers. They certainly would be essential in any precisely designed experiment which worked like the tumbler–soap bubble phenomenon. But why should we suppose that they have to play a role in John Dewey's thoughts while he is washing the dishes, that they are essential for this cognitive process also? The gas laws *might well not be in his thoughts at all*. He certainly is not, simply by explaining the phenomenon to himself or recognizing why it occurs, necessarily also calculating and confirming his result with known laws.

Hempel's claim that the explanation hinges on the gas laws is cogent to this extent: it is necessary to the explanation that a functional interdependence among the variables of pressure, volume, and temperature *exists*. However, the quantitative exactitude of the *laws*, that which may make them indispensable for, say, precision design, seems to have no central bearing on the explanation. Indeed, it seems that the phenomenon, the explanation, and the gas laws themselves, would not be what they are but for a certain *mechanism* that connects these variables. Dewey's brief "qualitative" description of this mechanical interconnection is distinctly more pertinent to an understanding of the phenomenon than any exact quantitative formula.

Someone might follow Hempel's lead and object to my pitting against the deductive model simple narrative explanations such as the one given for the "dip" phenomenon of the thermometer, or Dewey's explanation of the expanding and receding bubbles, by saying that these accounts merely fail to mention the various scientific laws which are "implied" in them. What could be meant by this? Carnap argues that all explanations are based on laws, whether explicitly or not. In one of his examples, someone seeks an explanation for why his watch is missing from the table, where he left it. The explanation is that "Jones took it." Carnap thinks even this is really a "law explanation in disguise." He says this:

> . . . "Jones took it" would not be considered a satisfactory explanation if we did not assume the universal law: whenever someone takes a watch from a table, the watch is no longer on the table.[18]

Carnap's claim is that a "universal" regularity involving an action and its immediate consequence (if these are really differentiable) is presupposed in the explanation. Perhaps this is true. But citing this "universal law"—action X always has "consequence" Y—does not improve on the explanation as it was; certainly *by itself* it would not satisfy the watch owner's query at all, and therefore cannot be the undisguised form of the given explanation.

It may be the tidiness and precision of laws that leads a thinker to value them as against "crude" or "merely qualitative" causal accounts, especially if there is pressure to separate scientific from ordinary explanations and to exalt the former above the latter. But let alone laws hidden in causal stories, is there any reason to think that more detailed and precise laws are *referred to by implication* in some fairly approximate laws, such as the laws (explanations, according to the "covering law" view) of classical, "macro" mechanics? Certainly pre–twentieth-century developers and utilizers of the laws of mechanics were not, in this practice, referring by implication to the laws covering more detailed regularities underlying all mechanical processes, such as the laws known as quantum mechanics, which were yet to be discovered. Thus it seems highly forced to claim that an account of a causal process is somehow a veiled reference to a system of equations. Why not instead, if one wants to insist upon further analysis, say that the causal process described is composed of more detailed causal activity? The difference in intelligibility is remarkable.

The strategy of saying that an explanation which does not mention laws might really be about hidden laws is a symptom of a fundamental problem with the deductive model, one that has already been remarked upon. The logico-empiricist writers, once they present an example of an explanation, then see a need to find the laws that are pertinent in order to explain or clarify the explanation. The laws that they look for are "generalizations" under which the thing explained can be "subsumed." To subsume under a generalization means that there is to be found in our knowledge a conditional proposition which describes circumstances under which a certain thing always occurs. In the examples we have looked at so far, however, the thing mentioned or "covered" in the generalization that is cited, apart from Carnap's example, is *not the thing-to-be-explained itself,* but some related quantitative regularity on which the philosopher can attempt to pin the essence of the explanation, something taken from scientific knowledge of measurable properties of material and its interactions. But if one throws off the pressure to find a law-based meaning of "explanation" by supposing that explanation in science is causal narrative, one no longer sees why this move toward laws should take place at all. Our critical resistance to this move has gained force as we have seen that this search for the law with the right connection to the "explanandum" has not managed to hit upon the essence of any one explanation.

The persistent, troubling thought about such a quest is this: Suppose that to find and cite an appropriate conditional proposition taken from a broader knowledge of such propositions really does accomplish explanation. Why not, then, cite the exactly appropriate one: "If a mercury thermometer is placed in hot water, the mercury level will dip briefly," or, "If ice which is floating in water melts, the water level will not change as a result." With these conditionals, one is sure of hitting the mark.

The trouble is, of course, that these are not specially scientific laws suitable for an account of specially scientific explanation. An even more conspicuous problem is that these statements merely describe the thing to be explained, they do not explain it. The more plausible course is therefore to seek more general facts of nature—hopefully, quantitative laws in use by science—from which these particular phenomena can be derived. If this can be done (the thinking goes) it will maintain the sense that what explains is something other than the thing explained, and also that the explanation is scientific. So one looks for the general laws that might seem essential to the observed phenomenon. But why should one think that such general laws, conditional propositions "subsuming" more specific laws, are any more of an explanation of the broader regularities *they* designate, than was the conditional proposition that was merely a description of the phenomenon? Does not a law, rendered in prose form, only amount to a conditional proposition describing an observed regularity, *and not to an explanation?* According to the regularist view of causation, a law is what in the last analysis is meant by a *causal* description; but even this doctrine does not render a law more plausibly an explanation, but only suggests that the explanatory "bringing about" is chimerical.

This leads to the question, what is to be gained by the insistent quest for laws at the heart of explanations? One might easily become captivated by the idea that the discovery and precise determination of order and regularity in nature, a process perhaps yielding things of beauty from a mathematical point of view, is itself explanation. But it is at least questionable whether such determination of the laws actually moves beyond establishment of fact to the explanation of fact. What leads one to think that it should? Is it for some reason desirable that it should? If it is desirable in order that quantitative-empirical physics may be viewed not only as an explanatory science, but as paradigmatic explanation, then this merely reinforces the prior belief, itself without basis, that it is *accurately* so viewed. Another possible motivation is the general aim of disclosing a formal logic within explanations, but this is peripheral to the project of an acausal theory of science, and is outside the present concern.

Moreover, laws can readily be seen as designating facts *to be explained*. I will take Archimedes' law as an example.

In Hempel's text, there is surprise at the discovery that the water level in a beaker of water does not rise as the ice in it melts, and this leads to a recollection of Archimedes' principle. Suppose instead that one day an amateur scientist who is unfamiliar with Archimedes' principle is tinkering with floating objects and with weight and volume measurements. Through "blind intuition" and more or less random tinkering, he hits upon the fact that a floating object displaces an amount of water with the same weight as itself. He might shout "Eureka!" But is this because he explained something? He was not *trying* to explain anything; he was only tinkering in his shop. He would have discovered something "nifty," a piece of mathematical symmetry that is a property of nature itself. He might not feel he had discovered anything that *needed* explaining, and this may be partly because he is pleased with it. However, one might have in such a way discovered this mathematical relationship, or just learned about it, and still not have an explanation of it—that is, one could still entertain the question why it is the case. What would *seeing* why it is the case involve?

One would think about floating objects, and the fact that some things float and some do not, and that certain things float in some liquids but not in others. A liquid will support an object if it has at least as much density as the object, just as objects lighter than air will rise. In supporting the object the weight of the liquid resists the force of gravity. The density of the object and the density of the liquid are in mechanical struggle. When the object "wins," it sinks; when there is equilibrium, the object is suspended within the liquid; and when the liquid "wins," there is buoyancy. Among floaters there is variation in buoyancy—some things float lower than others. Low floaters obviously displace more of the liquid. Thinking hard enough about this struggle of forces with density as the deciding factor could lead not only to understanding why Archimedes' law holds, but possibly even to the first intuition, formulation, and testing of it. If an object floats, there is an exact equality between two forces: the weight of the object, and the upward pressure of the liquid upon it. What might be the relation between the upward force and the amount of liquid giving way to the object? Could the force be equal to the weight of this amount of liquid?

The equality of weights is a fact about certain conditions, something one can either reason toward, or discover and subsequently come to understand why it is so. Thus the law itself is not something that explains, unless an artificial sense of "to explain" can be derived from its intrinsic interest as a precise piece of knowledge or even from its utility. Much less problematically, the law signifies a fact with a causal-mechanical explanation.

Why Theory of Explanation?

The discussion has shown how a law-based theory of explanation in science fares philosophically from a standpoint that has not acquired, or has set aside, any predisposition to think that it is outdated and mistaken to ascribe an essential role in scientific knowledge to causal narrative. This standpoint reveals logico-empiricism on explanation as resorting to a procedure which, having decided on acausalism through a description/explanation contrast, has already devalued narrative causal explanation by assigning it the status of "mere description." At the same time, seeking to uphold the simple dogma "Science explains" in full generality and particularly for the case of physics, it grasps at the possibility that truly scientific explanation might have its foundations in some celebrated cognitive equipment of physical science that falls under the broad heading of "laws." The implausibility of this move from a causalist standpoint is highlighted by the observation just made that in fact laws are by all appearances "explananda," things which might well *call for* explanation—in terms, of course, of physical processes taking place in experiments and other conditions of physical observation.

The literature of philosophy of science normally does not understand itself as limited in its sphere of concern to physics. Nevertheless, the standard view holds that to explain some natural occurrence is not to talk about its causes, but rather to cite one or more conditional propositions stating what happens or what is measured under specified conditions. From out of the blue, this seems perfectly absurd as a characterization of scientific understanding. Science has determined the true general explanation of the points of light and nebulae seen in the night sky, while everyday experience and speculation, it might be said, could not. Here is some specially scientific explanation. What could conditional propositions or equations have to do with it? If someone for some reason was ignorant about what he was seeing on a clear night, one would not explain it to him in terms of laws.

One source of support for this odd generalization of the image of physics over science is an idea known as "theory reduction," which is related to the positivist picture of science as a single, complete system or structure of propositions. It is thought that since the deeper physical explanation behind biology is chemistry, and the deeper physical explanation behind chemistry is physics, all scientific knowledge reduces to physics.[19] If physics is permanently acausal (and it certainly seems to be so on the surface), then at bottom so is science in general. But for this conclusion to follow it must *at least* be shown that knowledge in the domain of physics is really permanently acausal, that is, that there can be no scientifically sound and compelling causal understanding of its ultimate observations. My arguments aim to show in the long run that not even this petrified assumption is correct, and in the short run that it is in any case perfectly possible and indeed cogent to say that current physics is anomalous among the sciences in not providing fundamental explanations, such as (e.g.) biology provides with the theory of natural selection. This course of argument, even if it were to end up as an instrumentalist nihilism due to failure of causal ontology, avoids many of the problems and perplexities that arise in the standard approaches to philosophy of science.

A major problem with logico-empiricist explanation theory has also come to light in the discussion. It could only have been clear even to the proponents of the covering law model that an individual law describing a phenomenon or regularity does not by itself explain the phenomenon or regularity. Scientific explanation as law explanation cannot be a collection of individual laws. What answers this problem for the proponents is that the explanation by covering law refers and defers to more general laws which organize or "unify" them together with other "phenomenological" laws, and so on through higher orders of inclusion until the whole superstructure of science comes into view. The weight of science as a total system is thus placed behind the particular explanation. A clear statement of this view can be found in Karl Popper.[20]

But the more general laws are still merely empirical descriptions in the form of conditionals (or their quantitative renderings) in which the phenomenological laws take the place of the items of immediate observation that the specific law functionally interrelates: they are laws about laws. They therefore only identify facts of observation which are taken from different circumstances and placed into a relation. The general law is no more explanatory in character than the specific law, unless perhaps one can in following out each case through successive orders of generality get a "glimpse of the whole structure." And what could even this disclose except an ultimately inclusive law? Again, does a law explain? The very general law will presumably *have* a correspondingly broad physical explanation. The critique of "law explanation" theory applies also to the ubiquitous notion that mathematical theory is encoded understanding, since the former is only the formal philosophical explication of the latter. The notion that theories of physics disclose a mathematical structure of nature behind the perceptual appearances is another version of the same misguided reaction to the accomplishments of physics.

The major conclusion for the purpose of the present work is that the demand for productionistic explanation throughout science is undiminished by the logico-empiricist objection to this demand. A law-based model of explanation does not have the kind of independent success in supplying a meaning to "explanation" that would allow it to supercede causal stories in the explanation of phenomena. Laws of measurement, and observed phenomena generally, are from the naive productionist perspective "explananda" of natural science. The fact that knowledge in the domain of physics became confined to technical mastery of regularities is entirely inadequate support for a logico-empiricist view either of science or of scientific explanations. What this development does represent is a shift of focus entirely away from physical explanations, a fact that logico-empiricism merely evades and loses sight of.

A tradition of philosophers has shown that in general, it is possible to construe laws as explanation in a specially contrived sense while paying tribute to the traditional intellectual icon of formal deductive logic. The thrust of this chapter has been to make evident the artificiality of this procedure, especially as a conceptual foundation for natural science. This philosophical artifice in its recent developments fills a need created by neo-Humean theory of science. My view is that this whole enterprise is radically misguided, and that the need to overcome instrumentalism by somehow viewing physics in its present form as explanatory does not exist. Notwithstanding tendencies on the part of some philosophers to artificially separate concepts by exploiting distinctions, "explanation" and "understanding" roughly correspond in natural science, and both correspond to tracing the origins and penetrating the causal structure of facts or phenomena or events. Nothing in physics threatens this truth; if a science becomes entirely acausal in its mode of theory, it comes to lack a dimension of explanation and understanding. Despite a public persona that physics is on the verge of "wrapping it all up" at the final frontiers of knowledge, official puzzlement in the face of apparent basic causal inexplicability is profound and sustained.[21]

3

Philosophy and the
Structure of Causation

THE NEXT TASK IN THE CRITICISM of neo-Humean theory of science is to inquire into the foundational status of *discreteness* and *contiguity* for traditional philosophical analysis of the causal relation. These conceptual foundations have the effect of rendering fully occult the implication in "causation" that an event is traceable to another along some kind of transition that genuinely *connects* the events. If such genuine genetic connection is really a preanalytic conceptual illusion, perhaps one with sources in human psychology, this would give powerful support to the nihilist outlook that theoretical science is properly limited to laws of succession in directly observed events. At bottom this nihilist alliance exists in the postmechanist era of physics because these conceptual foundations and those of mechanism are the same, so that given these foundations causal explanation cannot escape mechanism for the same reason it cannot evade the Humean analysis. Thus in the background of the attempts to conceive explanation as a function of laws there lies the decision that any causal process that might be described in a narrative explanation reduces under analysis to *pure reliable succession in discrete events*, reinforcing the idea that scientific knowledge and/or explanation is really inductive knowledge of recurring patterns of succession in observed events (excluding intervening transmissions or transitions), summarized in the laws of nature. I seek to show in this chapter that the assumption that causal event-sequences must be cases of discrete and possibly contiguous succession springs from the same fundamental conceptual constraint that presents physicists with the stark options of *mechanistic* causal models or the abandonment of narrative explanation. (The strategy for constructively superceding the mechanistic causal ontology in physics is covered in Part Two.)

This conceptual reduction embodied in both Humean and mechanistic causation has some different but overlapping sources in physics and philosophy. It begins to operate, for example, whenever events and processes in general are preconceived through the use of a framework of space and time derived from the practice of *measurement*. Under this preconception, events and processes take place *within* intervals of space and time that form a pregiven scheme of extensiveness for them to span or stretch through as they arise, happen, and pass. This useful, practical conception quickly slips into indistinguishability from the fully spatialized picture that events and processes *occupy, either individually or side by side in a series,* the regions of space and segments of temporal duration required for their occurrence. Events are thus conveniently and almost irresistibly pictured as a *discrete*, that is, as if they were objects occupying separate or at most contiguous regions of a "four-dimensional space-time," whether or not they are, innocently of this picture, spoken of as causally connected. I am not disputing the usefulness of this conception of spatialized time and objectlike events, but it certainly makes trouble wherever it is construed ontologically, that is, as determining the content of physical explanations. An important pillar of scientific nihilism has been the assumption that this conception lies at the unavoidable endpoint of any analysis of succession in events. To demonstrate an alternative basic conception of events and processes and their relations in space and time is a central project of the present work.

One way of formulating the basic mechanistic-Humean doctrine might be this: If there is a succession in (differentiated) physical events purported to be a case of causality, and if it is not a series of entirely separate events, then it can only be understood as *ideal or perfect mechanical collisions* (elastic or inelastic), that is, cases of spatio-temporal contiguity. But this would be an abnormally concrete formulation. This chapter explores some of the original expressions of the neo-Humean view of causality that occurred in the early stages of the recent empiricist/nihilist tradition. Criticism of the discreteness/contiguity picture continues beyond this chapter.

The Positivist Conception of Science

Around the time that the famous paper of Hempel and Oppenheim appeared, philosophers of science were setting the stage for making laws the centerpiece of explanation, by reinforcing and developing the idea of a sharp distinction between description and explanation. This distinction grew out of a philosophical aim of positivism, that of "reconstructing" scientific knowledge as a *complete or completeable system of empirical propositions.* The impetus for a description-explanation contrast was partly the need to overcome instrumentalism and affirm the status of physics as an explanatory science. Description would be confined to description of phenomena as directly observed, while explanation of these phenomena would not be description of natural events giving rise to them, but instead would be a scientific process of formulating all relevant "observation sentences"—conditionals that describe and quantify phenomena and their observable contexts—and situating these within a superstructure of such sentences, this superstructure being (once fully realized) the "unified" system of science.

The background rationale for this was that philosophers of this tradition wanted to display science as knowledge meeting a certain standard of indubitability, namely, that it can be affirmed by certain investigative acts that seemed to comprise what may properly be called "making an observation." Observability meant that an event or object is potentially part of the field of direct perception. Science, properly purified of all "metaphysical" suggestions, had to be an organized system of propositions, all of which were such that they were subject to verification at any time in that they were statements of the fully sensible facts. The propositions to be retained as genuine science had to describe these observations and nothing more. A philosophical standard of "certainty" in knowledge, in keeping with traditional epistemologies, was to be maintained by not allowing any inference to causal features of nature that are permanently hidden from the confirmatory glance; such inference seemed to them inherently baseless conjecture. Strong support for this project appeared to come from Copenhagenist atomic physics, which adopted the view that physical theory in this area could only be a calculus of the observable; this was a kind of empiricism adopted as an explicit research philosophy in physics itself.

What positivists were faced with in their philosophy of science was the fact that modern physics, despite Copenhagenism and its largely pragmatic rationale, has "unobservable" entities as its primary subject matter. To reject claims for the actuality of whatever is named by "electron," "photon," "magnetic field," on the grounds that such claims are not verifiable by *seeing the thing named*, would mean casting out of genuine knowledge what appears to be a fundamental category of natural occurrences encountered in the course of science, in fact virtually the whole natural subject matter of contemporary physics. Nevertheless the positivists' program amounted to just such an intellectual housecleaning, or as they called it, a "clarification of scientific concepts."

They did after all have some important thinkers of the past from whom to claim descendancy. Hume had concluded that the very idea of "causation" could in any case only be a purely mental construct formed as a result of the "constant conjunction" of pairs of individual inputs through the senses, the "cause" and the "effect," occurring in successive order on repeated occasions. Causal connection, on this account, only means a "determination of the mind" such that, when a cause-event is perceived or thought, the mind will "pass from the idea of [this] object to that of its usual attendant," the effect-event.[1] On this view there is nothing in our knowledge of the world (independent of mind) corresponding to the idea of the production or bringing about of one event by another event. For Hume a causal connection would have to be something in addition to separate mental impressions, the evidence for which is merely the fact that these impressions consistently occur in immediate succession; it is therefore a construct according to this epistemology. The positivist appropriation of Hume's view for a theory of science combined this view with an existing tradition of *scientific antirealism*. For this tradition, the immediate givens of experimental observation had a role analogous to the Humean "sense impressions." Just as Hume had concluded that causal connections cannot form a part of our knowledge of nature, so for scientific antirealism (in its fully conscious form) any "unobservable" entities whose role might be, for instance, to transmit a connection from initial conditions of an experiment to its measured outcome as in the case of the variety of causal transmissions which are now most often called "particles," are not to be ranked among facts of nature disclosed, discovered, or illuminated by science; they are instead fictions that prove useful in scientific procedures.

In adopting scientific antirealism positivists claimed the physicist/philosopher Ernst Mach as a founding father. Mach had garnered a certain amount of prestige through the fact that his rejection of absolute space and time (another class of "unobservables") seemed to have been born out in Einstein's theories, the success of which confused the distinction between empirical-predictive models and physical explanations. Mach thought that in general explanations citing unseen entities such as fields or atoms tended to implicate physics in beliefs about an occult realm of entities, and argued that no one such explanation of a set of phenomena had more claim to truth than others that might be constructed.[2]

A comparison of two antirealist physicists of the same period, Mach and Wilhelm Ostwald, is instructive. Ostwald held out staunchly against the theory, already widely applied and generally favored, that material substances consist of ultimate, discrete, material parts, the "atomic-molecular hypothesis." Eventually Jean Perrin conducted a long series of experiments, all using independent procedures, which were specifically designed, it may well be said, to *detect* atoms. The results of this experimentation, amounting to a considerable convergence of independent pieces of evidence,[3] went against the skeptical view, so that Ostwald finally admitted that there are atoms and molecules. Especially if it serves as a goad for this kind of brilliant experimentation, the skeptical posture can be seen as a healthy influence, even though in this case it meant that someone had to admit to a lifelong error. But if in this case the skepticism was defeated *not by learning how to see atoms, but through the detection of their effects*, it would seem that a certain traditional prejudice about the perceptual (as opposed to the observational) basis of knowledge had also been defeated. It might be said that Ostwald the empiricist philosopher gave way to Ostwald the scientist. Mach, however, remained stubbornly antirealist about atoms, and it was Mach who, because of convictions of precisely this sort, was made a hero of positivist philosophy of science.

As I indicated in the previous chapter, the grounds for general antirealism about the generative background of phenomena can have little to do with, and should not be confused with, the reasons why hypotheses in science are considered as yet unconfirmed, where this is part of the process of scientific discovery in textbook accounts. As we will continue to see, antirealism finds a natural ally in the Humean treatment of causality.

Schlick's Humean Arguments

Herbert Feigl can be credited with a remark summarizing the positivist view of causality. It occurs at the beginning of his essay "Notes on Causality":[4] "The clarified (purified) concept of causality is defined in terms of *predictability according to a law* (or, more adequately, according to a set of laws)." In Feigl's subsequent discussion of the matter there is scarcely an argument for this position. The attitude in the writing is one of, "By now (in the positivist tradition) we have it right about causality. Let's move on." But he does provide some references to writings from the works of earlier positivists, which to him represent the reasoning by which this view was settled upon.

One of these works is Moritz Schlick's article, "Causality in Everyday Life and in Recent Science."[5] It is worth treating at length. Schlick raises the problem of the "post hoc/propter hoc" distinction: on what basis does one judge whether an event is the cause of another or just happens to precede it? He claims that this distinction marks our causal judgments, for example, when we consider which among several preceding factors is the cause of our state of depression, or when we try to determine whether a certain medicine is effective against a certain disease. His quest for the grounds of causal judgments leads him to a Humean view about the meaning of the causal relation.

The first part of his article, ostensibly concerned with "everyday life," discusses how in fact one would ordinarily go about establishing that a causal connection exists. In his language, he is considering how a causal judgment is "verified." This term should be understood in light of the verificationist thesis of positivism (to which Schlick was an important contributor) that for a statement of the form "the cause of this phenomenon is X" to have scientific validity and truth, all physical referents of the statement—the phenomenon, its purported cause X, and the causal relation (whatever this means)—would have to be observable in the narrow sense of directly perceptible; indeed, these all must be observed as interconnected or at least occurring together at once in a single observation, in order to confirm their factual "causal" association.

Schlick is thus seeking to explicate how talk of causes touches down in some way to observations, presumably in being ultimately derived directly from them. He presents some examples meant to illustrate the process of looking for and determining causes and causal connections through observation; but in addition, in accordance with his philosophical bent, these examples are intended to show how from such observations we gain a meaning for the very concept of "cause" (since to have physical meaning "the causal relation" must itself belong in some way to what is observed). What he calls the "causal judgment" is not a mere scientific decision about whether an inquirer has hit upon the correct cause of some observation; he thinks that the causal mode of understanding itself, its first possibility, springs from something about observations.

He tries out a few different possibilities for sources of the concept of "cause." The first candidate is closely related to Feigl's "predictability according to a law" and can be associated with Hume, that of *pure regularity of sequence in observed events:* "If C is *regularly* followed by E, then C is the cause of E."[6] As he illustrates it, if in a sufficient number of instances the use of a certain drug is followed by recovery from a certain disease, one will, he says, speak of cause and effect. He favors this candidate for a concept of the causal relation, and enunciates its virtues with gavel strokes:

> since . . . the observation of the regularity was . . . the *only* thing that was done, it was necessarily the *only* reason for speaking of cause and effect, it was the *sufficient* reason. The word cause, as used in everyday life, implies *nothing but* regularity of sequence, because *nothing else* is used to verify the proposition in which it occurs.[7]

But there is actually sparse basis in such examples for these impassioned claims. The truth in what Schlick says is that given a few repeated observations, one might well have sufficient reason (in the sense of quite good reason) to think that a causal connection exists between administration of the drug and remission of the disease. And this regularity *could* indeed be the only reason for so thinking, apart from a background of medical knowledge. Suppose, for example, that a spontaneous recovery often occurred that was unexplained, but only because it had gone unnoticed that in each case it had been preceded by the administration of a certain substance (for some other reason). At some point this regularity is recognized and the causal relation seen, so that the idea of a particular causal connection is eventually brought on by a repeated succession in observation (and a subsequent realization). But a medical researcher looking for a cure may well come by his causal judgment not by accumulating observations of a specific regularity at all. Normally, in fact, the case would proceed like this: there would be some substantial reason in pharmaceutical knowledge to think that a certain drug will produce a beneficial result, and trials would *confirm* that it does.

To say that the word "cause" as it would be used in connection with his example, let alone in general use, "implies *nothing but* regularity of sequence," is surely not warranted on the basis of an example of this kind. This is not to say that it "implies" anything else that is not already clear in its unanalyzed meaning. But if what the "nothing but" excludes is causal connection, one might well want to say there is more content to a "causal judgment" than accumulated regularity of succession, especially since in some cases one might even quite reasonably think after *only one observation* that a certain drug has a curative effect. There is certainly room for more to be meant by "causes" at this point in the argument.

But we soon learn from Schlick's discussion that he thinks it is *impossible* that causation could mean anything more than regularity of succession. It seems to him that if we say it is more than mere chance that E follows upon C, we must be saying that E will *always* follow upon C. The alternative to chance succession, he thinks, must be *necessary* succession, in which case an instance of E not following upon C would force one to revoke the causal connection. (Since Hume, "necessary connection" has been believed to be the "ordinary" notion as to the meaning of causation, the meaning that it has until philosophers expose this notion. I argue below that "necessary connection" is actually a philosopher's construction imposed upon causation.) But he reasons, this "always" is unverifiable, because only a finite number of instances of E follow-ing C can ever be observed, and therefore, according to the verifi-ability requirement for meaning, there can be no meaning for "cause" beyond the fact that there is a regularity in past experience which can be extrapolated into successful prediction, *if one assumes that this additional or alternative meaning would be necessary connection, which means, the absolutely inevitable succession of C by E.*

Schlick makes note of cases in which E does not follow upon C due to an intervening cause. If in one out of a hundred cases the cure fails, the physician will not decide the other ninety-nine cases were coincidences; rather, he will think that in this one case some-thing interfered with or counteracted the usual effect of cure. Though there is an exception to the regular succession, the judg-ment of causal connection is not revoked. This kind of case helps Schlick dispose of the idea that causation is necessitation. For Schlick the regularity analysis must apply to this intervening cause as well as to the original case of causation. For this analysis to hold, it must be the case that a causal connection is disclosed through an observed regular succession, which is then simply verified through repeated observations. It would seem that on the regularity view, one has to have accumulated some observations of the medicine failing to cure under similar circumstances, in order to have ac-quired the first basis for a "causal judgment." But Schlick is in serious trouble here, because this basic requirement of the regular-ity view is not fulfilled in the case of the counteracting cause. This is because the counteracting cause is *unobserved,* both at first and on repeated instances of its occurrence—even though its existence is strongly surmised after only one instance, which means that the concept of cause is already applied.

What is being called a counteracting or intervening cause here might be, for instance, that the disease symptoms in this case have a different cause, or that the disease-causing organism has a resistant strain, or that there is something about this patient's physiology that prevents the drug from working efficiently. In any case, it re-sists the regularity view for the following reasons: First, a causal judgment, forceful enough that it may well spark a special investi-gation, occurs after only one observation; second, there is no obser-vation of the counteracting cause, but only of its effect, namely, failure of cure; third, no number of occurrences of merely this same

observation—the drug is given, and no cure follows—would add to
or support an understanding of this as an instance of causality. It is
in fact its exceptionalness, far from an accumulation of like cases,
that makes a causal judgment about it so strong.

But what Schlick specifies as the observations which on his
account must give rise to our knowing and speaking of an intervening cause are not those of no cure following administration of the
drug. He refers, rather, to observations which go on "in the laboratory," rather than in the clinic, in the investigation of the intervening
cause. Does this rescue the regularity view?

Some research in the counteracting cause is carried out. It is
found to be a resistant strain of the agent of the disease; the microorganism taken from the body of the one hundredth patient is not
destroyed by the chemical in petri dishes (let's say). This "regularly
occurs," that is, a few experiments are done to make sure. This
allows the researcher to securely surmise the specific intervening
cause. However, on the regularity view the observed successions
which take place in these experiments could not be the source of a
causal judgment about the action of this cause preventing the action
of the medicine from curing the disease in the patient. These cases
of the microorganism surviving the drug could give rise at most to
a causal judgment about a process in a petri dish. The succession:
medicine given—no cure, which is the one about which there is a
judgment of a counteracting cause, takes place in a different setting,
and has quite different antecedents and consequents. So far there *is
no observed regularity* which could give rise to the causal judgment
about the clinical observation that there is an obstructive action of
one cause upon another.

In order to confirm the finding about the cause, disease organisms are taken from the body of the one hundredth patient and
injected into rats, and after they become diseased, the drug is tried
out on them. Cure does not result. The judgment of a particular
cause is confirmed. Does this arise from an observed succession,
say, those observations that are made in the course of this complex
series of actions? But this is not a causal process on anyone's analysis, but only a deliberate order in experimentation. How on the
regularity view would one make any connection between this
experiment and the clinical observations? What possible regular
succession could supply this connection? So far this view leaves out

some important judgments of causal connection. In fact, causal understanding permeates any such clinical and laboratory activity from start to finish.

To summarize, Schlick admits that a cause is inferred in a case where it is actually unobserved, a case of one cause obstructing the effect of another. His theory of causality cannot account for this. As succeeding chapters will continue to show, an identifying feature of acausalism in philosophy of science is a theory of causality that fails (in different ways) to account for the interaction of two or more causes.

There is an important assumption, one we have already met with in Schlick's article, that is traditionally put to use in arguing for the regularist view about causation: that if there is any other meaning of causation beyond reliable succession, it would have to be the absolutely invariable occurrence of the effect, given the cause. The traditional doctrine, also found in Hume, is that necessitation is a basic meaning of "causation" that one begins with, and which one would be left with if one were to try dropping the claim that regularity in observations is really the only physical meaning. For the positivist, even the claim that a certain effect *will always* occur given certain antecedent occurrences shatters against the demand for verifiability. "Will always" cannot, strictly speaking, be tested.

There are reasons for rejecting the concept of causation as necessitation or necessary connection apart from the verifiability criterion of the positivists. Immediately there is the question of what exactly it means. Is it somehow logical, or somehow physical, necessitation? What could "physical necessitation" mean other than "bringing about," or in other words, causing? If it were really necessitation, it might have to be some monstrous, unchallengeable cause, the unstoppable force. And there cannot be any *logical* necessity connecting one event with a different event. To say that a particular event or kind of event *entails* another is unsalvagable nonsense, confounding the domain of logic and meaning with that of physical actuality. What then does "necessity" mean in the context of discussions of causality everywhere in philosophy? "Under given conditions C, E is always brought about"—this is certainly unobjectionable; perhaps even "will always" can be allowed, as long as any suggestion of mathematical or deductive certainty is

avoided. But to then add "of necessity" is to import some extraneous and probably inappropriate conceptual content. "It cannot be
otherwise than that E follows C"—this thought catches us off
guard, but it has already slipped out of the proper element of a
discussion of causality into untenable claims that certain logical
constraints exist; and it is shown false as a general doctrine about
causation by the existence of counteracting causes.

Also, talk of the "necessary conditions" for an event merely
refers to the truth that such-and-such conditions always in fact sur-
round the event, that it does in fact arise out of these conditions in a
traceable way, or in other words, that these conditions are *the* physical context for it (certainly this language cannot be construed *logically*). There may be nothing in this relation between a phenomenon
and its causal context other than a potentially traceable physical
transition, along with the expectation that this physical connection
can be analyzed, to whatever level an investigation achieves, into a
continuous structure of connected events and connecting processes
(we can ignore "quantum discontinuities" for the time being); that
is, perhaps physical causation reduces only to *genetic transition*. "So
what does 'genetic transition' amount to, that is, reduce to?" My
answer, the details of which are elaborated in Part Two, is that
genetic or causal transition (a) successfully and adequately identifies causal connection; (b) is not further reducible *as such*; and (c) is
always subject to analysis in the sense that actual instances of
physical causality can be analyzed into genetic transitions along a
gradation of levels. To investigate the latter is to do physical science
in a sense that challenges contemporary assumptions, which is not
to say that such an investigation has nothing to do with philosophy,
but only that it is not a matter for critical-philosophical scholarship.
Without carrying the argument any further here, I will voice the
intuition that philosophers' quandaries about causality are partly
traceable to the persistent intrusion of the concept of necessity in
various functions.

Allowing that necessitation is not workable as the alternative
to observed regularity for an analysis of causality, the question remains, might there be some other options? Schlick is aware of this
question. He considers and rejects the logical entailment of the
effect by the cause, and then presses on to test further possibilities
for a basis for causal judgments. I will retrace this whole path in his

considerations. He first gives an example that he thinks will show the source of the notion that causation is more than regularity, whether logical entailment or something else:

> If a surgeon amputates a man's leg, he will know beforehand that the man will be one-legged afterwards. Nobody thinks that we have to wait for a long series of experiments in order to know that amputation results in the loss of a limb. We feel we "understand" the whole process and therefore know its result without having experienced it.[8]

(By "experienced" he of course means "observed" from one standpoint or another.) He spots where the idea that causation is logical necessitation might come from in this example. In fact, he points out, the logic has "nothing to do with causation," but only with what "amputation" means. The tautology in the statement, "amputation causes the loss of a limb" might create the illusion that here "cause" means logical entailment.

Then Schlick says something more, something very curious, about this example. He generously tries to trace this causal understanding to something different than regularity, something called our "comprehension of the process" (emphasis added):

> We usually believe we understand this connection, because we think we comprehend the process, say, of a saw cutting through a bone; the hard particles of the steel are *in immediate contact with* the soft particles of the bone, and the latter somehow give way to the former. Here again we have *contiguity in space and time*, which appears to flatter our imagination, but apart from that we have again nothing but a sequence of events which we have often observed to happen in a similar way and which we therefore expect to happen again.[9]

Here Schlick is again exploring possible alternatives to his view by considering the possibility that "we comprehend [causally] the process" beyond the mere discrete succession of events. His description of this process as we perhaps comprehend it is a strange microanalysis that I cannot imagine scientists giving as an expression of their understanding of it, for all their concern with mini-micro dimensions. The point is to show the process as a detailed sequence of events, with the aim of throwing us back on the regular succession analysis. He claims to understand the intellectual psychology of his opposition here: because we have this fineness of detail, so that all the gaps in the process appear to be filled in, we think we have a special comprehension of causality. But in fact, he claims, the events in the causal chain are *at most merely contiguous* and no more show connections than would any less detailed event-sequence. His argument is flawed by the fact that the event-sequence he speaks of, particles making contact with particles, is a strange contrivance.

Schlick does not live up very satisfactorily to his intention of considering, through this example, a possible meaning of "causation" that is not logical entailment but has more than that vanishing Humean identity, regularity accumulated in the observing subject. Instead of a "particle" analysis of the amputation, would not a causal understanding be better expressed by saying, "A sharp saw, wielded by a strong arm, will cut through flesh and bone"? But Schlick will want to say, as he does following the quoted passage, that this assertion would be merely a summary of past experience, whether or not one had ever witnessed or undergone the operation:[10] "For aught we know we might some day come across a bone that would resist any saw and that no human power would be able to cut in two." This is the verification criterion at its best. He speaks here of the stubborn contingency that keeps one from being able to verify the "will always." Here it serves not only to ward off the logical entailment view, but also against what he calls a "comprehension of the process" that is more than regularity. But if I were to say I understand the *mechanism* involved here, his imaginary possibility does not speak against me at all, since I understand *it* mechanically as well: in this imaginary case the bone is too hard for the saw, so that the usual mechanism cannot proceed.

It is hard to say at this point whether a detailed description of the cutting action of the saw's teeth would support a regularist view, which denies event-connections, as opposed to a productionist view, which affirms them. He does try to make this same argument with a case that is perhaps more manageable and is certainly less abrasive, the old one of a drug curing a disease. He provides a physiological sketch of the cure process, with the aim of showing it to be a mere sequence of contiguous events, and he therefore leaves out any terms of causal connection such as "affects," "acts on," "produces change in," etc., that might be found in such an account if it were given without any such philosophical aim. This caution in language makes his account sound slightly peculiar (emphasis added):

> The drug, e.g., is injected into the veins, we know that it *comes into contact with* the blood particles, we know that these *will then undergo* a certain chemical change, they will travel through the body, they will *come into contact with* a certain organ, this organ *will be changed* in a certain way, and so on.[11]

And from this series of events, perhaps contiguous, but in any case merely successive in Schlick's description, we can derive no different meaning for "causality" than we could from observing numerous instances of use of the drug followed by cure; the detailing of the process just gives more cases of regularity of succession which are known from "experiences in the laboratory."

It is important to see what Schlick is trying to get us to think when he describes this causal process as a series of distinct events, simply following upon one another. On the one hand he claims to be unveiling the source of the idea that causality can be understood as something more than regularity of succession, namely the detail and depth to which an explanatory account might proceed. But, he argues, no causal analysis can proceed beyond events in *contiguous* series; it can only find more detailed events at most contiguous to one another in space and time. Thus if we were to speak, as we well might, of the drug infusing some tissue and acting on it in such a way as to change its chemistry, we would in the final analysis only be referring to the fact that the drug moved into the tissue and that the tissue subsequently changed. "To act on" designates an illusion of connection which cannot actually be found by the analysis of a process.

A causal "chain" as he sees it is like some bricks lined up end to end, perhaps actually in contact, but capable of no interconnection. Each event in the chain, like a brick, is a quite discrete particular. "Connection" between them might at most mean that they are touching, contiguous. Either there is only one event, as when there is one long, unbroken brick, or if there are more than one, then the closest possible intimacy between them which can hold up to analysis is that of contiguity, being directly adjacent to one another. If some of the events . . .

> ...are still separated, we have to look for new events between
> them, and so on, until all the gaps are filled out and the chain has
> become perfectly continuous in space and time. But evidently we
> can *go no further*, and it would be nonsense to expect more of us.
> If we look for the causal link that links two events together, we
> cannot find anything but another event (or perhaps several).
> Whatever can be observed and shown in the causal chain will be
> the links, but it would be nonsense to look for the linkage.[12]

It is cogent to claim that whatever we find to fill in the causal
chain will be events. Whether perceptibility confers knowability is a
question I treat elsewhere. What needs investigation here is
whether there is any reason to conclude that our understanding of
successive events (in general) and their relations in space and time
in constrained to conceive them as side-by-side and perhaps contiguous. This would amount to a mosaic theory of events, according
to which individual events can only compose into a more extensive
process the way objects combine into larger objects, the way bricks
compose a house. On this theory a causal "chain" has links but no
linkage.

A consequence of the mosaic theory of events is that when
talking about a process such as the physiological action of a medicine against a disease, terms such as "affects," "brings about
a change in," "acts on," etc., which seem to speak of event-
connections, actually thereby mislead by suggesting a chimerical
connection that dissipates with correct thinking. But it seems possible at this point that the philosophical attack on the connectedness
understood in such language may be based on an uncritically accepted general theory about events.

Hume generally refers to the members of the cause-effect relation as "objects," and to this relation as one of contiguity in space
and time. The following passage summarizes his analysis of causation and contains his concluding definition of "cause":

> When I examine with the utmost accuracy those objects, which
> are commonly denominated causes and effects, I find, in consid-
> ering a single instance, that the one object is precedent and con-
> tiguous to the other; and in inlarging my view to consider sev-
> eral instances, I find only, that like objects are constantly plac'd in
> like relations of succession and contiguity.[13]

In view of Hume's classic account of causality, the mosaic theory seems well embedded in the empiricist tradition. It involves the assumption that an analysis of a causal process into its constituent events will ultimately come to a stop at a mosaic of (at best) contiguous parts. That we are not bound to this assumption might be shown if we can apply a genuinely alternative picture of event relations. (I make no claim as to whether the neo-Humean tradition is correct in its understanding of Hume; I am only investigating the more recent philosophies.)

If we are trying to conceive how in general events might compose into a causal process, and we want to supply these composing parts with genuine connections, we might try supposing that a causal process is analogous to the natural spectrum of colors, with distinct regions of color representing its constituent event-parts. Different colors in the spectrum are connected by continuous transitions. This analogy tends to support the supposition that language such as "brings about" need not mean mere side-by-side succession, but might possibly mean precisely the connection of one event with another. It shows that *connected difference* is at least a general, conceptual possibility. In other words, the example of the natural spectrum suggests that there may be alternatives to the mosaic analysis, though the analogy with events and causation is of course limited.

Suppose we make use of the analogy and try to analyze a spectrum further, just as Schlick suggests we analyze further the causal process of a drug remedying a disease. A spectrum has the obvious analysis into a series of distinct colors, so that further analysis will be interested in the transitional regions joining these successively ordered regions. Though it would normally not be the case in an actual spectrum of colors, these transitional regions might, in some complex case of spectral variation, be themselves organized spectrally. Therefore what the further analysis will find will be either (a) more distinct and connected regions (a more detailed spectrum), hence more transition points for further analysis, or (b) if the analysis has proceeded to the level at which there are no longer more or less unvarying segments (particular colors), there will be regions of continuous variation. Clearly the transitions in the spectrum resist any reduction to contiguity between mosaiclike parts. "How does this show connected difference? It only shows that lines of contiguity through the spectrum would be arbitrarily assigned." But this is precisely the point: the lines would be imposed and have no place in the description of the example. The question whether there is necessarily a contiguity of the parts of space through which the spectrum extends I will not treat at this point; this is not an example of causality or events and there is no warrant for pushing the analogy to this extent.

How might it be decided whether one or the other mode of analysis of a causal process, into discrete parts or into transitionally connected parts (as in the spectrum model), is correct, or more correct? On the side of the mosaic model, it seems one can speak about the causal examples in terms of separate, successive events, as in Schlick's description of the disease treatment process, without using causal language such as "makes," produces, "acts on." One can, in other words, leave out any mention of causal connections and still have a detailed description. But this is the case with a spectrum of colors just as it is with a mosaic of objects; one can speak of the ordered succession of distinct colors without mentioning the transitions connecting them. Thus the analogy with a spectrum would show how a discrete analysis is possible though the connections are not undiscoverable. It would allow causal connections (whatever they are) to belong to events and still show why the traditional empiricist or phenomenalist treatment is a possibility: simply because though genuine connectedness is not at all disallowed, what would thus be connected are also distinct and separately identifiable, and so can be treated as discrete by ignoring the transitions.

It would be helpful to have an example of two causally related events with their relation clearly displayed, rather than relying solely on models and analogies. A hammer driving a nail is a case of a "productive process" given by Salmon (cited in Chapter 5). Suppose we analyze a single hammer blow into two events: the motion of the hammer and the penetration of the wood by the nail—two distinct object-motions. The first can be said to bring about the second. Is the productive connection on display in the example? It is clear that if the hammer blow is so analyzed, the two events are not merely contiguous. The motion of the hammer is the only one that even occurs independently of the other; when the nail moves into the wood, the hammer moves along with it. Thus there is a region of overlap or intersection of these events in space and time, which includes the whole of the nail motion, but is only a part of the hammer motion. The terminal penetration of the wood by the nail with the hammer following it can be seen as the product of the antecedent hammer motion. This account of overlapping events is compatible with saying, "the impact of the hammer makes the nail penetrate the wood." This language is also compatible with describing the hammer blow as a single event, consisting of an antecedent process and a terminal phase.

But instead of the connected events which the ordinary causal language, taken innocently, seems to speak of, Schlick or Hume would presumable want to give this analysis: There are two merely contiguous events. First, the hammer proceeds up to the head of the nail, then, as a separate event, nail and hammer proceed together beyond this point-instant of contact, with the nail penetrating the wood. Thus nothing specially identifiable as "productive connection" is to be found at all. So far as this apparently complete analysis is concerned, there are only these two side-by-side events; "the bringing-about of one by the other" is merely a normal but misleading way of interpreting our experience and expectation.

If this traditional empiricist analysis of a casual process, caught up in the mosaic account of event relations, is possible in this example, then surely it could be opted for as an analysis of *any* series of events, with some minor forcing of the description. But it is important to recognize that whenever this is done, a decision has already been made between two alternatives: events occupying separate, at most contiguous, spatio-temporal segments; or events differentiated by transitions.

What comparative strengths do these two possibilities have? On the side of connecting transitions, this would create no obvious difficulties about the source of causal talk and of our natural understanding of causation. The mosaic view, on the other hand, would make the origin and sense of the causal language as a language of connections mysterious, perhaps leading to an explanation in terms of purely mental function, as in Hume. On the side of the mosaic view, one can say that it gives in a sense a clearer picture in that the events are unambiguously marked off from one another. An advocate of this view might critique the other possibility as follows: "Connection in the hammer example is supposed to be represented by overlap of events. But why should spatio-temporal overlap of the events necessarily mean causal connection? Perhaps it is only an indistinctness belonging to this particular analysis, so that the parts blend into one another." This indistinctness of boundary is precisely what is needed, according to the analogy with the color spectrum, to identify connection between what is differentiated. But it seems one can provide no proof, against this objection, that the overlap or interconnection between hammer motion and nail motion represents *bringing about*, causation in its fully naive suggestion. Perhaps this shows that our understanding of the happening of causation, here no more than "parts blending into one another," is more primitive than our reflections, so that at this stage of the inquiry it can only be roughly identified as overlap in regions of space and time, analogous to a Venn diagram. This resort to a flattened analysis is susceptible to the mosaic picture of events, but even so it thwarts the claim that no candidates for a physical "connection" are to be found in the hammer stroke example.

The clarity (or rather, sharpness and tidiness) which comes from picturing events as if they were discrete objects seems to have no advantage in terms of accuracy or completeness of description. And it is bought at the expense of the ordinary way of talking about the example. The analysis into hammer motion event—nail motion event discriminates cause and effect more naturally than the analysis that insists on a boundary of contiguity at an instantaneous slice of space and time.

It is obviously not conceptually impossible that events should be joined by continous transitions. Why can it not be similar to the case of a musical composition in which sections of the piece, differentiated by mood or key, are connected by a *segue* in which the music does not stop?

Finally it should be said that a natural spectrum of colors and a musical *segue* are only illustrations of possible differentiating relations, and cannot ultimately be relied upon to provide an understanding of the causal relation. What they do is throw back upon the neo-Humean the burden of showing why the causal relation must not be like these, but rather must be like bricks laid end to end. After all, the language of causality bespeaks *productive connection, the engenderment of one event by another,* in whatever sense is given in the particular case. If this were a mere rewording of the expression "causal connection" it would be unhelpful; but it is more than this: it is a concept of differentiating transition that *leaves out contiguity*. This is not an interpretation of "causation," but a provisional understanding of its primordial meaning.

Hilary Putnam objects to the claim that "causal connection" is philosophically adequate without a reflective interpretation; he says that if it were, this notion would have to be "self-identifying," which he apparently thinks is nonsensical or impossible.[14] I do not wish to deny the importance to philosophy of the question of causation, but as I see it, Putnam makes the traditional mistake of viewing the problem of causation as a problem of supplying a basic meaning to the causal relation. It is not that "causation" is an occult and unknown *meaning*; the truth is only that the primordiality of this relation resists the traditional artificial analysis that first collapses its structure to accord with the model of discreteness/contiguity and then perhaps does it a semblance of honor by talking about "necessary connection." The discussion of explanation as causal narrative has suggested that a causal sequence is a sequence involving connection through transitional process such that one event or phase in some way engenders another; causation in general may have no further precision of meaning beyond this. I propose as a provisional response to the problem of causation an abstract and general concept, to be further developed in Part Two in application to actual cases of physical causality, which is essentially contrary to the mosaic concept: differentiating transition in events. The exact way in which the parts of a causal process on this conception are identified and differentiated will in due course become a focus of interest. Part Two goes beyond the general critique of the contiguity assumption and argues that one of the main sources of problems in thinking about causation is the belief that a causal relation is *essentially* a relation of *succession*,[15] which leaves out a whole structural aspect of physical causality. For now the claim is only that solid reasons are lacking for denying causal language its native implication, a thesis opposed to a variety of disastrous reductions.

Positivists on Indeterminism

In the second part of Schlick's article, dealing with causality in "recent science," he makes use of some developments in physics that were new and dramatic at the time the article was written to make a different argument for the regularist view. He takes as his focus the apparent *indeterminism* that is observed in advanced experiments with light and other radiation.

This feature of experiments concerns the specific atomic locations at which radiation interacts with matter, or the exact timing of such interactions. There are only equations yielding *probabilities* for specific times or sites of interactions; there are not equations which predict with certainty sites or times for particular interactions, nor is it likely that such deterministic laws will be found. As a result of extensive investigation and analysis, it is generally accepted that this is a genuine indeterminism, unlike the randomness in the outcomes of a deterministic process such as coin tossing, which is an unpredictability resulting from limitations in knowledge of details about the particular occurrence. I will limit my discussion to indeterminism about atomic sites of interactions, ignoring indeterminism about times.

Schlick's analysis of the everyday notion of "causation" had reached the conclusion that all it could really mean is that certain regularities are observed which can be formulated into laws embodying predictive knowledge. In the second part of his article he claims to find "striking confirmation"[16] of this view in quantum theory (in which the probabilistic equations for indeterministic effects occur). Here is how he begins his discussion of this.

> As is well known, the quantum theory in its present form asserts that a strictly deterministic description of nature is impossible; in other words, that physics has to abandon the Principle of Causality.[17] *not the same*

The Principle of Causality is another name for determinism, that anything that happens has an antecedent cause. It is indeed true that physics has abandoned the principle insofar as it would lead a scientist to look for a cause for why an individual interaction in a radiation experiment occurred just where it did and not at any other place. The breakdown of determinism in *such points of detail in scientifically studied and measured events* is what Schlick can cogently mean in saying that a "strictly" deterministic science is no longer possible.

But then it seems rhetorically heavy-handed to call this the abandonment of determinism (let alone causality) by physics.

Schlick seems to realize this:

> ...for all ordinary purposes of science and everyday life the deterministic attitude not only remains justified but is the only one compatible with our knowledge of nature.[18]

He recognizes that determinism is not at all defunct. But in order to make his case that the discovery of indeterministic effects by atomic physics has impact on the general meaning of "causality" in such a way as to support his analysis of the concept, he does need to make the claim that because of this discovery, physics abandons "causality" as it has been traditionally applied in science, that is, unaffected by the Humean critique. For him causality as such is inseparable from the Principle of Causality, that is, from determinism. He asks, "What does physics mean when it thus denies causality [by 'abandoning' determinism]?"[19] When physics gives up on the expectation of "strict" determinism it "abandons" determinism, and furthermore, "denies" causality. These are hasty and confusing conclusions that need examination.

Schlick himself seemed to admit that despite the highly specific and minute element of unpredictability in certain experimental contexts, it remains true that for nearly all practical and explanatory purposes the world is deterministic. What therefore could be meant by saying that because of this indeterminism causality should be "denied"? If there were a general indeterminism, that is, if to some degree regularities were not reliable, were all probabilistic, or there were no regularities, it might be said that to get things right would be to deny causality. But one can avoid these murky imaginings, because there is a concrete, contextualized case of indeterminism at hand (if one adopts the generally accepted view in physics). What does the fact that certain effects feature indeterminism say about "causality" *as applied specifically to those observed events bearing this indeterministic detail?*

Assuming that a realist view about the "unobservables" of physics is sustained, then these *effects* are *produced by the kind* of transmission that is called radiation. The effects are individual localized interactions with matter whose sites in detection material only lend themselves to probabilistic prediction. "Denying causality" here could only mean determining that there is no discoverable cause for why the unit of radiation produces an effect at one specific site rather than another. What is thus stated in our "productionist" description, speaking of radiation as producing effects, is that *within a certain context of the causal production of events* there are specific features of the events, their exact locations, which appear to have no discoverable cause. According to this account, "denying causality" is a wildly inappropriate characterization. What is appropriate is to say that a certain presupposition *about causality* is being challenged: the presupposition that it must always involve a thorough determinism.[20]

The clearest point Schlick is making with his "denial of causality" is that since there are no mathematical models to predict that an interaction will occur at a particular site and not at another, and since knowledge of specific causes is denied where predictive knowledge is lacking, quantum physics at least *fits with* the claim that causality is nothing more than predictability, as Feigl said. (Surprise! Neo-Humean doctrines fit with the decisions in theoretical physics which inspired them.) Schlick writes:

> What *is* the real reason [for rejecting determinism]? None other but that it is found impossible to *predict* phenomena with perfect accuracy. Within certain well-defined limits it is impossible to construct functions that can be used for extrapolation. [This] proves that the physicist in his actual proceeding has adopted just that view of causality which we have been advocating.[21]

And later:

> We found that recent science confirmed the view that causality must be understood as meaning "possibility of extrapolation," because we found that this was exactly the sense in which the word is used in quantum physics.[22]

In an era of physics in which theory is encompassed by predictive algorithm, it is only natural that "possibility of extrapolation" came to be the technical meaning of "causality." But the productionist point of view has kept us wary of the idea that the discovery of indeterminism by quantum physics meant the denial and abandonment of causality (on the naive understanding). It was noted above that a productionist reading of this case of indeterminism (if that is what it is) is also possible. Schlick's case here amounts to the trivial point that where no further "functions" for prediction can be "constructed," this is accurately referred to as the breakdown of determinism—thus according to science causation *might* reduce to prediction.

What is true of physics is the absence of the possibility of deterministic, and not only probabilistic, prediction in certain cases. But if "possibility of extrapolation" means the existence of useful predictive mathematics, then this possibility did not fail physics. Quantum theory is a brilliantly successful mathematical equipment, with many important applications, though its use involves an element of probability in prediction. So there is really no failure of the "possibility of extrapolation," but instead highly successful "extrapolation" which is not fully deterministic. Of course physics did not suddenly find itself in the midst of senselessness and absence of regularity, rather, there is no lack of quantifiable and utilizable regularity.

As mentioned, in Schlick's thinking causality and determinism are inseparably wed, in contrast to my suggestion above that a challenge to this presupposition is present in the facts:

> Lack of determination means pure chance, randomness; the alternative "either determination or chance" is a logical one, there is no escape from it, no third possibility.[23]

Note the tone of "necessitation" that lurks in this concept of "determination." The claim here implies that there is no room for *indeterministic causality*, for a generative process with certain statistical outcomes, which is how these cases of indeterminism would be understood within our commonsense assumption that the observed interactions are caused or produced (by radiation). But it should be recalled that Schlick thinks that the only viable meaning for "causality" is "consistently observed succession," in other words, what can be formulated into laws by and for science. And there are probabilistic laws at the very center of this discussion. So really Schlick should recognize a meaning for probabilistic or statistical causality, even on his analysis of the concept—namely, the statistical regularities designated by the laws of quantum theory.

Schlick cannot see that there may be certain *causal processes* (however this is understood) with detailed facts about their outcomes following only statistical laws. He does not recognize the possibility of a statistical regularity occurring within a causal and otherwise deterministic context. Even more sophisticated contemporary discussions still proceed as if firm bonds join causation not only with determinism but also with necessitation.[24] (A different view might allow an understanding of nature as necessitously productive that is not given over to an outdated determinism.) This study is working toward a positive conception of specifically indeterministic causal structure.

Once the possibility of statistical regularity is admitted, there can still be two views of causality: that it means nothing but regularity in observation (in these cases statistical), or that there are *connecting* causal processes, with some purely statistical outcomes, underlying these regularities. The existence of indeterminism does not argue for one or the other of these positions. There are evidently such things as indeterministic regularities, formulated into the laws of quantum theory, and their explanations may be physical, causal processes (on a productionist understanding) with indeterministic features. If there are no clear signs of an available description of such a physical process which maintains the genetic continuity in events required by narrative causality, this is not already a logical constraint against such a description. It might only mean that a new stage of physical inquiry is in the offing that would seek to disclose and understand the actual physical nature of transitions to indeterministic effects.

Philipp Frank provides more circumspect discussions of the relations between causality, predictive mathematics, and indeterminism, both in a work already cited,[25] and in a contribution to the ambitious positivist tract *International Encyclopedia of Unified Science*.[26] For Frank the positivist outlook, a combination of the regularity view of causation and a tentative antirealism about scientifically "postulated" entities, is taken as a matter of course. Frank states the "law of causality" this way: If all the variables in the initial state of a system are known, the values of these variables in a subsequent state of the system can be predicted.[27] Based on this definition, he claims that "the question of whether the law of causality has survived in twentieth-century physics cannot be answered by a simple 'yes' or 'no.' "[28] His reasoning is based on the fact that, on the one hand, physicists talk about the computational devices of quantum mechanics as though they referred to a set of mysterious physical variables of a system which evolves deterministically, though on the other hand, as regards observations these devices predict only the probabilities for particular sites of individual interactions. The predictive mathematics, on a predominant interpretation by physicists, designates a "probability wave" which, if it can be interpreted as a physical condition antecedent to measurement (a possibility I dispute), would be said to evolve deterministically.

In the earlier work, the contribution to the *Encyclopedia*, Frank notes that no definite conclusion regarding causality can be drawn from quantum physicists call "state variables." Since they do not appear to correspond to Newtonian masses, that is, essentially inert objects present in space and subject to mechanical interaction, he could see no clear way to compare quantum mechanics with deterministic Newtonian mechanics in order to decide whether the "strictly" applied law of causality—determinism—still applies in this domain. In the later work he also claims, as quoted above, that this question has an ambiguous answer.

The approach of the present work is not inclined to ascribe any physical interpretation to "the evolution of a probability wave," and hence gives this conception no voice in favor of "strict causality" in physical events. Instead it views the equations of quantum mechanics (together with these "state variable" models) in the same instrumentalistic way as other mathematical theory. This avoids any confusion and ambiguity about whether there is indeterminism in the detailed effects of radiation. The possibility is allowed that indeterministic events covered by probabilistic laws can occur within a context of generative physical process.

My suggestion about indeterminism could be put in the terms of Frank's formulation of the law of causality: the law would be amended to allow for the possibility of an ingredient of intrinsically unpredictable detail in effects, that is, of specific outcomes which *do not have* causal traces extending back into the given antecedent state of the system, within an otherwise deterministic context. To acknowledge indeterminism in the atomic domain and, further, to suppose that indeterminism might surprise our prejudices and prove possible to incorporate within narrative causal explanation, goes along with the intention to pursue physical explanation beyond the confines of mechanism. The specifically indeterministic aspect of causal structure is described in detail in Chapter 8, second and third sections.

Hobart's Defense of Hume

Another work whose conclusions Feigl claimed to be representing with his pronouncement about the purification of the concept of causality is R. E. Hobart's article, "Hume without Skepticism."[29] This article brings up for criticism an idiosyncratic theory of causality that seeks to overcome the Humean reduction. This is the view of A. N. Whitehead that "causal efficacy" is something like a primitive organic feeling that underlies all our experience. It is not an a priori condition of perception, but a form of perception, though different from "sense impressions."

Hobart's discussion focuses on a text of Whitehead's that disputes Hume on causality. The expression "presentational immediacy" is Whitehead's term of art for the usual "sense data" on which, according to empiricist philosophy, all knowledge stands. In the quoted text Whitehead presents his idea that the feeling of causal efficacy is a different mode of perception, one overlooked by Hume.

> In the dark, the electric light is suddenly turned on and the man's eyes blink. . . . The philosophy of organism [Whitehead's philosophy] accepts the man's statement, that the flash *made* him blink . . . [Hume] first points out that in the mode of presentational immediacy there is no percept of the flash *making* the man blink . . . Hume refuses to admit the man's protestation, that the compulsion to blink is just what he did feel. The refusal is based on the dogma that all percepts are in the mode of presentational immediacy.[30]

Hobart, defending Hume and positivistic empiricism, disputes Whitehead's alternative view as it is presented through this example. Hobart's subsequent defense of Hume against Whitehead is highly turgid, but his clearer thoughts can be found scattered in the text. He tried to argue that Hume . . .

> . . . is not denying causation or derivation, or what we mean by the word "made," he is analyzing it . . . he says that the real production of one event by another, in the common human sense of these words, may be proved to consist in what he specifies.[31]

We do not begin to learn what this is until the next page:

Hume says that the flash makes us blink in that, given the circumstances, the blinking does inescapably follow upon it; that to
make a thing happen means to occasion it, or be the occasion of it,
unconditionally—without chance of another sequel—and could
not intelligibly mean anything else. The flash makes us blink,
that is, when it occurs we blink forthwith as a matter of course
and cannot help blinking. . . . The flash calls the blinking
into existence; that is, when it calls, the blinking comes into
existence.[32]

This is a Herculean effort to portray both pure necessity and
pure regularity of succession, and at the same time be claiming to
explicate our ordinary idea of "to make happen." Hobart is correctly representing Hume as thinking that causation in its ordinary
meaning is necessitation or "necessary connection," and also we
again see the caution about language which tries to eliminate causal
action terms in order to suggest bare succession. Real murkiness
and confusion results here from a deliberate warping of language.

Hobart does not give Whitehead's example an accurate or adequate treatment. For one thing, he ignores Whitehead's point
about the feeling of compulsion as a candidate for a "percept."
Leaving aside this claim of Whitehead's, this example of causation,
if that is what it is, is clearly not the necessitation of one event by
another. In this example, when the man says, "the flash made me
blink," or "it caused me to blink," or "I felt the compulsion to
blink," he need not mean that it was impossible for him not to have
blinked. Suppose a friend of his had a camera, and wanted to obtain a candid shot of the man groping about in the dark—an artistic
technique, let us say. The friend says, "Now the flash will surprise
you, but whatever you do, do not blink." On Hobart's view, either
the man cannot comply with his friend's request, or if he can, then
the usual mechanism, producing the compulsion, is now somehow
gone, so that there is nothing for the man to overcome.

The flash-blink mechanism is a kind of reflex, or near-reflex. A
more accessible example might be the knee-tap reflex, which one
might be able to overcome, though one would certainly have to
practice (suppose one were trying hard to obtain a medical excuse
from a doctor). If one could succeed in holding one's leg rigid, this
would be to struggle against a *felt* impulse to jerk one's knee (one

would know it if the doctor hit the wrong place), not to somehow cancel or nullify the impulse. If the compulsion to blink can be overcome, then there can be the flash, yet not this particular effect of blinking. To overcome the impulse or compulsion to blink does not, however, *eliminate* the causal action of the flash. That one has to exert control to prevent the effect shows that something happens to one, something arising directly out of the light flash.

It is important to notice what it is about this discussion of causal examples that suggests abandoning the usual empiricist approach. Notably, these reflex or reflexlike examples encourage the avoidance of thinking in terms of necessity. And the discussion shows how, though it is always possible to give the neo-Humean account in terms of mere regular succession, this account might well be leaving something out of these cases. Making sense of these examples favors the un-"purified" understanding of causal language and its implication of event connections—for a productionist view of causality. What is conspicuous in the flash-blink example, and I will have occasion to bring this up again, is the possibility that a certain causal process might in principle have one of its usual effects *not happen*, and still be the same particular process, the same otherwise causally efficacious actual occurrence, that it normally is. Necessity of succession is thus ruled out, and causality is attributable to the case even when *the usual succession does not occur.* I do not use the example to support Whitehead's theory about our knowledge of causality (which I will make no further attempt to explicate), but as an illustration of interacting causes, which, as we have seen and will continue to see, in a class of cases about which the acausalist tradition is highly vulnerable.

Is "Productionism" Anthropomorphism?

A type of objection in the Humean tradition to my proffering a "productionist" alternative to the regularist view must be addressed. The objection may be either that this word says nothing more than that whatever is designated a "cause" has the *power* to produce something, or that "production" is derived from the "feeling" of action or muscular exertion; in either case, the objector would claim that the ordinary language of causation is anthropomorphic.[33]

My answer is that "production" does not necessarily imply an *attribute of power had by something,* nor is it necessarily derived from human action.

One might conceivably say that a certain drug has a power to cure a certain disease (as a loose figure of speech). But when *antecedent events* are designated "causal" and adduced in explanation, namely, the mechanical flows and interactions of chemicals leading up to recovery, then neither this event-sequence as a whole nor any of its part-events is anything about which one would say it has a power. An antecedent causal event is in any case not the kind of thing that can "have a power"; the relevant "attribute" is simply the fact of *production.* The causal event produces by *happening.* How and when could "it has a power" be applied to it? Causality in events is not a priori a matter of latent properties of causative entities.

Efficacy through action by humans and animals is only a special case of "production." This word is really just part of a complex of more-or-less interchangeable causal expressions, such as "bringing about," "making happen," and "giving rise to." As in Hobart's usage, "production" can be used to speak in the indefinite case about a causal process or sequence of events. In my view, causal language just bespeaks the fact that transitional connections of a certain kind exist joining events, so that human action would be only a particular context for such connections. "Power," on the other hand, is most likely the projection of human skills, capacities, and capabilities upon a case of causality—and is sometimes used in a loose fashion as a stand-in for the identification of a cause (as in "the power of action at a distance"). The next chapter contains my own critique of the contemporary use of "powers" and similar terms to fill in the knowledge-lacuna of generative processes in physics, in which I concur with the Humeans that *these* terms are fatally anthropomorphic in such contexts.

Conclusions

I hope to have demonstrated in this chapter that the philosophical conclusions about causality framed by Humean traditions are not without a viable—and generally better—alternative. The two main props of the neo-Humean view are not sufficient to deny that causal *connections* belong to the actual structure of natural events. One of the props is the idea that causation, if it means more than regularity of succession, must mean some kind of necessitation. This had very little basis in terms of what can be clearly and cogently meant by "necessitation," and in any case, this alternative meaning of "cause" was defeated by counterexamples. The necessitation concept was used by Schlick as a foil for arguments toward the regularity concept, in that once set up as the only other option, it fell easily. Hobart tried to fuse the necessitation concept together with the reduction to discrete succession. But a cause neither necessitates an effect nor is a necessary condition for it in any philosophically compelling sense. If causation has an essence, it is *active differentiating transition* in which events or stages of a process arise out of other events or stages in a sense not further specified for the general case.[34]

The other prop is the assumption that the analysis of any process into its parts can only turn up a disconnected, at most contiguous, series of events. There is the alternative to this picture that a causal process may be somehow analogous to a spectrum of variation, in which case though the process has distinct parts, to analyze these successive parts as discrete would be incomplete; there are also graduated transitions. A look at the causal example of a hammer blow to a nail showed that an analysis into contiguous events is neither the only option nor the best analysis. It should be noted that while there is certainly the general possibility of continuous transition joining events, and intuitively this might indeed take the form of "causally connected difference," the study has not so far addressed the question whether there is an essential spatial and temporal background of any causal succession that reduces to contiguity or discreteness in its relations (though such a claim smacks of artificiality stemming from the broad utility of abstract concepts).

My claim for this chapter is that the viability of productionistic explanation is upheld against the general philosophical critique of causal connection. The traditional competing analyses of causality, whether they conceive connection as having something to do with necessity or attack connection as such, are probably missteps of thought. The second leaves no place for genetic transition between events, while the first misconceives and entirely misses this essential structure. Certainly, in light of their association with philosophical quandaries, and because of the central importance of an accurate and coherent concept of causality for physical ontology, adoption of any of these would have little to recommend it. Our mood at this point is, Enough of these notions! One might think that if a breakthrough could be made in the disclosure of some fundamental *cases* of physical causality in the domain of physics, this might shed light on the general question as to the status and significance of the causal relation in and for knowledge of nature. Thinking about it carefully, one should be heartened by the fact that mechanistic models *both* fail to describe these crucial cases *and* have traditionally failed to elucidate causation. Part Two explains how this whole complex question is constructively resolved not at a purely philosophical level, by pasting a concept onto the causal relation, but at the level of physical explanation, by disclosing the ultimate structure of actual physical causality.

4

Causal Realist Projects, I

THIS CHAPTER AND THE NEXT are concerned with a diverse group of contemporary philosophers who may be called "causal realists," though I do not think it can be said that their philosophical positions amount to what I have labeled "productionism," nor do they move outside theory of science and seek any new scientific explanations. Their goals, nevertheless, include one or more of the following: (a) to show that the Humean account of causality is incomplete; (b) to defend the physical reality of at least some unseen generative factors in physics; (c) to furnish the latter with some kind of general, positive characterization.

My interest in these writers lies in showing how within the standard contemporary outlook attempts to think and speak constructively about physical causality in physics lead only to new forms of acausalism, even when the philosophers adopt positions opposed to traditional acausalist conclusions such as the Humean analysis of causality or entity-antirealism. This phenomenon is one effect of a conventional assumption that philosophy of science is limited to theory of science and can do nothing more in the course of its concern with science than serve to clarify or at most interpret concepts, laws, models, and theories already in use by science, and has no options for autonomous exertion toward the development of explanatory concepts and strategies. As a consequence of this assumption causal realism suffers from a lack of bearings as counternihilism. What is shown in these two chapters is that to carry causal realism beyond a purely critical approach within and toward theory of science is to immediately encounter the ontological lacunas of physics, and given the assumption of a radical separation in the roles of philosophy and science, the philosopher cannot legitimately proceed further. But some causal realists do succeed within the critical approach at overcoming certain standard philosophical determinations regarding causality and physical reality.

The causal realist arguments to be studied are divided into two categories. Those that propose general concepts under which to subsume causal occurrences outside the sphere of direct observation in an attempt to in some way further their understanding, I will name "constructive realisms." Where philosophers steer clear of such substantive suggestions as to ways to talk and think about these occurrences, and limit themselves to defending their physical reality in unambiguous opposition to antirealism, I will call the arguments "critical realism." Nancy Cartwright gives both kinds of arguments, and they will be treated separately.

One mode of constructive realism that is discussed in both chapters is the popular analysis of the physical field as in one way or another consisting of "tendencies," "propensities," or "capacities." The discussion makes clear that whatever value strategies of this type may have, their results cannot amount to concrete identification and description of actual events, as is necessary if the aim is ultimately, as I think it is, to answer a need for some fundamental causal explanations. All contemporary *constructive* realisms labor under some form of this same limitation: without a fore-conception of the subjects of their "realism" as physical, causal occurrences, they lack a necessary orientation. Much of this results from merely taking over from technicalized physics the attitudes to be adopted toward the fact that the nature of these occurrences so far thoroughly resists a basic comprehension. Causal realism so far remains a sort of satellite system around a tradition of acausalism, kept in motion by a vague antinihilist yearning.

This problem of conceiving field events as something less than actual does not occur with the critical realist focus in texts of Nancy Cartwright[1] and Ian Hacking.[2] These arguments claim that the life-activity of experimental and engineering science, working with the "entities" in question, involves the recognition that modern experiments and technology *detect* and *put to use* "concrete causal processes,"[3] which are named and differentiated in the familiar "particle" terminology of physics. The critical realist claim is that antirealist arguments cannot stand up to the preepistemological realism of a working physicist in direct encounter with electrons, protons, and the like. Cartwright and Hacking show that causal attributions framed in the "particle" terminology arise from the context of experimental activity, rather than being explanatory constructions or "theoretical entities" (though Cartwright retains this expression and entertains her own brand of antirealist skepticism).

The critical realist strategy of drawing attention to the activities of applied physics is at some points a refreshing change of perspective away from the traditional fixation on mathematical theory (laws). Cartwright's critique of entity-antirealisms that are based on consideration of theories leads her into attacks on what can be called "theory-realism"—roughly, the claim that the mathematical predictive models that physics invents and adopts, which are conventionally thought to be elemental "descriptions" of the world, are also "true," that is, describe the world accurately and correctly. Her theory-antirealism takes the form succinctly expressed by the title of her book, *How the Laws of Physics Lie*, and it recognizes in its own way the truth of instrumentalism. Her claim about the absence of "facticity" or "truth" on the part of certain laws turns out, at least in the case I will presently examine, not to be only a skepticism that somehow attaches to a formula, but to be an antirealism about the *forces* whose quantities certain laws specify as a function of other measured quantities.

Whether certain laws are true or false, or whether this is even an appropriate question considering the kind of thing a law is, is not an immediate concern here. In the first section I will consider one of Cartwright's arguments for the "falsity" of certain laws, because of its significance for questions about causality and entity-realism. It occurs in her discussion of the composition or interaction of causes. Her claims about an example of such a circumstance bring on an instructive dispute with Rom Harré.

Composition of Causes

There is an inverse-square law quantifying the gravitational force exerted by bodies upon each other, and there is a corresponding inverse-square law for the force of electrical attraction. The "field" of any object always has some element, perhaps immeasurable, of each of these "field" components. In cases such as the moon's attraction toward the earth, the electrical force is negligible, and in the case of electrons and protons in atoms, gravity is negligible. But for the absolutely precise quantification of the actual force exerted by any body upon another, the two laws would have to be applied in combination. It seems therefore that one can never measure the *pure* forces of gravity and electrical attraction individually at work.

Cartwright concludes from this that these two laws taken separately do not designate "real, occurrent forces."[4] For her a force is something that really occurs only with "the actual behavior of objects."[5] She thus construes "force" as, in Rom Harré's words, "an alternative term for mass acceleration."[6] If an object is accelerated, there is a force, and the precise measurement of acceleration is the precise measurement of the force. Taking the observations of gravitational and electrical forces to ultimate exactitude, and considering exclusively these observable motions, each law by its term "force" refers only to an *ideal* acceleration of a body in which only this force and no other is operating. Given this approach, what is thus referred to in the individual law is nonoccurrent, that is, it is merely a construct of the algorithmic procedure. It is very important for understanding the course of Cartwright's thinking to see that, for her, (observable) motion is the *sole reality* with which the concept of force connects.

She expresses her claim as a difference with Lewis Creary. Creary thinks component forces are real as causal influences (transmitted, in some way, from one object to another), and that two such physical influences can have as their combined effect an observed, resultant motion.[7] But Cartwright contends that there is only a real "resultant" force, given by an actual motion, and quantified by the vector addition of two functions whose individual terms of "force" signify... well, perhaps *something*. What might it be? She suggests a possible answer, though it is one with which she herself is not in the end satisfied: perhaps the component forces are the "causal powers that bodies have."[8] Here she attempts, for the sake of argument, to make use of a distinction which she acknowledges to have been declared illicit by Hume: that between power and the exercize of it. Powers "exercized" (she suggests) might be the "occurrent" forces shown in mass accelerations, while the forces individually signified by the laws of gravitational and electrical attraction are the "unexercized" powers had by bodies.

"Unexercized" must mean that such a power is an unactualized potential. But then how could it be something actively contributing to a resultant force? Cartwright claims that this suggestion about the powers of bodies "makes sense of the story about vector addition."[9] But potentials cannot themselves be the realities represented by vectors which are added into an actually occurring force. An actualization of a potential would have to be an interruption of the latent condition, not an example of it. What occurrence would the *addition* of the potentials represent; what would "composition" into a "resultant" refer to? With merely an unexercized power, *nothing* occurs; there is no individual action, hence no *inter*action, of causes. The resultant occurrence would have to arise spontaneously from the combined latency of inactive potentials, rather than being *generated* at all. This is not to give a causal interpretation to individual laws, nor would it seem to be helpful to the general aims of causal realism.

But this failure of the "powers" interpretation serves Cartwright's argument that laws "lie." To explain a physical force by a "power" has a notoriously occult feeling, and Cartwright reasons that to say that a law "describes" only a power had by an object which is not put to effect is really to say that the law "describes" no actual occurrence. The procedure of adding vectors to obtain a resultant force, the "vector addition law," however, does "describe reality," since the force yielded by this compound algorithm "really occurs" as the motion of a body.

As mentioned, in this argument she is taking issue with Creary's "causal influences," which, on his view, are evidenced in the force of gravity and the force of electrical attraction. Creary gives a commonsense reading of physics, at least to this extent. But Cartwright thinks that a realist about a particular law—one who holds that the particular force really occurs as a particular mode of "field"—has the burden of pointing to the specific "effect" that the law "produces," by which she means an *exact*, observable prediction derivable from it alone, a criterion not met by the individual laws of these forces. She says that Creary's influences belong to a class of suggestions that tries to keep the separate causal laws in something like their original form, and simultaneously to defend their facticity by postulating some intermediate effect which they produce, such as a force due to gravity, a gravitational potential, or a field.[10]

And later:

Creary's influences seem to me just to be shadow occurrences which stand in for the effect we would like to see but cannot in fact find.[11]

She thinks gravitational and electrical forces (in the sense of "fields," or interspatial background occurrences) are stand-in hypotheses with no concrete evidence. But her antirealism about these influences does not arise from traditional epistemology. She does not disbelieve in them because one could never in principle watch such an influence, whatever it is, happening, could never perceptually "catch it in the act" of transmission. In fact, for her *some* "intermediate influences" or "theoretical entities" are evidently real things:

I see no reason to think that these intermediate influences can always be found. I am not opposed to them because of any general objection to theoretical entities, but rather because I think every new theoretical entity which is admitted should be grounded in experimentation, which shows up its causal structure in detail.[12]

Her position seems close to the practice in physics of considering gravity radiation (the gravity "wave" or "particle") hypothetical, strictly speaking, until such time as it is experimentally detected in the same manner that other forms of propagating radiation ("particles") are detected. Her position is that a force should not be believed in merely because it is a term in a law quantifying it under given conditions. In the end she does not seem happy with a realism about laws based on causal powers. Thus she ends up with a kind of antirealism about certain individual laws, as the title of her book indicates, and this turns out to mean an antirealism about the specific generative processes giving rise to the forces mentioned in the laws. And from the point of view of particle physics, one might take her to be skeptical about the graviton. The question remains open how exactly her reservation is related to a vestigal scientific agnosticism about whether such a cause would show itself in a different experimental context yet to be produced. If Cartwright's view is a realism about radiation or "particles" and an antirealism or skepticism about fields, this would fly in the face of the apparent fact that a unity of reference underlies these terms: they seem to be alternative names for a single natural class, though it includes a wide spectrum of entities or processes that generate a variety of effects under a variety of conditions.

Rom Harré takes on Cartwright and makes the opposite claim about composition of causes.[13] For him the composite forces are realities. He wants to say they are real in that they are "tendencies," "propensities," or "dispositions," which only manifest themselves insofar as there are no countervailing "tendencies" which nullify their effects. By this, he thinks, an actual effect (meaning an observed effect) can be said to be the result of interacting influences— the tendencies—yet neither influence has just the observable effect it would have if it acted alone, just as a tendency may be not manifested or be only partially manifested. This "tendency" analysis is popular with philosophers of science and originated with physicists themselves,[14] but from the point of view of this work, though "tendency" may be in some way a useful designation, it has intrinsic limitations when used in a constructive realism.

Harré's counterargument based on the "tendency" characterization of individual forces does not quite manage to engage Cartwright's argument. Either a tendency is activated or it is not. Cartwright's argument would apply the same way to tendencies as to Creary's influences: even if what are thought to be interacting are tendencies, one prevailing in most cases over another, her argument is that neither force *as quantified in the quantitative law* is activated in any real case (and as we will see, Harré essentially shares this view of the ontological status of the law). The component force as a tendency would still be no real occurrence. Can a tendency ever be a real occurrence? The point I will try to make presently, and once again in the following chapter, is that whatever the practical linguistic utility of these terms may be, "tendency" and the like cannot define the conceptual strategy of a causal ontology for contemporary physics. There is only a stand-in notion here, a sort of safe harbor or resting place for thought.

Ontology of Latent Properties

Harré realizes that a tendency by itself is not a thing, but can only be an attribute of a thing:

> Tendencies are the dispositions of individuals. *This* cannonball has a tendency to fall, *this* electron has a tendency to accelerate toward an electrically charged plate, and so on.[15]

This sounds a lot like Cartwright's suggestion about causal powers. Harré claims that tendencies are what are "described" by laws, and that the tendencies are "dispositional" properties of objects. He agrees with Cartwright that individual laws do not "describe" actual events, but he disagrees with her in that he thinks her antirealism about forces concluded from this

> is the fallacy of 'actualism.' It prejudges the way that the laws of nature engage physical reality. According to the arguments of this chapter, laws of nature do not describe sequences of actual events, but rather the tendencies that produce them. Tendencies are almost never exactly realized because of countervailing tendencies.[16]

Tendencies, for Harré, are what allow one to be a realist about laws of nature—that is, about the terms of "force" that occur in them. And, from the quotation preceding the last, the tendencies are properties of the objects which exhibit motion due to forces. But from his way of speaking in this last quotation, it is as though tendencies are to be understood as active, productive, *yet somehow nonactual* causal factors which bring about actual events—presumably observable motions. He wants to advocate a "modest" realism. But about these "modestly real" things, the forces of attraction signified in laws (and apparent in phenomena): are they latent powers of objects, or are they actual productive factors, that is, physical events? "Modest realism" seems to designate an unresolved ambivalence.

"This cannonball has a tendency to fall." What could this be saying about gravity? The cannonball will fall if we let go of it. Can we say it has a tendency to fall when we let go of it? No. When we let go of it, it falls. If we do not let go of it, it weighs quite heavy in our hand. Is its weight its tendency to fall? Weight is mentioned by

Harré as an example of a tendency.[17] But why call it a tendency? It simply weighs this much in the earth's gravitational field. Suppose it is floating in outer space away from other objects, and is thus weightless. In this case the "tendency" would be unactivated; like all tendencies, it only manifests itself under certain conditions. But now weight appears to be a manifestation, not a tendency. "It has a tendency," where this refers to a definite property of the object, suggests that it occasionally exhibits weight even where there are no other objects to fall toward.

Suppose the cannonball is on display in a science exhibit. It is atop a slender iron pedestal. A guide explains to onlookers, "Here we have an illustration of the force of gravity, in that this cannonball has a tendency to fall."

Some onlookers are puzzled. "Is something wrong with the pedestal?" "What are we supposed to be looking at?" One of them thinks hard and asks, "Is the pedestal deliberately designed so that the cannonball tends to fall off and illustrate the force of gravity?"

"No," the guide patiently explains, "the exhibit is that this cannonball, just as you see it, has a tendency downward."

"Oh. You mean the thing *weighs* quite a bit," says one very bright onlooker.

"I believe that," says another. "And we had better be careful standing around it if it has a tendency to fall off."

Does anything warrant saying that the electron and the cannonball have tendencies, just because they respond to certain environmental conditions? They *do* respond, if only by a minute stress or deformation when they are blocked in one way or another from actually moving; they do not *tend to* respond. "The quantitative law signifies a tendency for the occurrence of a given measured quantity which may be, and always is to some extent, modified by a different tendency." But what could this be saying other than that one physical cause partially counteracts another physical cause? How could a tendency itself be a physical cause at all, let alone one that acts "regularly and constantly" (Clarke)? Or does the object with the tendency "wait" for a certain appropriate circumstance, and then express this particular tendency, that is, to the degree that exactly corresponds to those conditions? If "tendency" retains any of its original meaning, does it not lead down a path of anthropomorphism or teleology?

Could one say that the cannon ball on the pedestal represents a power in check? A tendency in check? The "in check" implies an opposing force, so that "power" and "tendency" become synonyms for "force." In this usage they become, in other words, not latent properties, but active and actual occurrences, "exerted" forces.

"Tendency" (as a latent property) does not find a fitting application in forces of attraction (or repulsion), because no one has ever known a case in which, for example, two masses failed to exert a force on each other gravitationally in the expected way (according to the law of gravity), regardless of other forces at work that may determine what motion takes place or does not take place. The cannonball can fall off things, or be dropped, and it has measurable weight. This is all that can be meant, unless one is resorting to teleology, appropriating a meaning from the tendencies of persons, which is unsuitable for basic reasons.

Suppose then that "tendency" refers not to properties of objects, but to properties of unobservable transmissions through the space between objects, or in other words that it characterizes the *field*, as might be expected of a venturesome contemporary realism (and Harré gives some reason to think he does intend this). This is the way that the philosophical physicists Werner Heisenberg and Henry Stapp have proposed to use the terms "*potentia*" and "tendency." Can "tendency" and similar terms play some role in a concept of unobservable causes in physics?

We have already heard Harré admit that there are not tendencies by themselves, but that there needs to be something which has them. The next step, then, is to try thinking of a tendency as an attribute of particular causal events, rather than of objects. A problem with this is already apparent. Suppose there is an event or process that brings about something observed. What could be gained by way of an explanation of the observation, by supplying this event or process with a tendency? Is not its causal action all that needs to be described in the explanation? If it has a tendency to do just the thing causally attributed to it, then it is the *activation* of the tendency that is the causal-explanatory event. Citing *this* causal event would be what explains. What role could the tendency as a latent property play in the explanation? (One never quite knows whether "tendency" and the like are intended as ontological designators, or whether such terms voice a fundamentally nihilist outlook that blends Bohr with Heisenberg and views the field as pure potentiality.)

In any case, it is difficult to conceive how a latent property or potentiality can even be attributed to an event (as opposed to an object) in the first place. What stage or part of the event is the latent property? If it is some stage or part of an event, then it is itself an occurrence, not merely a potential for an occurrence.

If a cause is a tendency, then (roughly) at some times a certain thing will be brought about, and at other (comparable) times it will not. This is how the concept of tendency seems to serve Harré: sometimes gravity is active, that is, it "prevails" in producing motion, while sometimes a different force is responsible for the motion, and this is not the motion that gravity would produce in this circumstance. Obviously there is nothing capricious about the effects: the force whose effect is not the overriding one is thought of as lying latent, awaiting a different sort of circumstance in which it *would be* activated or manifested. However, if there is only an unactivated tendency, then there is *no causal event at all,* just where Harré wanted to claim that in the example taken from Cartwright there are two "modestly real" interacting causes producing a resultant force. Cartwright and Harré were both motivated to talk about powers and tendencies, due to consciously or unconsciously confining the class of real, active events to observable motions. I will again take up the critique of this procedure for thinking about interacting causes, after bringing to light a different source of "tendency"-type analysis.

In a paper coauthored with John Dupré,[18] Cartwright makes a claim for causal "capacities," an idea which seems indistinguishable from causal powers, except that the "capacities" seem to be attributed to physical events rather than to objects. This paper is an earnest and creative attempt to establish the reality of concrete physical causality in physics in a way that is directed against the Humean reduction of causality to regular succession. "Capacities" are what there is in addition to regularity in observation:

> The Humean tradition downplays capacities, and conceives of them as no more than misleading ways of referring to lawlike regularities. We want to reverse this idea: it is better to think of lawlike regularities as misleading ways of referring to the exercize of capacities. If we try to tailor our causal claims to

match the regularities we see in nature, we will miss a good deal of the causal structure.[19]

One of these assertions calls for criticism right off. Lawlike regularities are physical facts; they are not ways of referring, misleading or not. A *law* might be said to refer to something, namely, a regularity. But what could be misleading about it? It is just a formulation of a factual relationship found by observation.

The author's basic aim is one that others have also taken up in their own ways: to supplement the account of causality with what Hume left out. It was seen in Chapter 3, fourth section that Whitehead tried to specify what Hume left out by talking about a different mode of perception. Rom Harré would fill in the causality picture at one level of explanation with causal mechanisms, and at another level (beyond the observable) with powers, tendencies, or dispositions. The ideas of Cartwright, Dupré, and Harré in this regard are attempts at counter-Humean causal realism about physics. I too have made such an attempt: my analysis has indicated that the Humean (or at least the neo-Humean) view leaves out event-connections due to the unwarranted presupposition that events in succession necessarily have the spatial and temporal character of discrete objects in a row. A claim of this chapter is that there is no reason whatever to doubt the reality of physical processes underlying the physical "regularities."

What is a "capacity" for Dupré and Cartwright? Ordinarily one would say that *something has* a capacity of some sort. Normally, in fact, the capacity would be attributed to a person. The definition of "capacity" in *Webster's Dictionary* that is relevant here is "legal qualification, power, or fitness." One might, though, try to think of a capacity as an attribute of material things; for example, a quantity of explosive might have the capacity to move a certain mass of rock if skillfully placed (though this may stretch the term), or an acre of forest might have a certain capacity for atmospheric restoration. The intent does lie somewhere along these lines:

> A basic assumption of this paper is that things or events have causal capacities: in virtue of the properties they possess, they have the power to bring about other events or states.[20]

But a capacity, for Dupré and Cartwright, is not a simple property of things or events, but is rather a *property of a property*. They characterize the Humean view as follows:

> Causality is supposed to be a relation which holds between events, a relation which holds in virtue of the empirically distinguishable properties that events have; the relation consists in, or at least is marked by, the regular association of these properties. . . . This does not work, we want to argue, because the right sort of connections between capacities and properties do not exist. Capacities are carried by properties. That is, you cannot have the capacity without having one of the right properties. But the same properties can carry mixed capacities, and so the true complexity of the situation cannot be revealed by the associations of properties.[21]

As quoted before, they talk about the "exercize of capacities." Apparently then, the point is to say that a property of some causal factor may or may not exercize some particular capacity which it possesses. But this idea has the same problem as Harré's "tendencies." If the exercize of a capacity is the causal event that explains an observation, then what is the explanatory role of the capacity itself? A capacity "just sits there" in whatever has it, it is not itself an occurrence whose adducement explains. Only a power or capacity *put to effect* can be a causal occurrence, and to cite a power or capacity cannot further the explanation of what is brought about. And I have made the point above and in Chapter 3 (in the section entitled "Is 'Productionism' Anthropomorphism?") that events per se do not "have capacities" or bring things about "by way of" capacities. They happen, and in so happening, they bring about.

In the dispute about composing causes, the motivation for talking about tendencies and powers was to account for causes interacting with one another while tacitly limiting the category of actual events to observable motions. The motivation of Dupré and Cartwright for talking about capacities in this paper is not this, but instead, to further their aim as stated in the title of the paper: "Probability and Causality: Why Hume and Indeterminism Don't Mix." The authors seek to show that, while Hume's analysis of causality may have had great force in the context of Newtonian science, in the quantum era it has proved too simple, just as the Newtonian framework itself has proved too simple. The Humean account, they claim, cannot survive the transition to a fundamentally probabilistic or statistical causality. That this claim harbors a correct insight will be shown by the broader results of the present work.

The idea in talking about probabilistic causality in terms of "capacities" is that one capacity might sometimes be activated, producing a particular determinate outcome, and sometimes a different capacity will be activated, thus accounting for fundamental indeterminism in measured events (and "tendency" and "power" can be put to this same use). I will call this the "capricious causality" interpretation of indeterminism. It is modeled after a feature of the behavior of persons, who may or may not activate or exercize their tendencies and capacities. Its essential claim is that *there are causal agencies or factors which at times are active and at times lie latent.* One cannot predict with certainty when one or the other is going to become effective. Moreover, it is *in the latent condition* that they are the "causal realities" in some sense that underlie observed phenomena, for example, radiation effects. It is as though to ask about the nature of their actualizing activity by which they bring things about is not to the point! But the question, *What happens* to bring about . . . ? cannot be answered by a latent condition.

I propose a contrasting provisional interpretation of indeterminism, to be further developed in this work. It avoids anthropomorphic usages and does not have the fatal limitation as causal explanation that has been seen in tendencies and capacities, that of citing latent properties rather than actual events. The initial proposal is to say that within certain contexts of causal processes of transmission and interaction there are attributes of terminal or interaction events, such as their exact location or timing, which *have no specific causal correlates* in the antecedent conditions. This says that there is precisely no explanatory factor for these particular determinate outcomes present in those conditions, let alone a latent and capricious one. It maintains that indeterministic features of events are *causeless* features (though events bearing them are physically engendered or produced).

The question of causeless variables warrants digression into a different philosophical topic involving events and causes. First, I have just described atomic indeterminism in a way that avoids speaking of causeless *occurrence*; but supposing for the moment that

I were necessarily at some point talking about causeless occurrence, this is certainly a conceivable thing in general. I would argue that in fact there are commonplace examples of causeless events. Despite the persistence of the issue of "free will" versus determinism in Anglo-American philosophy, the detailed intentional behavior of humans and animals is uncaused (though there may be any number of reasons or motivations for it). The cause of my action would have to be either me myself or something other than myself, for example, my body independently of my volition. If I am the cause of my action, then I do something to bring about my action, which begins a vicious regress. Suppose something else is its cause—say, for instance, the brain and central nervous system, a popular candidate. But to suppose that brain function directly determines behavior is to forget the "structure of intentionality" that belongs to action as well as to thought and feeling, the "of or about something," the "directedness toward . . . ," the involvement with life and the world; my brain is something other than, indeed categorically different from, myself, my person, and does not have the ontological status to be the agent of my action.[22] Nothing except myself, in fact, can be the basic impetus of *my* action, and since it is regressive to say that I am its cause, it has no cause. This digression into the "free will" issue has an ulterior motive. Here and in two other asides in Chapters 7 and 10, I take the opportunity to enframe my general promotion of causal explanation by pointing to philosophical contexts in which *wrongheaded* involvements with causal explanation occur due to the spell exerted over philosophy by special scientific orientations and procedures. In the instance just mentioned, the trained reflex to reduce phenomena to subjects of special sciences in order to have a "scientific" approach, instead of looking the matter straight in the face, generates the highly fashionable conviction that all so-called mental phenomena are "actually" physical events in the domain of neurophysiology. There is never enough examination of the science-inspired notion that thoughts, for instance, are occurrences quite literally inside our heads; according to this a well-developed fetus, since it has a brain, might in principle entertain thoughts, take up attitudes, or form beliefs (what would these be of, toward, or about?).

But I need not depend on this controversial defense of free action, which is likely to raise fears that I am myself reverting to teleology about nonliving domains, because I have not actually claimed that in atomic physics there are causeless events. The inter- action events are caused; they are the products of antecedent trans- missions. What are causeless are particular spatial and temporal variables attributed to these events, such as (in cases I have dis- cussed) their particular locations within detection material (or the exact timing of events in cases of radioactive decay). Once the possi- bility of *causeless quantitative specifics of terminal events* is accepted, the assigning of attributes of capriciousness to antecedent events appears to be the introduction of quasi-causes, when what is meant by "indeterminism" is that there is an absence of certain specific causes at an ultimate level of detail in measurement. The site of atomic interaction and the site at which a bird alights on a wire each independently refutes *universal* determinism, keeping in mind the extreme difference of context.

The other kind of motivation for talk of "tendencies" and "powers" came from cases of two causes interacting in a single observed event. Cartwright's and Harré's views on this turned out really to be the same. Both end up concluding that instead of saying *"two causes interact to produce an effect"* which is not exactly the effect that would be produced by either cause acting alone," one must say instead that the two individual causes are not any kind of actual occurrence (they are at most tendencies or powers). If a causal fac- tor does not bring about just what it is supposed to bring about or normally brings about, then (the thinking goes) one is entitled to skeptical antirealism or "modest realism" about this specific causal factor. But then *what does the interacting* in the case of composition of two causes into a single effect? Harré gives the cause a reality-in- latency as a "tendency," but this has been shown to amount to acausalism. Cartwright in *How the Laws of Physics Lie* does not care much for the "powers" interpretation of these cases.

Their position (partially unwitting in the case of Harré) against the actuality of the individual causes, based on the absence of their "pure" effects, is an unwarranted conclusion. Consider an example from the previous chapter: Suppose I experience a bright flash, but I suppress my blinking response. The flash certainly had its usual effect on me; that an effort was required to suppress the reflex shows this. The nonexistence of a specific terminal effect in no way negates the reality of the causal event.

The thinking of the realists Cartwright and Harré is in a sense empiricist and acausalist, since in the last analysis they both limit real events (as opposed to tendencies) either to events that can be observed, or to those generating events whose pure effects, that is, effects not modified or counteracted by other causes, can be observed. Conspicuously absent in both authors is any thought of "realism" in the sense of causal explanation outside of mechanism, though Cartwright cogently insists that whatever the status of our knowledge about it, radiation is "concrete causal process." Cartwright's antirealism about certain fields, which Harré unwittingly goes along with in part, is based on thinking that, given two causal sequences of events A and B, the prevention of a particular outcome of A by B renders the antecedent events of A subject to skepticism (or "modest realism"), even though two contributing causes are cited in this description.

Physicist Henry P. Stapp is a well-known figure among those physicists who pursue explanations in the "quantum" domain outside the mainstream Copenhagenist positions (the nontechnical aspect of his work in this area seems to me outstanding for philosophical sanity and a sense of reality). In personal correspondence to the author he has criticized the above attacks on tendency ontology. Stapp feels that "tendency" appropriately characterizes cases of purely statistical regularity. It captures the fact that observed effects are not purely "helter-skelter," that there *is regularity*. At the same time it is fitting where there is not deterministic causation. A specific effect will not necessarily occur given its antecedent conditions, fully specified, but it will have a tendency to occur, a tendency which is given an exact quantitative value by a probabilistic equation. He gives this example:

> Consider a device that can be described . . . as a particle gun (accelerator): it periodically ejects a "particle." The gun is aimed at a set of three detectors, and quantum theory predicts that the probabilities for detection are, say, 1/10, 2/10, and 3/10 in the three detectors, respectively. In this case the "particle" is conceived as a "tendency" for the various alternative possible detection events. One does not say that the tendency is activated. The tendency is present and it either produces an actual event or it does not produce an actual event. The law describes the tendency for an actual event to occur: it is the essence of

indeterminism that there can be tendencies for an event to occur without the event actually occurring. If every tendency actually produced the unique corresponding event then it would not be a "tendency" but rather a deterministic cause.

This is a very instructive presentation of how the idea of "tendencies" takes hold in the minds of physicists, from whom it is adopted by philosophers of science. It is strikingly close to what Harré says in a passage quoted above: "laws of nature do not describe sequences of actual events, but rather the tendencies that produce them." There are two important things to notice about Stapp's presentation of his thinking.

(1) He says that "the tendency is present and it either produces an actual event or it does not produce an actual event." Though the tendency is "present," something (as it were) embedded in the experimental circumstance, it is not, or is not quite, an actual event, and hence is radically set apart from being an observed event. Here again actual events seem to be equated with observed or observable events. But Stapp is not skeptical about the existence of something on grounds of unobservability. The claim for a tendency is a claim for *something existing and present which is inherently antecedent to a measurement and hence cannot be observed.* But what the existing thing is that is not an actual event (or an object), is hard to conceive.

It might be best to think of Stapp's "tendency," not fully occurrent but also not pure, nonactual potentiality, as something *quasireal.* In earlier correspondence he speaks of the events preceding localized interactions as "more tentative (or more like propensity) than the terminal event, which *seems more actual*" (my emphasis). This shows the origin of the tendency idea in the experience of physicists such as the original quantum theorists with their "probability wave" and "*potentia*" interpretations of the theory.[23] They are faced with an object of investigation (radiation, or "particles") which does not have the combinations of measurable attributes that belong to material objects (this is explained further in Chapter 8). Whatever one makes of "particles," there is not a bit of matter in motion with a definite course toward a terminal site. In the words of Merleau-Ponty, the causal factor or process is apparently at no time "constrained to a unique and fixed location, to an absolute density of being."[24]

Such an object of study is, in short, outside an ontology of nature in which fully localized matter and spatial localities are fundamental. Hence the suitability of thinking of it as somehow not quite fully real, as deficient in being. The thought is that prior to the observable and hence fully real interaction (with matter) is the potential for, or the tendency for, this determinate event. It should be noted that this view of the matter does not involve an epistemological prohibition against claims to knowledge beyond the observable (though it may ally itself with such epistemologies). It rather *asserts the existence* of quasi-actual or fuzzily actual things. Neils Bohr famously held the view that there is no reality to be comprehended in the "quantum" experiment apart from what is "described" by the mathematical models. But again, it is the encounter with something outside the working ontology—or should it rather be said, on its outer fringes, where causal factors become differentiated in character from localized, present objects—that leads to these conclusions, and not a general philosophical empiricism.

(2) In Stapp's account the idea of a tendency is directly tied to the mathematics of probability. The "particle" is a "set of tendencies," one for each of the quantitative probabilities for an event in each of the detectors. "Tendency" is a *direct interpretation of the meaning of a probabilistic equation*, whose solution in a given case is a quantity of probability. Just as Harré said, the equation (law) "describes" the tendency, something waiting in the wings of actuality, its likelihood of causal action subject to statistical calculation—thus the tendency is signified and identified by a quantity. A physical referent identified in this way is derived directly from a law, radically avoiding description of the "particle" and its causal activity.

"Tendency" is a usefully appropriate word for situations with uncertain outcomes which have a certain degree of predictability: "My car has a tendency to stall at intersections." Hence it may serve as a convenient linguistic response to the encounter with effects in experiments that are predictable purely in a statistical sense. But whatever purpose is served by such a linguistic adoption by a physicist is sharply distinct from the aim of "productionistic" explanation in this area. The latter aim, as I see it, is that of sustaining a concept of physical causality suitable for a renewed, full-fledged aspect (1) theory; not surprisingly, it has very different assumptions from those prevailing in physics. "Tendency," where this refers to a state of affairs antecedent to an interaction, is not something that can serve in this descriptive project, because, as stated earlier, it is not itself a causal event; it does not have the form "such-and-such happens." The important distinction, it seems to me, is between the need for a ready-to-hand utility of language and a concern for a genuine explanation. The goal of the latter should not be simply an apt characterization of the observed facts—statistical regularity— but a description of a causal context for indeterminism, one which makes this phenomenon fully coherent by giving it a place in our understanding of natural processes.

Stapp himself has indicated in correspondence that he considers the concept "tendency" of transitory scientific value in the pursuit of a better understanding.

The Success of Critical Realism

The contemporary issue of scientific realism/antirealism is primarily about a certain class of "unobservables," the established "particles" of physics; these are varieties of radiant transmission that are emitted at one end of a typical experiment and detected at the other. A prominent contemporary view that is straightforwardly antirealist is that of Bas Van Fraassen. The following passage shows how Van Fraassen's antirealist arguments are framed in terms of what can and cannot be *seen* and in terms of *particles* (taking this technical term of physics uncritically as though it referred to something have certain characteristics of material objects). He is discussing the tiny luminous streaks which can be seen in a device called a Wilson cloud chamber, a kind of detection device for radiation.

> The theory says that if a charged particle traverses a chamber filled with saturated vapour, some atoms in the neighborhood of its path are ionized. If this vapour is decompressed, and hence becomes supersaturated, it condenses in droplets on the ions, thus marking the path of the particle. The resulting silver-grey line is similar . . . to the vapour trail left in the sky when a jet passes. Suppose I point to such a trail and say: 'Look, there is a jet'; might you not say: 'I see the vapour trail, but where is the jet?' Then I would answer, 'Look just a bit ahead of the

trail . . . there! Do you see it?' Now, in the case of the cloud chamber this response is not possible. So while the particle is detected by means of the cloud chamber, and the detection is based on observation, it is clearly not a case of the article's being observed.[25]

The distinction between detecting and observing is thought to make room for scientific skepticism about what is identified and/or described by theory. That the theoretical entity is in the nature of the case unobservable—in the specific sense in which one says that to see an object is to observe its presence and be able to point to it—means, for Van Fraassen, that its reality (the correctness of the theory) cannot be established as a matter of science, a doctrine characterizing both positivism and logico-empiricism.

Cartwright and Hacking give important critiques of the doctrine that physical reality knowable with a special standard of certainty is limited to what can in principle be observed in the sense of directly perceived. They respond to its claim from the standpoint of the activities of experimentation and observation, prior to any epistemological reflection.

Ian Hacking repositions our standpoint on examples from physics, as against Van Fraassen's approach. He asks us to take the standpoint of the experimental physicist, not merely as an observer of phenomena but having himself a role in the *production* of effects to be observed.[26] Hacking supplies an example from his own experience in which as an experimental researcher he is trying to detect a certain entity which the mathematical models predict to exist. The procedure is to make use of a device which emits a certain kind of known "particle." By aiming the emission at an object made of a certain material, he hopes to produce specific effects that would indicate the presence of the hypothetical entity. Under the conditions of such experimental activity, which is the source and reference of theory, if someone now says about that which is emitted by his instrument and which is thus itself a tool of research, "It may be nothing at all," this should bring on a question whether this remark arose from a mentally sound condition. Hacking's argument that entities put to use by a physicist can only be considered actual contrasts strongly with Van Fraassen's procedure of conjuring an image of a conjectured and theoretical causal object, where skepticism was made to seem appropriate.

An important feature of Hacking's realism is that he does not leave the reader with the idea that physicists have some one definite picture of electrons (and the like):

> ... for different purposes we use different and incompatible models of electrons which one does not think are literally true, but there are electrons, nonetheless.[27]

What we know electrons to be, on Hacking's view, are complexes of causal properties or causal powers, which are understood only partially but well enough to put them to use for a variety of specific purposes.[28] Our knowledge of electrons is really knowledge of their various effects as part of the technology of experiment. This recognition speaks against thinking that they are "theoretical entities" in the sense of products of the physicist's explanatory imagination.

Hacking's experimentalist standpoint on antirealism is much like the case I made use of in Chapter 3, of a carpenter wielding a hammer. It is about as cogent in either case to claim that these are not cases of using some actual and effective means, some mediating physical causality, to bring about a result. An experimenter who has set up in his emitter just the species of radiation he needs, and directs a beam of it at an object to look for specific effects, can only insist, against antirealism about "unobservables," that he is putting to use a real causal process. It is the same commonsense assuredness with which one can say even in an ordinary life context that light is *some* form of causal transmission across space.

Of course, the antirealist is likely to say about Hacking's argument, "*What* stuff emitted by the instrument? To talk about particles being emitted is to prejudge the whole issue." But the real force of Hacking's argument comes from placing the question of realism in the light of the relevant real-life context, instead of supplying, as in Van Fraassen's text, only the simple framework of a casual observation plus a "particle" that is treated as if its very existence were a causal conjecture.

Van Fraassen speaks as if "theory" meant a conjecture about the existence of something quite definite: a subvisible object. It is likened to a jet airplane seen at high altitude, except that it is useless to try to see it by squinting one's eyes or using an optical instrument. Whatever is observed is thought to necessarily be an object of an ostensive gesture: "There it is!" I would point out that this is a theory of observation, one that defines an observation more narrowly than, say, Whitehead's view that what are observed are events which present to the observer not only themselves in their immediate perceptible content but a "relatedness" beyond themselves. Hacking does not invoke the narrower presumptions either about the character of the "unobserved" entity or about what constitutes observation; according to the causalist intuition as I describe it in Whitehead, Hacking might if he wishes claim that the entity is actually observed as a mediating and generating factor in his experiment. Hacking's positrons radically elude the tidy criteria of the narrow theory of observation, if only because for him the "particle" as a moving object is one useful model which is "not literally true." This recognizes the impossibility of concrete representational-mechanistic description, avoiding pictures that mislead thinking. The suggestion Van Fraassen derives from the phrase "unobservable entity," by contrast, is just that of a subvisible object, as if it were in effect moving around behind an impenetrable barrier.

Cartwright in her arguments toward realism about radiation does not go so far as to describe in detail the standpoint of the experimental producer of effects. She simply calls upon the prereflective realism that the scientist has about experimental observations, and calls attention, for instance, to experience applying quantum theory in the designing of lasers. About the cloud chamber example treated by Van Fraassen, she says this:

> In explaining the track by the particle, I am saying that the particle causes the track, and that explanation . . . has no sense unless one is asserting that the particle in motion brings about, causes, makes, produces, that very track.[29]

And later:

> If there are no electrons in the cloud chamber, I do not know why the tracks are there.[30]

Her claim is simply that for specific observed events under closely specified conditions, there are the characteristic causes named by the term "electron" bringing them about. Her language of "particle in motion" might suggest that she pictures this cause as a subvisible object, as does Van Fraassen. But her claim does not require this picture. A causal realism need only involve the claim that some specific kind of process, agency, or entity, named "electron" by physics, produces the track. It may be identified solely by its exact experimental context: it makes *this* kind of track in *this* kind of medium and has *this* kind of emitter as its source. There is no need for causal realism to picture the individual electron *in any particular way whatever*, and this holds also for science exploring the causal possibilities of the electron and making use of them. "Something made the track" is the essential claim of the causal realist; *this* raw, picture-independent truth rests on a contact between the scientist and reality that is out of reach of skeptical arguments.

As I have suggested also in the case of Hacking, an important connection links these remarks of Cartwright's with the founding approach of Whitehead's physical theory. Whitehead distinguishes two primary sources of knowledge in natural science: "cognisance by adjective" and "cognisance by relatedness."[31] A brief and partial explication of this idea outfitted with the terms of the present discussion will be adequate for now. What is the essential constitution of a fact of scientific observation? Traditional accounts in philosophy concerning perception and observation have placed heavy emphasis upon "sense qualities"—the coloredness of a thing, a certain texture, a resemblance to a jet trail—and whatever else comes closest to conforming to the concept of an "impression," or more currently, a "sense datum." But what Cartwright's remark should tell us is that there is at least one kind of *observed character* that is remote from these, namely that of *being produced*. The cloud chamber track as arriving in and tracing through a medium is an event one of whose observed characters is an internal reference to antecedent events, whether or not we are given to know any "adjectival" characters of the latter. We "cognize by relatedness" that the track is *something brought about*. The looks-like-a-jet-trail residuum is an "adjectival" character that has for its first explanation some scientific hypothesis: it is suitable to ask, *What* brings about this distinctive structure, color, etc.? But on Whitehead's view the event-character of the track simply as a spatially and temporally concrete and particular happening refers beyond itself to a structure of relationships in its background (and ultimately to a whole of nature); "the event is essentially a 'field.'"[32] The track comes about as a transition embedded in an indefinitely extended background of physical structure. This includes a physical factor of immediate antecedence.

The significant difference in Cartwright's treatment of the cloud chamber example as against Van Fraassen's is that Cartwright keeps in view the real context of technological activity, a context of developing knowledge involving cloud chambers and other equipment. As part of this scientific legacy there is a linguistic tradition of identifying, in a technical language, causal agencies and sequences of events presumed to lie behind the observations. Van Fraassen, on the other hand, takes this traditional linguistic response to scientific experience to be wholly a product of the scientific imagination, describing a concocted world of the inherently imperceptible (and hence unknowable, according to traditional prejudices). He ignores the historical experience through which the various forms of invisible process—light "waves," electricity, "particles"—came to be identified. He presents the cloud chamber example as if scientific observations occur without a context of knowledge, as though one had wandered into a laboratory, come upon a curious container, and observed a little streak within it that looked something like a jet trail. About this distinctive datum, one might then make the ad hoc causal supposition, as a convenient explanation, that a tiny object moved through the chamber and made the track. And the accounts of the matter by physics, which are broadly and unreflectively referred to as "theory," certainly suggest this, if taken at face value. But physics also says that it is impossible ever to see it. What gives us the warrant, then, to take this causal account as more than a convenient fiction, a parable metaphorizing the track, which can be used to communicate its character to others?

What Van Fraassen recounts as the theoretical story is a technical account, not a description of events in ordinary language. The current state of science does not include descriptions of the actual sequences of occurrences classed as "particles" and the specific types designated "electron" and so on. Except for the knowledge based on superposition effects that such events have a cyclic structure analogous to a wave, this domain of nature is simply a large lacuna for narrative understanding, an ontological question mark. Nevertheless Van Fraassen writes as if it is completely clear *what the thing is* whose reality we are to be skeptical *about*: an object, like a meteor entering the atmosphere. His notion "unobservable in principle" implicitly accepts the term "particle" as if it *did* mean what it does in ordinary language. "In principle" means for him that there is no physical means by which one could see the electron. To argue that one might see the jet but never the electron generates the thought-picture of a subvisible object, even though his very point is that unobservability in this case actually has nothing to do with the physical limits of microscope and microimaging technology. To see through this confusion means realizing that the direct ostensive gesture *"There* it is!" makes no more sense in connection with the electron than does *"There* goes a light wave!"

So what he is saying amounts to this: Physics tells a story about an infinitesimal object, but no one can securely claim to know it exists. The most basic problem with this is that if one considers a few simple experimental results (described in Part Two, Chapters 6 and 8), one finds that the suggestion that the electron is a tiny object, a suggestion derived from the language of "particles," is simply mistaken. As a causal occurrence independent of its observable effects it is a form of radiation that will produce superposition effects and therefore is wavelike in character, and also the track at the atomic level is really a series of localizing interactions rather than a continuous transition through space; one must take this knowledge seriously and not ignore it for convenience as is so often done. Despite the thought-pictures that are generally employed, the physical referent of "electron" is not a localized object, but a field or field element.

Suppose one were to say to the entity-antirealist, "Nancy Cartwright is convincing when she voices the conviction that *something* brings about these tracks, leaving aside any picture of the cause as a moving particle. Would you not be content to say that we do not know, perhaps cannot know, *what it is*, rather than saying that whatever it is, we cannot know that it exists?"

"It is the same with any inherently hidden cause, in my view."

"Your argument, then, has nothing to do with the seemingly irremovable obscurity of the electron's nature. You make no use of the fact that as a 'quantum' entity, its complex effects seem to put it beyond the reach of any ordinary understanding."

"That is correct. However it is understood or not understood, it belongs to a causal hypothesis constructed by science, a construct with a useful application, but without any possibility of confirmation. My argument is not that of the Copenhagenists, that the quest for a causal understanding of atomic interactions should be given up in light of its difficulties; this is physics coming to the truths of empiricism by its own route, not by that of the philosopher."

"I'm glad that is clear. And ionization also is an unobservable change in an atom, so that your skepticism must apply to this occurrence as well, and not only to the existence of a moving charge. If so, then your full-fledged claim is as follows: For all we really know, it may be that nothing (as opposed to something inaccurately or incompletely understood) brings about ionization; it may be that nothing brings about condensation of droplets; and ditto for anything which as you say *results in* the silver-grey line which we see, but which cannot itself be seen."

Van Fraassen takes the position that the success of science means only that theories are empirically adequate, not that they are true. The intent of this is the usual one of empiricism, notable in Mach, to argue that one cannot infer that an explanation is correct when other explanations exist or may exist which explain equally well, unless our favored explanation ("there are particles") can be independently confirmed, which requires an observation. He uses the word "theory" in a way suggesting that the target of his skepticism is some one clear, definite hypothesis about observations, one which presumably could be evaluated in comparison with other, competing hypotheses. Now, some cases of scientific problems do fit this model. Recently some extraordinary features were revealed on the satellite Miranda of Uranus by the Voyager spacecraft. Wide-ranging causal hypotheses were put forth, but the matter was for a time unsettlable.[33] Van Fraassen seems to view the cloud chamber track as just like this, except that one explanation, the "charged particle," is settled upon for some such reason as simplicity or convenience, a reason that falls short of establishing the unique correctness of this "explanation."

But as we have seen, if the electron is really a "theoretical entity," then "theory" here does not mean some one, distinct, causal hypothesis, as opposed to some other, about the cloud chamber track, but the working language and understanding, both background and immediate, of physicists, which is through and through causal. After all, the only necessary claim of causal realism is that *some distinctive physical factor* brings this track about. Cartwright's and Hacking's strategies are to promote this claim, whose source lies prior to the traditional epistemological foundations, and certainly prior to theory in any of the usual senses.

The critical realist arguments are remedies for thinking of physics as a mythology of objects too small to be seen, objects whose existence is postulated to explain particular observation data. When physics is seen instead as a real-life activity of experimentation and invention, the entities it designates through the various particle names are recognized as context-specified species of causal activity that are investigated and put to use. The lay person can usefully refer to ordinary experience with light transmission to fully appreciate the point. Knowledge about entities in this broad class consists of their differentiation and naming in the course of technical exploration and exploitation; "particles" are not only "unobservable" according to the narrow criterion that ties observation to ostensive perception; they also operate fully incognito, that is, without being described or identified ontologically by existing science.

Whatever its errors and successes, the discussion between antirealism and its critics belongs to the tradition of philosophy of science descended from physics, and as such, ironically enough, never entertains the thought that the entities in question have a physical structure that is simply *unknown*, but *might* be known. Though the importance of this thought for the issue of realism should be clear, to take it seriously would seem from a theory of science point of view an indulgence in pointless speculation about a possible advance in physical science whose shape cannot be made out.

Causality and Forces

Setting aside all traces of acausalist/empiricist doctrine, either about the limitation of knowledge to the observable or about the primitive status of narrative natural science: Is a philosopher of nature warranted in expecting that potentially comprehensible causal occurrences, inherently outside of and antecedent to direct perception (as of objects), are designated by the term "fields" and by the term "forces" intended in the sense of the action of "fields" upon matter? If so, how would inquiry about the force or field itself be related to knowledge of the law or laws of its observable effects?

Suppose one observes that two objects of a certain type attract each other when they are placed in proximity, without having any idea at all about how the exact quantity of force might depend on other measurable conditions. One might naturally say that a force is exerted, even that something exerts a force, on these occasions. Now suppose one diligently investigates and formulates the law yielding the quantity of force as a function of other quantities. How would this new piece of knowledge affect what one is able to say about the force? There would certainly be no guarantee that by this determination one had gained any insight into *why* the force occurs or what produces it, though there may well be clues about this in the newly discovered quantitative relationship. But knowing the law would just as certainly have no negative effect on one's ability to say that a force is exerted by one object upon another or by a mediating factor upon both. Primarily one would now have at hand a procedure for determining the quantity of force, and one may have grouped a set of observations (say, planetary and lunar motions) under a single mathematical principle.

Now suppose a long tradition develops around this law and its applications, perhaps as a component of a more inclusive system of laws covering a wider set of attraction/repulsion phenomena, in other words, a science advances primarily by means of this and other laws and becomes identified with them. The force is cited to explain certain facts, and the law becomes the main focus of scientific concern with the force. When the force is discussed in science books and science classes, the law is brought forth and explained; the things to learn about the force are first of all an equation and how to use it, and perhaps the history of its formulation. Students may now be unlikely to raise the question, "But what is this force? What causes it?" They would be likely to feel somewhat foolish asking such a question, and to receive gentle, patronizing correction for their naiveté from instructors. But a thoughtful and independent-minded student might well find himself or herself pondering this question. Indeed, there is no reason that the naive question about what sort of natural process could be at work in this force should be any less viable than it was in the context of a crude observation that the force occurs, though it could easily have gotten lost in a knowledge-snowstorm when science became focused on the possibly much more fruitful course of mathematical analysis.

However, the question will meet a more basic obstacle than this cultural and institutional one. Those who have sought to answer the question of the cause of gravity have found themselves proposing contrived and ultimately unsuccessful mechanisms. What physical process, even ignoring the requirement that it be otherwise unobservable, could have the characteristics to draw together two objects whenever they are in proximity? The etherial or corpuscular explanations that had occasionally been put forth in the past, for example, for gravity, found little or no scientific favor, and for the most part inspired little confidence even in their authors. And from our standpoint today it seems absolutely clear, at least to this writer, that if there are causal explanations of the forces they are not mechanistic explanations (this is both a deeper mystery and a decisive clarification). To explain a force of attraction would be tricky enough in any case, but today it must be framed in terms of the possibility of empty space (or nonmechanistic) events, wherever this might take the inquirer. Even though there is no valid reason to say that gravity has *no* physical cause (indeed, physicists now talk about the gravitational field as a form of radiation), the question seems no better oriented than it has ever been. How in the first place can the basic form of an empty space event be conceived? There is an absence of what Cartwright calls a "background knowledge" for the question to have bearings, so that one's response to the question may well be to declare *"hypotheses non fingo"* with redoubled emphasis. Rom Harré, in fact, says this:

> At the end of any explanatory regress we must perforce shift from causal mechanisms to causal powers. . . . Gravity may . . . be a referent of last resort, explanatorily. To explain the behavior of falling bodies by reference to gravitational potential may be to cite a basic causal power.[34]

How about by reference to a gravitational *field*? The answer is no doubt the same. The last stand in the explanation of forces is a power or "disposition." But a point I have been making is that a power by itself does not explain any physical phenomenon, and this is true whether it is attributed to the objects involved, or to intervening events, or is not given any specific subject. If explanation in such cases really stops at a power, then basic phenomena of physics are permanently unexplained. But according to the generally accepted hypothesis of gravity radiation, the expectation of basic causal explainability in the case of gravitational attraction has exactly as much force as the expectation that optical phenomena have a basic causal explanation involving a specific form of propagation across space. I would wager that Harré would not bring the explanation of, say, illumination to a stop at a reference to a "power." The critical realists, at any rate, would insist that concrete, antecedent causal process is involved.

The quest for a genuine answer to the question of the cause or physical origin of a force should prepare the general ontological question about the "field," that is, the question (if its primitive impulse can be recovered) as to *what it is*. The complete ontological formlessness of the "field" in the present state of knowledge (assuming mechanistic models really did collapse) is only betrayed by the common intellectual adaptation of thinking that the field is somehow described by laws. But the general requirements, at least, of an explanatory description can be outlined by referring to the phenomenon itself. An essential causalism along a Whiteheadian (and naively commonsense) line would begin with the point that to observe a force of attraction in operation is to observe an occurrence with a causal structure, specifically that of an effect tracing somehow into the intervening space of the objects. "Field" means that the objects are embedded in an environment (in some sense) of physical activity (of some kind) through which they interact with one another, and that the mediating activity is integrated with the structure of the locality of each object. There is a strong suggestion in certain discoveries and models of physics that a field is in some way to be identified with the space itself in which an object is situated. Since this cannot mean that because its field is inseparable from it the object is fixed to a particular locality or "carries its space around with it," it may only mean that the presence of the object imparts to an active spatial environment a specific character. The field is not merely a potential for an effect on a test object, but a dynamic physical background, just as the radiation transiting the field is a form of propagative activity and not, for instance, a "probability wave." Any explanation should avoid all conceptual prejudices derived from experience with simple mechanical processes, since it is now clear that the field is simply not comprised of mechanistic activity, but of some other form of physical activity.

Attitudes fostered by practical science with its dependence on simple linguistic models and mathematical formalism should not lead one to reject out of hand the intuition that operating conceptions of causal structure may have remained unnecessarily limited, closing off explanation in spite of all the wondrous advances of physics. This raises the possibility of achieving a causal-narrative explanation of forces even though this might seem to require examining our notions as to the essential content of physical explanations. Part Two proposes a possible constructive course for a distinctly philosophical approach to furthering understanding.

Conclusions

Our examination of the dispute over composition of causes had the upshot that when a single observed event is the result of two interacting causes, this situation contains no warrant for saying that either cause may be unreal or can only be identified as a latent property, a tendency or power. The empiricist doctrine of "observability" as the boundary of science operates here in a confused way. And it was made clear that for a causal explanation, in any case, simply citing the potentiality for such and such cannot substitute for describing actual causal events. The field is not a tendency, propensity, or "disposition."

What Harré thinks in claiming to concur with Cartwright that laws do not describe sequences of actual events represents an interesting confusion. It is certainly true that a law of nature does not describe a sequence of events intervening between correlated observations. *It is not a causal narrative.* However, the implication in his thinking is that when a law quantifies a force as a function of other measured variables, no real causal occurrence corresponds to this designated "force" (it is only a "tendency"). This is quite different from saying (correctly) that a law does not tell a story about this generative occurrence, that is, does not answer the question of *what it is*. As to what the basic forces of nature actually are in this causal–ontological sense, science has yet to uncover this. This certainly does not mean that the law of gravity does not identify by its variable "force" a specific physical occurrence apart from a consequent motion.

The critical realists supplied the insight that a specific pre-observational cause does not have to be a "theoretical entity," identified only as a permanently hypothetical explanation.

All of this speaks against the ubiquitous idea that because certain processes that are adduced as causes for observed phenomena happen in empty space or in the "subatomic realm" and are therefore not the sort of thing one could watch happening—in short, they are not object-motions or changes in objects—this is a reason to doubt their status as active, physical actualities. Composition of physical forces, for instance, only makes sense if some depth of antecedence is ascribed to the forces as occurrent causal activity: apart from the dominant generating field itself, the acceleration produced in a test body has an additional component of the effect (perhaps extremely minute) of a different generating field; what else is meant by "composition"? When "force" is another name for the field, forces *bring about* motions (if not prevented from doing so); they are not identical with these motions. Nor are they—assuming one takes the "breakdown of mechanistic models" at all seriously—some subtler motions. The technical character of the "particle" terminology can be a source of antirealism, not to mention the opacity of these entities to understanding; but this "lab talk" nevertheless does designate species of physical causality.

I submit that this general confusion is a sign that the struggle for a philosophical approach to "scientific realism" has pushed its own central questions to the periphery of the discussion if not wholly outside of it. The real task of causal realism is to look toward and prepare for the disclosure and exploration of a new region of possibilities for understanding within natural science; investigation so far into the aspect of nature in question has systematized its observable manifestations with astounding ingenuity while standing before it in a wonder that has given way to dwindling expectations amid increasingly puzzling results. At this point the need is for a broad explanatory brushstroke: a background physical story, compensating at last for the permanent loss of the "ether," which would supply a physical context for understanding empty space "fields" in general. Its exposition would have to reject any regression to naive mechanistic models, aware of the subtle deceptions of linguistic convenience.

5

Causal Realist Projects, II

THE LOGICIAN and philosopher of science Wesley Salmon published a work in 1984 with the captivating and optimistic title, *Scientific Explanation and the Causal Structure of the World.*[1] It has two major parts: (a) a theory of explanation in the tradition of logico-empiricism, and (b) a general theory of causality covering two basic types of causal occurrence—transmission and interaction. The theory of explanation purports to improve on the deductive theory of Hempel and Oppenheim. According to their "standard" doctrine (Chapter 2), an explanation is the deduction of an event from a general law. Salmon hopes to improve on this basic model by providing a place for probabilistic laws, such as those by which physics responds to the encounter with indeterministic processes, which could not yield specific "explananda" events deductively. I will not attempt to determine whether this project is successful; my interest is in the theory of the basic types of physical causality and the proposal for a constructive realism that comes out of this theory.

The importance of Salmon from the standpoint of the present work is that he straddles, or tries to straddle, the acausalist and productionist positions, or in other words, he tries to further both causal ontology and the standard empiricist program of philosophy of science. On the one hand, he believes that the world has a "causal structure," which for radiation science means that the gaps, so to speak, in sequences of events subject to direct observation are filled in by continuous, intervening transmissions. On the other hand, he strives to further and perfect the traditional acausalist project of a law-based model of explanation, on the Humean assumption that scientific knowledge, including knowledge of causal connections if there are any such things, is built up through the exploratory disclosure and formalization of regular associations in observation, and he formulates the logic leading from such accumulated experience to a "statistical" (as opposed to a deductive) structure of explanation.

Salmon's constructive realism is a proposal for how to talk about intervening or underlying processes, for instance in physics, as continuous and causally connected. As it turns out, however, his commitment to the tradition of Hume subverts this aim: his account of causality unwittingly continues to exclude the relation of production connecting events or stages of a process. To leave out this relation, in the final analysis, has the usual result in the neo-Humean tradition, that of reducing knowledge in the domain of physics to quantitative knowledge of regularity in observed and discrete events, effectively exiling generative activity from all physical theory and philosophical explications. His intention, however, is to combine two views about explanation: first, his "statistical" model in the tradition of the law-based model; and second, what he calls the "ontic conception" of explanation, according to which to explain a phenomenon, in a sense to be upheld as legitimate and important, is to talk about natural entities or goings-on in nature, rather than about laws and their logical force as conditional propositions. He attempts a hybrid of empiricism with causal realism or causal ontology.

The three sections in this chapter show, respectively, (a) that Salmon's attempt to affirm the reality of transmissions connecting observed events does not succeed, and in its failure provides a stark illustration of traditional reductions of causal relations; (b) how Salmon exemplifies the traditional view, contrary to that of the present work, that theories of physics explain and that the role of philosophy of science is to analyze the meaning that "explain" must have in these cases; and (c) how Salmon's attempt to think about "probabilistic" process ends in a "propensity" analysis, whose basic failing from a productionist point of view was made clear in the previous chapter. One merit of Salmon's work is that by attempting to uphold the importance of *connecting* causal processes for explanations, he brings the consequences of logico-empiricism (a tradition he continues) to a new clarity. Another important merit is that causal ontology is pursued by investigating the uses of "event," "process," "production," "propagation," "interaction." In addition to being a heartfelt attempt at conceiving a causal realism, the book contains a good discussion of an example from science of a convergence argument for entity-realism (see note 12, this chapter), and some other valuable items that lie outside the present concern.

The Theory of Transmission

A theory of physical causality must take into account both transmission through space and its interactions with other transmissions and with objects. Salmon adopts the natural starting point of considering a variety of cases of transmission and interaction in general. In so doing he brings into use, for his own purposes, some fundamental distinctions. He attempts first to establish a distinction between *propagation*, or transmission, and *production*, illustrated by cases of interaction such as one object colliding with another. This distinction appears to be taken quite sharply, as if propagation could not be production nor vice versa. The other distinction, unveiled later in the discussion, is alleged to hold between "process" and "event," with "process" roughly corresponding to transmission and "event" to interaction.

With these special distinctions and usages he aims to form a conception of transmission as continuous process intervening between separate interactions. For Salmon (and we will come to see why this is), only a "process" can unproblematically be said to be continuous in the sense required by causal connection, while succession in "events" cannot. The propagation-production distinction apparently serves to reinforce the difference so conceived between transmission and interaction. In the following passage he tries to use the propagation-production distinction to sort out these two forms of causality.

> There are, I believe, two fundamental causal concepts that need to be explicated, and if that can be achieved, we will be in a position to deal with the problems of causality in general. The two basic concepts are *propagation* and *production*, and both are familiar to common sense. . . . When we say that the blow of a

hammer drives a nail, we mean that the impact produces penetration of the nail into the wood. When we say that a horse pulls a cart, we mean that the force exerted by the horse produces motion in the cart. When we say that lightning ignites a forest, we mean that the electrical discharge produces a fire. When we say that a person's embarrassment was due to a thoughtless remark, we mean that an inappropriate comment produced psychological discomfort. Such examples of causal production occur frequently in everyday contexts.[2]

After thus identifying cases of "production," which he will later call "interaction," he then turns to cases of transmission:

Causal propagation (or transmission) is equally familiar. Experiences that we had earlier in our lives affect our current behavior. By means of memory, the influence of these past events is transmitted to the present (see Rosen, 1975). A sonic boom makes us aware of the passage of a jet airplane overhead; a disturbance in the air is propagated from the upper atmosphere to our location on the ground. Signals transmitted from a broadcasting station are received by the radio in our home. News or music reaches us because electromagnetic waves are propagated from the transmitter to the receiver. In 1775, some Massachusetts farmers—in initiating the American revolutionary war—"fired the shot heard 'round the world" (Emerson, 1836). As all of these examples show, what happens at one place and time can have significant influence upon what happens at other places and times. This is possible because causal influence can be propagated through time and space. Although causal production and causal propagation are intimately related to one another, we should, I believe, resist any temptation to reduce one to the other.[3]

There are some nice examples here, leaving aside the problematic assertion about memory as transmission. In these two paragraphs Salmon provides some cases of causal production and causal propagation. But does his sorting out of these examples really show that these are neatly separable categories of occurrence, as he suggests? Not only are they "intimately related"; it would be hard to think of two words that come closer to meaning the same thing, or at least, that are so generally overlapping in their examples. Take the cases of water surface waves and sound waves. Why are they commonly called propagation? Because antecedent stages of the process give rise to, produce, succeeding stages. That is how a ripple intrudes upon and activates waters previously still. It produces a disturbance on the advance. A wave is productive of itself; it is a reproduction or propagation.

Or take one of Salmon's examples: electromagnetic waves, also commonly termed propagation. Today such "waves" are no longer thought of as vibration in material which is everywhere present. Still, their effects show frequency and phases, hence in some sense successive reiteration, despite the fate of the waves-in-the-ether conception. There is no reason at all to suppose, just because these are not oscillations in and of some substance, that this successive reiteration is not any kind of productive succession, so that the use here of both the term "propagation" and term "production" would be inaccurate or inappropriate. They are still called "waves," and according to some basic experiments must in fact be wavelike.

If one wanted to differentiate the usages of "propagation" and "production" in a general way, one might say something like this: Propagation is at least a kind of production which is reiterated through successive events in space and time (without treating the question of whether the events need all be of the same type). A case of production which contrasted with ongoing propagation would then be one that terminated in an effect, which then produced nothing further (or nothing relevant to the account being given). I will now study as a possible case of such a terminal effect a favorite example of Salmon's.[4]

In a darkened Astrodome a spotlight shines on the ceiling. The beam of light (transmission) produces, we would say, the spot (interaction). Now set the spotlight turning; the spot moves around. *This* process of the spot moving around does not produce any further effect apart from itself, as does the light transmission process, which produces a spot. Because of this Salmon wants to call this movement of the spot a "pseudo-process."[5] This seems inappropriate. There is no reason not to call it a real process.

"The beam is now turning figure eights on the ceiling."

"Yes, but do not imagine that this is a process; it is only a pseudo-process."

I will not adopt this expression, but will take Salmon's intent to be that the movement of the spot is not a causal process, but a pattern in terminal effects. One can understand his point; spot *movement* may or may not itself produce any particular physical effect at the point of interaction with the ceiling. Salmon explains the difference between causal and noncausal or "pseudo" processes in general as follows. If one inserts a red filter into the beam, it will alter the beam, and the beam will pass on this alteration to the spot on the dome, which becomes red. Salmon calls this "mark transmission." The beam of light is altered "structurally," and this new structure is transmitted to the wall. But, he argues, the movement of the spot is not capable of transmitting a mark, since if one holds the filter up and rotates the beam through it and on past, the spot will momentarily change to red, but this change will not be passed on to successive positions of the spot. Since it cannot transmit such a change, spot movement is a "pseudo-process."

But since the light transmission and the movement of the spot are quite different kinds of process, why should it be assumed that the very thing that leaves a mark on one kind through portions of the process successive to the marking, should similarly leave a mark on the other? To be free of this assumption, we should ask whether there is any way in which the spot movement process could transmit—pass on into further effects—some kind of modification of itself. Suppose there is a computerized device in the Astrodome which is equipped with a light receptor. It is designed to read patterns of light movement reflected from the ceiling and translate them into outputs specific to those patterns. Different ways in which the spotlight is moved—ways of "marking" the spot movement process—produce different responses from the computer. This would be a kind of "mark transmission" appropriate to this kind of process, and within its capacity. For that matter, what about a certain cycle of movement of the spot leaving in our eye a retinal image of a figure eight? Spot movement is marked by the way the spotlight is moved, and this pattern in turn marks our vision.

Another one of Salmon's examples of a "pseudo-process" is a moving shadow.[6] But imagine a situation right out of the movies, in which the shadow of a feared assailant betrays the fact that he is creeping around the door with a gun in his hand (by holding the gun in the air, he "marks" the shadow in an unmistakable way). Here a moving shadow has quite an impact on the possible victim. The sudden appearance of a shadow can startle us, and a passing cloud will produce definite effects in light meters and solar cells. The motion of a shadow is certainly a real, physical occurrence. One might object that the *effects* in these examples are effects, respectively, of the *appearance* of a shadow and of a *cloud blocking the sun*, not of the motion of shadows as such. Certainly the motion of a shadow is not a *productive* process in the sense of a propagation, since stages of the motion do not give rise to succeeding stages.

In Salmon's favor, it is possible to view certain occurrences as *termini* of causal processes, which do not themselves give rise to any further effects within our horizon of interest. However, we have seen that these same product occurrences are not in their nature excluded from the role of transmission to further effects. Thus Salmon's "pseudo-processes" are not a category separate from causal processes, as he suggests. From our discussion it looks as though there could be a productive process that is not a case of propagation, but not vice versa. It does not therefore appear likely that transmission and interaction as different kinds or aspects of causality can be explicated in the manner Salmon proposes, though some possibly useful overlapping distinctions have come to light.

Salmon proposes a *sharp* distinction between transmission and interaction in a characteristically empiricist conception appropriated from Russell and Reichenbach: the "at-at" theory. According to this theory there are some separate events, interactions, and there is their occurrence in succession, the latter being the whole meaning of "transmission." For example, there is an interaction at the site of a red filter inserted into the spotlight beam, and there is an interaction identified by a red spot on the dome ceiling, and their *succession in time and space constitutes transmission*. The entire description is of an event at one point succeeded by an event at another point, hence the name of this theory. As Part Two explains, Salmon here exemplifies a standard artificial reduction of the structure of causal processes to linearity and successiveness. This procedure for the analysis of transmission and interaction is influenced by a tradition of thinking about the implications of the special theory of relativity for causal relations.

From the standpoint of the present work, the "at-at" theory is recognized as an explicit acausalism about transmissions, leaving them wholly out of the account as independent connecting processes with their own mode of activity. Note that it does reinforce

Salmon's propagation-production distinction: only the interactions are cases of production, while propagation just means they happen first at one place, and then at another, in an orderly and calculable succession. But an aproductive process—that is, one that is only "propagative" according to Salmon's distinction—cannot re*produce* itself (*really* propagate), and it also cannot, as aproductive, *do* anything by way of what should be called its role in an interaction. Thus on his view there can be no causal action of transmission producing an interaction event; there can only be an unproduced, hence causally unexplained and unexplainable, "interaction" event. This is the same symptom that Cartwright and Harré had in their discussions of composition of causes: excluding from the account (in one way or another) the independent activity of the individual, preobservational contributing causes.

Interestingly, Salmon nevertheless sees that, especially from the point of view of physics and the history of experimentation, there is in the spotlight example "continuous intervening process" termed electromagnetic propagation.[7] Given his "at-at" theory, how is it that he can speak of such a thing? He thinks he can still claim reality for the intervening transmission by a special tailoring of the term "process" as against "event." This second basic distinction is explained as follows:

> One of the fundamental changes that I propose in approaching causality is to take processes rather than events as basic entities. I shall not attempt any rigorous definition of processes; rather, I shall cite examples and make some very informal remarks. The main difference between events and processes is that events are relatively localized in space and time, while processes have much greater temporal duration, and in many cases, much greater spatial extent. In space-time diagrams, events are represented by points, while processes are represented by lines. A baseball colliding with a window would count as an event; the baseball, travelling from the bat to the window, would constitute a process. The activation of a photocell by a pulse of light would be an event; the pulse of light, travelling, perhaps from a distant star, would be a process. A sneeze is an event. The shadow of a cloud moving across the landscape is a process.... Among the physically important processes are waves and material objects that persist through time. As I shall use these terms, even a material object at rest will qualify as a process.[8]

Salmon is trying to establish, for his own special purpose, that there are cases where one of these concepts—event or process—applies but the other definitely does not, and vice versa. But it seems clear that this is not established by these examples. Why can't a baseball colliding with a window be called a process? If one were studying the shattering properties of the glass, capturing the event on film in order to slow it down and examine it, one would be likely to call it a process. A scientist could surely give a fairly detailed story about the activation of a photocell by a pulse of light; the word "process" would not be inappropriate for what is described. "A sneeze is an event [and not a process]"; but a sneeze could be called a process if it were a subject of physiological or aerodynamic study. And a cloud shadow suddenly moving across a meadow might be called a process in certain contexts, but as it is it makes a perfectly good event.

There is one possible case in which "event" applies but "process" does not, that of an instantaneous event. But it seems likely that these words will thoroughly overlap in their potential applicability to occurrences with temporal extension. One can gain from these examples and discussion a sense for how these two words are actually differentiated in their usage. It is *not something factual about an occurrence itself* that makes it a process rather than an event or vice versa; this depends on what aspect of the occurrence is relevant to the purpose at hand in talking about it. If one is viewing it as something that comes to pass at a particular time and place and is gone, one will call it an event. But if one is interested in the composition of the occurrence, what parts or stages it has and how it goes through them, one will refer to it as a process.

The notion of a nonoverlapping separation in usage of "process" and "event" for different kinds of occurrences, leaving aside these fatal objections to it, serves Salmon's purposes well. His "at-at" theory leaves transmission without any description independent of interaction; it is just succession in interactions and *as such* is distinct from interactions themselves. Transmission is a sequence of events, but not one that is identifiable independently of interactions. But for Salmon this does not prevent it from being something intervening between interactions: his strategy is to speak of it as a "process" that is not a sequence of events but rather is somehow featureless and uniform, an *undifferentiated continuity* of "process" stretching between the observed events. By cleanly separating "event" from "process" he can, consistently with the "at-at" theory, refer to intervening transmission as "process"—thus, as it were, giving it being through the concept as a proposal toward a constructive realism. Salmon seeks to outline a "process ontology," which calls to mind Whitehead, while at the same time insisting on traditional empiricism.

In pursuing the reasons for this separation of concepts one is also led directly to a historical reason for this whole procedure: Hume. Here is Salmon in some closing remarks to a chapter:

> Throughout the discussion of causality, in this chapter and the preceding one, I have laid particular stress upon the role of causal processes, and I have even suggested the abandonment of the so-called event ontology. It might be asked whether it would not be possible to carry through the same analysis, within the framework of an event ontology, by considering processes as continuous series of events. I see no reason for supposing that this program could not be carried through, but I would be inclined to ask why we should bother to do so. One important source of difficulty for Hume, if I understand him, is that he tried to account for causal connections between non-contiguous events by interpolating intervening events. This approach seemed only to raise precisely the same questions about causal connections between events, for one had to ask how the causal influence is transmitted from one intervening event to another along the chain. As I argued in chapter 5, the difficulty can be circumvented if we look to processes to provide the causal connections.[9]

The same objection to conceiving a causal process as a series of events is also expressed in closing remarks to an earlier chapter, during which he achieves a remarkable clarity:

> It is tempting, of course, to try to reduce causal processes to chains of events; indeed, people frequently speak of causal chains. Such talk can be seriously misleading if it is taken to mean that causal processes are composed of discrete events that are serially ordered so that any given event has an immediate successor. If, however, the continuous character of causal processes is kept clearly in mind, I would not argue that it is philosophically incorrect to regard processes as collections of events.

At the same time, it does seem heuristically disadvantageous to do so, for this practice seems almost inevitably to lead to the puzzle . . . of how these events, which make up a given process, are causally related to one another. The point of the "at-at" theory, it seems to me, is to show that no such question about the causal relations among the constituents of the process need arise—for the same reason that, aside from occupying intermediate positions at the appropriate times, there is no further question about how the flying arrow [of Zeno's paradox] gets from one place to another. With the aid of the "at-at" theory, we have a complete answer to Hume's penetrating question about the nature of causal connection. For this heuristic reason, then, I consider it advisable to resist the temptation always to return to formulations in terms of events.[10]

To speak of transmissions as serially ordered events, he thinks, tends to make one think of them as broken up into discrete parts which do not have the continuity required of causal connection. To try to reestablish connections by means of intervening events, also discrete, is to obtain at most contiguity, not connection. We recognize here the mosaic theory of events, which says that events must compose the way objects compose. But Salmon brings this Humean thinking to a new height of clarity. The belief he expresses here is that if a process *has parts, it must be discontinuous and disconnected.* This means that in order for causal connection to obtain, the "process" must be *completely uniform in space and time, with no differentiation into separately identifiable stages or parts.* There is no "question about how [the transmission] gets from one place to another"; this restates the view that *transmission* or *propagation* is not to be described, independently of interactions, as productive activity.

The central claim Salmon is making with all these distinctions turns out to be a stark version of the neo-Humean position on causality, under the heartening revision that the reality of causal connection can, it seems to him, now be identified. The claim is that *there can only be causal connection where there is complete uniformity and featurelessness in time and space along the route of "transmission."* As a causal realist he does not intend to be saying that there is *nothing* connecting the events. There is rather a real, *featureless* intervening "process." Being featureless allows it no independent description, beyond saying that it occupies, in some special sense in which a process can be said to "occupy," all points of a space-time interval. This can be called an object-picture of a process (with no differentiable event-parts), and it is easy to see how it is also exemplified in a "material object at rest," "occupying" the space and time of its enduring existence.

Concerning the spotlight example, Salmon points out that one can make a conditional assertion, saying that a "mark" imprinted on the process *will be passed on by it*, and this tells him that the process is "really there," filling the space-time region. For example, if a white card is placed in the beam of light at any point past the red filter, which "marks" the beam, "we will find the beam red at that point."[11] The transmission is thought of as occupying its temporal and spatial interval as a featureless simple objectlike entity, a "process," simply lying there in this interval, waiting for a discriminative testing and measuring of its effects, but *itself utterly without distinguishing character*. (This is close to the physicist's way of speaking about the field as a "potential" for an effect on a test object.) Thus Salmon would uphold the causal reality of the "unobservable," and also attempt something like a description or a step toward description (constructive realism) with this idea. But note that this success comes at the cost of actually eliminating and exiling from the account something that is known about the particular transmission in this example. Electromagnetic transmission has all the marks of a wavelike process; it is some kind of propagative reiteration of cyclic phases. It is therefore a process composed of a sequence of differentiable parts or events. If Salmon's claim is that in order to be causal it must have no differentiated parts, this not only ignores, but denies, the one thing that science does know about the causal "structure" of this genus of transmission, denies its known physical character. But the traditional flattened analysis of transmission and interaction accomplished this feat already. Given the reduction to discreteness and contiguity, the problem of causal connection arises with equal force for the distinction between transmission and interaction as it does for different events of a single transmission.

Some general remarks are in order about Salmon's appropriation of Hume's analysis of causality for application to intervening processes in physics. Hume's epistemological starting point was sense impressions, which we can understand as the mental reception of the images of objects qualified by the senses. It is one thing to say, given this starting point, that there are no productive relationships to be discovered among these object-images. It is quite another thing to say that given *any case whatever* of events in serial order, "productive connection" must already be without meaning, or is unavoidably barred from the description. In the present appropriation of Hume the matter at issue is a class of occurrences such as light transmission which are not subject to direct observation and can precisely never be "phenomenal," can comprise neither cause-events nor effect-events given a sense data epistemology. Speaking generally, it seems clearly injudicious to impose a particular thought-picture, such as that of featureless "process" or the underlying discreteness/contiguity picture, on physical causality per se. What could we presume about the latter? After all, it appears to have escaped the very mechanistic format that has been fundamental to accounts of physical processes in physics, and this format has not been replaced with another. It has evidently exceeded these simple and habitual pictures of events, processes, and space and time relations that come into play in philosophy of science as well as in physicist's practical conceptions. Since we are without any actual conception of the "wave" in the case of light transmission, the event-mosaic could not be more artificial, and there is no basis for the idea that productive connections would be a problem.

The point of departure for criticism of the empiricist view of intervening causal transmission is simply to point out, as I did in Chapter 3, that it is not a conceptual impossibility that things may be differentiated from one another and yet might also be *connected* in a perfectly sound sense of the word. The example of the color spectrum serves to establish the possibility, though in a concrete conception of physical causality the space and time relations will be different. Especially for causal occurrences whose accurate and concrete description is the central task that must ultimately be faced in physical ontology, it seems a serious and perhaps disastrous mistake to be ruling out connected differentiation in advance. But the problem of causation is not merely the failure to acknowledge the possibility of a differentiated continuity. More basically it is the crude question, What is "bringing about," anyway? An analysis of the question within the simple format of succession in events may turn out something like this: If the engendering activity of an event at some point passes into activity identified with the engendered event, is not this point of passage assigned quite arbitrarily, collapsing the differentiation between the events? What constitutes the individual identity of "events" having specific times and places of occurrence, in distinction from the more extended regenerative "process," if (so far as reflection on causation is concerned) the events are essentially instances of derivation, of "arising out of" and "passing into"? This matter of identity and difference in causation is a theme of Part Two.

Much more successful than the strategy of talking about "process" exclusively of "events" is Salmon's argument for realism from convergence of evidence.[12] The Perrin experiments argued definitively for the correctness of the atomic-molecular model as a matter of science, whereas the earlier successes of this model had not been so definitive. Salmon expands the argument by pointing to what is widely regarded as sweeping evidence for the "atomic" structure of physical reality in general, evidence such as the quantization of the mathematical analysis of the field in units of Planck's constant, a quantity signifying the irreducible "atom" of physical causality. But in carrying the convergence argument beyond the evidence for ultimate constituents of matter and into discussion of the structure of the field, philosophical problems arise that this generalization of the argument covers over; for example, the fact that "particles" is actually only a convenient model for what is yet undescribed remains a potent subliminal impetus for antirealism, if the atomic/discrete analysis is allowed to stand (and the mere widespread use of a term hardly amounts to a convergence argument for realism). The preparation for physical ontology adopts an independent perspective on the issue: the atomic/discrete interpretation of the field, indeed any interpretation in which contiguous relations are foundational, weakens realism by failing to acknowledge that if an actual description of field events is indeed possible, it can only be sought outside Humean-mechanistic conceptions.

Explanation as a Theme: A Mark of Basic Acausalism

In his chapter section entitled "Explanation in Quantum Mechanics," Salmon gives an illuminating exposition of some problems of comprehension faced by physicists. Besides this valuable exposition

there are some vague positive suggestions about particular problems, whose evaluation would require going into those problems.
But in his final estimation he does not feel he has gone far at all in
the matter. In the closing words of the book he cites a central unresolved mystery of quantum physics:

> In his address, "On the Present State of the Philosophy of
> Quantum Mechanics," delivered at the 1982 meeting of the Phi-
> losophy of Science Association, Howard Stein remarked, "The
> problem of reduction of the wave packet . . . remains what it
> has always been: baffling." If he is correct—and I feel fairly
> confident that he is—in claiming that we still do not under-
> stand so fundamental an aspect of quantum theory, it is hardly
> surprising that nothing approaching an adequate theory of
> quantum mechanical explanation has yet been offered. In any
> case, to provide a satisfactory treatment of microphysical expla-
> nation constitutes a premier challenge to contemporary phi-
> losophy of science—one that lies beyond the scope of this book
> and beyond my present capabilities.[13]

I want to make note here of how Salmon's language in this
passage situates him squarely in traditional philosophy of science,
and contrasts with the way I would speak of this same problem of
understanding, namely as a problem of how to causally explain a
phenomenon. He sees the problem as one of a "theory of quantum
mechanical explanation" and of the philosophical "treatment of microphysical explanation." Thus he represents the philosophical
need here as something like *understanding a case of explanation which
already exists in and as a certain achievement of science (quantum theory).*
He is continuing the traditional *issue or theme of explanation*, in
which theories of scientific explanation are propounded, and for
which the Hempel and Oppenheim paper is monumental. This tradition begins with the flat presumption, "Science explains," then
asks the question, in view of contemporary acausal physics taken as
a paradigm, "How does it do so?" This thematizing of explanation
is an outgrowth of scientific nihilism. My own view is that science,
in general, explains, and it does so with accounts of causal activity in
nature, *except for relativity and quantum physics,* which achieve dramatic
technical advances in interaction with certain domains of nature without corresponding advances in explanation. The essential content of
theory in these instances is something other than explanation.

The established results of the researches of quantum physics do not include what I would call explanations, but they do include some quite "baffling" facts. As I discussed briefly in Chapter 1, there are physicists who try to actually explain these facts. They comprise a dispersed group of "quantum" ontologists, working outside the mainstream and in a purely speculative explanatory exertion, so far without effect on practical research (as far as I know). Their specialty is that of quantum measurement theory, in which the perplexity mentioned by Stein is the special concern. This field arose and is sustained out of *dissatisfaction* with the Copenhagenism associated with Niels Bohr, which is the most conspicuous basis in physics itself for neo-Humean theory of science. In other words, this antiorthodox field exists because the baffling character of the quantum measurement event is unaffected by the successes of quantum mechanics. It does not appear to me that any of the existing quantum ontologies meets the productionist criteria for an explanation; I made general comments on them in Chapter 1, and more specific critiques occur in Chapter 8, first section. But my view is allied with these attempts in holding that to supply an explanatory dimension to physics is the business of a pursuit which is independent of the mainstream technical/theoretical procedures of physics, and that explanatory attempts can have validity independently of possible experimental confirmations and applications. A general strategy for a productionistic "quantum" ontology is proposed in Part Two, Chapter 8.

By clinging to assumptions of logico-empiricism, Salmon places himself in the position of saying that the baffling result of physics mentioned by Stein is a crucially important example of an explanation, one that philosophers at some point have the task of *explicating as such*. Since successful and achieving science is assumed to be successful explanation, one's task is then to discover how it is that existing quantum mechanics explains what everyone seems to agree it does not explain. Beneath the surface rationales it is originally because certain ontological questions in and of physical explanation remain in a condition of profound neglect that philosophers have become occupied with the "problem of explanation."

"Probabilistic" Process

According to the passage last quoted, Salmon does not think his project of constructive realism (or "ontic" explanation) has gotten very far with regard to the explanation problems of contemporary physics. However, he does offer a philosophical strategy which he apparently thinks is an achievement in this matter. Quantum physics encounters what appear to be irreducibly indeterministic effects. Hence certain mathematical models in use (which only correlate observed interactions and cannot be regarded as descriptions of connecting transmissions) yield probabilities as solutions. Salmon cogently demands that a contemporary theory of causality face this ineradicable indeterminism. Along with his "statistical" model of (law-based) explanation, he seeks to add to his conception of a connecting causal process the feature of being *probabilistic*, thereby supporting the idea of "causal structure" even for the strange "microcosmic world" of contemporary physics. But his proposal is not only about the indeterministic processes of physics. He wants to show that probabilistic processes are not something unique to this domain, but form a class including ordinary cases. This purports to be a general, probabilistic theory of causality.

He gives as an example the way a pond ripple affects a toy sailboat. He says about this,

> ... there is a certain probability that the boat will capsize when the wave reaches it. The probability will, of course, vary with the distance between the boat and the point of entry of the rock. For certain distances, the probability of capsizing will be unity [the boat *will* capsize] and for others it will be zero for all practical purposes, but for intermediate distances, the probability can reasonably be supposed to have intermediate values.... [14]

Ordinarily, when it is said that there is a certain probability that something will occur under given conditions, it is meant that given a number of cases of these same or similar conditions, the results show a variation which can be calculated as a statistical distribution. Though not in quantum mechanics, in ordinary cases such as this one it is assumed that this variation is due to variations in the conditions which were not accounted for in the description.

Thus to give more concrete meaning to this probability calculus about the sailboat, something like the following context should
be supplied. A group of boys has designed a toy sailboat for rac-
ing, and made it particularly slender and with a large sail. They
wonder how well it stays upright in waves. They experiment by
dropping rocks of various sizes into the water at varying distances
from the boat. After this they test a competing design in the same
way; one appears to be better by a small margin. They make a
rough probabilistic judgment: this one is less likely to capsize. If
one wanted to work into the example the calculation of specific
probabilities for capsizing from, let's say, average pond ripples,
one might try making the boys serious toy sailboat engineers with
marketing ambitions. But the example is already fairly contrived.
Calculation of probabilities does not seem to suit this example all
that well.

But what Salmon wants to suggest by talking about a quantifiable probability of capsizing is something very different from this.
For one thing, the probability is not something about the boat at all,
but something about the wave. The probability is in fact treated by
him as if it were a thing that either *rides along with* the wave, or *is*
the wave itself. He says about this and some other examples,

> The basic causal mechanism, in my opinion, is a causal process
> that carries with it probability distributions for various types of
> interactions. . . . In these cases, one causal process is propagating
> through space-time, and it carries probabilities for different sorts
> of interactions if it intersects with another process of a particular
> sort.[15]

Continuing, Salmon's thinking displays the same propensity that
was found in Harré:

> It seems to me altogether appropriate to refer to these probabili-
> ties as *propensities.*

He uses this language to suggest some definite embedded content
in the wave, just as Harré did in regard to objects. A subsequent
remark is quite clear:

> Causal processes transmit energy, among other things, but they
> also transmit propensities for various kinds of interactions under
> various specifiable circumstances.[16]

Energy is a measurable attribute of a wave. Salmon clearly intends to be saying that the propensity is, like energy, a property of the wave, so that inasmuch as the wave is transmitted, so is the propensity; and also that a propensity is the same thing as a probability. Is a propensity something that can be transmitted from place to place? Is a probability such a thing? If we can agree that these are properties of waves, it would seem they could be transmitted, just as my propensity to absent-mindedness goes wherever I go.

But a wave on the pond cannot be said to have a propensity to overturn the sailboat. It either will or it will not, and the exact conditions obtaining determine whether it will or will not. To say it has a propensity would have to mean that identical waves under identical conditions at times capsize the sailboat, and at times, unpredictably, do not. But this is not the case. The wave will capsize the sailboat if it is large enough to do so. A certain design of sailboat, on the other hand, can have a tendency to fall over in waves.

What about the claim that the wave "carries probabilities"? This claim was derived from occasions on which one says, "There is a certain probability that. . . . " This talk can only be informing about a statistic, computed or computable on the basis of a compilation of cases with waves of varying heights. One would only assign a probability to the event of capsizing if there were considerable unknowns among these conditions in the case at hand, say, if one threw a rock over one's shoulder into the vicinity of the boat. But given an actual wave under particular conditions, a probability for an effect is not a fact about that wave.

Let us take another look at the physicist's notion of a probability wave, since this appears to corroborate Salmon's strategy and is surely one of its sources. The physicist has a case of purely statistical predictability in a class of effects. He devises a formula that can be used to determine the correct probability for each of a set of particular outcomes under given conditions. There turns out to be a group of such formulas that are very different in form yet fully equivalent in their predictions. The idea takes hold that one or more of these formalisms describes or otherwise identifies a kind of physical wave, which seems to fulfill the hope for "realism" in physical theory. Naturally this is taken up by realist philosophers of science. But suppose for a moment that the aim of a realism is causal ontology, as appears to be part of Salmon's intent. To have at one's disposal any one of these formulas, or the whole set, is not to have a description of any generative physical condition or process. If one of them, say, the Schroedinger equation, is so construed, this selection from among the group is completely arbitrary; but this well-known problem is only a sign of something deeper, which is that this approach has attempted to hybridize different modes of knowledge. To call for a narrative explanation is to call for some determinate events: *what happens* to bring about . . . ?—but neither a distribution of probabilities nor a computational procedure yielding such a distribution is an advance in understanding of the kind called for. The "probability wave" is merely a law, a statistical calculus *of interactions*; if it "describes" anything, it is not the propagation phase of radiation. It could of course be a piece of evidence toward such a description. The basic reason why citing probabilities and propensities is not an auspicious procedure for causal ontology is this: If an event A engenders a succeeding event B either directly or by way of intermediate events, than it either precedes event B in time or is contemporaneous with event B as the process of its generation; it either *has happened* or *is happening* to bring about B. A physical cause is necessarily *determinate*, that is, an actual happening, not a probability or potentiality.

As for Salmon's probabilistic account of causality, probabilities and propensities are not things that can ride along with waves. Also, it is not clear how causal realism benefits by running together the different meanings of "probability" as applied to indeterministic and deterministic process, though of course there is deterministic randomness (and, more recently studied, deterministic "chaos"). Such running together of cases is certainly encouraged by the general focus on laws and predictive models, because statistical calculation bridges across the two kinds of randomness.

Conclusion

We have seen that the attempts Salmon has made toward an ontology of causes in physics reach some clarities but have basic inadequacies. Rom Harré, another constructive realist, is quite clear about ontological possibilities other than the "modest realism" of talking about dispositions and tendencies. He sees that causal transmissions in atomic experiments, about which the real difficulties for understanding arise, cannot be described on the basis of an ontology of particles moving against a reference grid,[17] and he says that he is far from knowing how to proceed toward a different ontology which might be adequate for these cases, and can only state an unargued preference for David Bohm's speculations about them.[18] Harré senses the need for a fundamental shift in conceptions of physical events, advancing beyond the elements of moving particles and local reference space, but stops well short of exploring the positive possibilities of such a conceptual advance. Salmon's constructive realism reaches about as far as does that of Harré, to propensities and featureless process. We must conclude that the true objective of constructive realism, to produce a causal ontology suitable and adequate for the domain of physics, is all too dimly perceived. Each constructive realist strategy studied has shown in its own way its roots in scientific nihilism. The position of the present work is that the aims of constructive realism should be formulated in a more auspicious manner by adopting a general point of view radically different from the neo-Humean outlook which goes by the name of empiricism. Its claims would be: (a) that contemporary physics is lacking an explanatory dimension; (b) that the missing explanations—narrative accounts of natural productive activity—might be produced when the special skills of the philosopher are allowed an autonomous role in scientific explanation, in sharp contrast with our recent traditions; and (c) that to make progress in the matter requires a transformed approach to thinking about causality.

PART TWO

Physical Ontology

Introduction

PART ONE WAS PERVADED by the suggestion that what is actually indicated by the long hibernation and reputed death of "natural philosophy" (in the sense of causalist science of radiation and forces) is that its new, revitalized prospect must pioneer its own basic strategy, completely distinct from the technical and mathematical procedures of present-day physics. This is the initial (idiosyncratic) response of physical ontology to the suspension of causal explanation in physics. The obvious next stage is to carry out this venture and actually produce some contributions to the understanding of physical phenomena. This whole point of view is quite alien to the standard philosophical response to twentieth-century developments in physics that informs the discipline known as philosophy of science, which is to elaborate an acausalist theory of science. But physical ontology sees the contemporary situation of physics as a phase of dormancy in a continuing quest for genuine physical understanding, a dormancy that is at bottom only a suspension of constructive ideas in the face of starkly enigmatic findings. On the standard view, no such constructive possibility is foreseen or expected, and the idea that the quest might be pursued outside of physics itself is deeply resisted.

Apparent fundamental impasses for physical explanation have certainly been brought to light by advanced investigation. I claim that these are nevertheless not *technical* problems of specialized physics, but questions for the natural philosopher which, despite a great deal of interest and sometimes unbridled speculation, have never been adequately formulated. The assumption that they must be specialized problems properly dealt with by physicists is in my view simply a natural and understandable prejudice that ignores historical fact. Remember that physics *abandoned* the causalist quest because the mechanistic physical concepts that were presumed basic to causal models failed to apply in the long run and the science was channeled into technical and mathematical imperatives. Inasmuch as natural philosophy in its revived and revised form conceives itself, in keeping with its older tradition, as a different mode of inquiry from that which constructs mathematical models and predicts new entities for experimental detection, it moves onto a turf that today's physics has forsaken for the reason that the basic repertoire of meanings for physical description proved inadequate. What distinguishes physical ontology from the contemporary science of physics is best shown by means of examples, and Part Two presents a series of such examples. Their remoteness from the kinds of things physicists say and think will be obvious. Many signs point toward reconsidering certain contemporary assumptions about how philosophy is related to science and what kinds of questions are legitimately pursued by each.

The procedure adopted here is to investigate a group of physical questions by scouting the limits of those aspects of the questions accessible to the nonphysicist. With luck any genuine blunders will be cases of stumbling beyond the boundary, and hopefully of surviving and moving on; I make no excuse for errors of a philosophical nature. The differences between physical ontology as I conceive it and the methods of physical analysis employed by physics can be sharply illuminated by comparing the two modes of inquiry in their approaches to the issue that has recently become the preeminent perplexity of quantum physics. A reader who is at all familiar with philosophical problems of quantum physics may well have felt his credulity strained since the beginning of this book, because the problem of *nonlocality* seems so completely intractable to any resolution through causal explanation. I will provide a background sketch of this problem, followed by my own response to it, displaying crucial idiosyncracies of my approach.

Einstein was never satisfied with quantum theory, and in a famous debate with Bohr he held that it must be an "incomplete" theory. He argued for this position by calling attention to an implication of quantum theory that came to be called nonlocality, which he thought could only be an incorrect physical prediction. With the help of Boris Podolsky and Nathan Rosen, he devised a thought experiment involving two "quantum" entities (e.g., electrons or photons) that become paired with one another through a certain procedure and then travel away from one another. What quantum theory then predicts is that regardless of the degree of spatial separation they reach, certain measurements on one of these entities will depend statistically upon outcomes of measurements performed on the other, *without any lapse of time* separating the measurements. It would be as if some influence could be "transmitted" instantaneously across a distance, regardless of the extent of spatial separation. This appears to conflict with Einstein's relativity theory, whose predictions had been well confirmed, and which establishes the velocity of light as an upper limit to the velocity of causal transmissions.

In the decades since the Einstein-Bohr debate, strong evidence has accumulated on the side of quantum theory and against Einstein's criticism. John S. Bell showed through what is known as Bell's Theorem that this prediction of quantum theory could, in principle, be tested. Experimenters found ways to actually do so, and the results have given great weight to the conclusion that quantum theory is *correct* in its prediction and that therefore Einstein was wrong in believing that this prediction, which he thought to be an impossibility, showed the limitations of the theory.

It thus seems highly probable if not certain that "nonlocal" connections do in fact exist or occur, since they are (a) predicted by quantum theory, which otherwise predicts correctly throughout its domain, and (b) corroborated by Bell's Theorem and some experimental confirmations. Whatever the exact nature of this phenomenon (and I will assume it is or could be a phenomenon), it is already clear that it poses a unique challenge to explanation in the tradition of narrative causal explanation which seeks to describe generating or mediating processes in the form of continuous transmissions wherever there are effects to be explained. It might appear that any notion of a mediating causality in this case evades completely our available notions of a continuous causal transmission, and consequently it might seem that prospects for narrative causal explanation are radically blocked. But actually, the nonlocality result helps light the way, poignantly clarifying the required course for a shift in ontology of the physical. This is because, as I see it, the project of generating explanations outside of mechanism that I have discussed by way of anticipation requires the development of a way to understand causal transmissions whose descriptions are not enframed by any system or systems of localized space, and it turns out that this strategy is also a precisely suitable basis for the explanation of "nonlocal" features of events. Nonmechanistic and nonlocal *forms of physical causality* turn out to be different aspects of the same background physical explanation.

Before explaining these last claims further I should deal with the objection which says that any causal account of nonlocal connections would find itself in opposition to relativity theory, with all of its intellectual weight, because of its stipulation that any causal transmission is limited to a velocity equal to or less than that of light. I have no less than quantum theory on my side in supporting the factuality of nonlocal connection (though I interpret this somewhat idiosyncratically as *causal* connection), but the theory of nonmechanistic causality is quite independent from quantum theory. Do I then presume to take on Einstein on my own? Remember that I consider Einstein's special and general theories of relativity to be wholly aspect (2) theories, and thus *not in competition* with aspect (1) theories because they belong to a different mode of knowledge altogether. Moreover, in the chapters to follow I propose an alternative explanatory framework to the space and time models constructed by Einstein; this alternative is an interpretation and extrapolation of Whitehead's approach in formulating his alternative to Einstein's two-part theory, an alternative enjoying roughly the same level of empirical confirmation. Physical ontology thus has no reason to strive for conformity with the space-time models of Einstein; furthermore, as Part Two is devoted to explaining, the fundamental constraint on *local* transmissions to velocity c is not something that presents an obstacle to physical ontology once its general positive strategy is accepted.

Next I should point out that the problem of comprehending nonlocality does not appear to be only that of how to fit a faster-than-light or infinite velocity into some current physical theory. The upshot of nonlocality really seems to be the *irrelevance* of spatial separation, regardless of magnitude, to a certain action of one physical existent upon another (and also, incidently, the irrelevance of physical barriers). The discovery here is primarily that differences of locality impose neither character nor constraint on a certain physical interconnection. The expression "nonlocal" is entirely fitting; it is not merely a problem of how to bring into the account a *local* transmission with a high enough velocity (I am not sure an infinite velocity in a local transmission even makes sense, but I leave this to specialized minds).

Explanation outside of mechanism is already conceived as nonlocal in that it proposes to escape all reduction to the elements of local motion: localized background space and present objects. From the viewpoint of this strategy (assuming it is workable), it appears that the usual impasse into which considerations about nonlocality lead is actually a conceptual bind. The raw results of investigation into the Einstein-Podolsky-Rosen problem could be described as the recognition of the existence of *nonlocal causal connections*; in fact the phrase "nonlocal connections" has come into the discussions. But despite this clarity about the nature of the given problem, it is clear that physicists have been *unable to form any concept* of nonlocal causal connection (normally the phenomenon is simply characterized as acausal or noncausal). Neither faster-than-light transmission nor transmission with infinite velocity actually fills the bill, since infinite velocity is a mathematical abstraction that is not amenable to narrative explication, and in any case neither of these, as velocities, specifies any escape from the idea of *transition through successive points in a local system of space*, and therefore neither can describe cases of nonlocal causal connection. Yet it is these possibilities that are called upon for a resolution of the problem (if such is deemed possible at all). The physicist can only work with these concepts, even though they are shown insufficient by the very problem at issue, just as the terms "particle" and "wave" continue to be used even though mechanistic explanation is left behind. To me, these are all indications that a shift needs to take place in the root concept of a physical propagation in order for understanding to be furthered, but that the character of such a shift has never revealed itself. In an unfinished work, Maurice Merleau-Ponty described this situation in recent physics as follows:

> Truths that should not have left its idea of Being unchanged are—at the cost of great difficulties of expression and thought—retranslated into the language of the traditional ontology. . . .[1]

This remark is highly significant for Part Two. The central thesis to be argued can be framed in Heideggerian terms: it is that what can be called the "ontic" question about light, that is, the simple question, what is light transmission? as it might have been asked in full innocence by a seventeenth-century optical experimenter, has in the long run turned out to involve a less easily contemplated "ontological" question, namely, how is the *scientific conception of physical being* to be brought into general conformity with nature as manifested in advanced experimentation? Saying that the new truths are "retranslated" means that even though a change in ontology would be needed if understanding were actually to keep abreast of the new knowledge, the operating conceptual models nevertheless retain the Cartesian ontology in which *res extensa* are foundational, if only in the form of idealizations such as a reference system (grid) of spatial position. The Einstein-Podolsky-Rosen thought experiment, for example, could only make use of the simple picture of "things flying apart to disparate localities." Speculative as well as practical procedures have continued to make use of the conceptual model of fully localized aggregate elements, even where the applicability of this model in actual physical description has radically broken down. If physical or causal explanation is constrained to this class of conceptual model for lack of any alternative, this means that as far as any narrative physical understanding is concerned causality is bound inextricably to locality. My position (in accord with Merleau-Ponty and also with Whitehead) is that it is just this unexamined constraint on concepts applied to understanding physical events that prevents explanations of the "quantum" phenomena in general from being given. The prevailing technical language and applied models comprise at most a substitute or stand-in for what is really needed: a background physical explanation that would make a first sense of radiation phenomena *over the whole range of their features*, from the wavelike structure of propagation to the element of nonlocality. This same scientific situation in which

"difficulties of expression and thought" arise in the midst of continued linguistic-conceptual recourse to an outdated ontology also
obtains with the accounts of cosmic origins and the ultimate dimensions of the universe, though these basic problems of understanding
are not as widely recognized as those associated with the "quantum"
explorations.

The argument of Part Two is that the way to break the spell of
this traditional Cartesian idea of physical being is to develop some
concrete conceptions of actual physical process suitable to the various perplexing phenomena, conceptions that reach beyond the
seemingly immutable connection between causality and locality, su-
perseding de facto the assumption that localized space and occupying entities are fundamental to any conception of a physical event.
It is projected that the nonlocality result of quantum physics is not
an isolated fact uniquely requiring this ontological shift as a specially fitted solution, because this is by no means the only fact that
calls for breaking the causality–locality link. This unlinking is also
precisely what is called for, I believe, by the general breakdown of
mechanistic models for radiation, by the truly "baffling" phenomenon of the "collapse of the wave packet," and by a series of other
physical perplexities. In short, "breaking the causality–locality
bond" is one way of identifying the basic strategy for the general
theory of radiation that takes shape in subsequent chapters, a
theory accounting for the basic array of its manifestations. It means
consciously rejecting the thought-picture suggested by physicists'
use of the term "particles," a thought-picture that tends to guide
thinking by default despite full knowledge of its radical limitations,
namely, that of discrete entities present in space and susceptible of
rest or motion. If I may venture a provisional positive description,
the paired entities with nonlocal connections are instead *wavelike*
pulsations of a certain type which remain interactive throughout
their histories. The concrete physical basis of this possibility will
be made clear by the central hypothesis to be developed in these
chapters.

The differences between a technical discussion of this issue by
physicists and the preceding discussion should be exceedingly
clear: I said nothing about how quantum theory and Bell's Theorem
produce the relevant predictions or suggest the experiments, nor
have I even described the phenomena produced in the laboratory

(which would have been highly technical and outside my competence). What I did instead was promote a certain direction for ontological research by showing that if this direction is feasible, then a phenomenon of nonlocal connection, whatever its detailed observational content, does not inherently preclude causal explanation as everyone thinks. I limited my claim to pointing out that the apparent causal intractability of nonlocality rests on the causality–locality bond, that is, the assumption that any intelligible physical and causal event necessarily involves localized transition through space, and that the projection of a neo-Whiteheadian ontology is precisely the denial of this assumption. By the end of the book I will have shown how, in the course of this ontological project, nonlocality can be accounted for together with a broad array of physical perplexities through a common background hypothesis. The above minidiscussion of nonlocality sets the tone for the rest of Part Two, although in cases other than this one I do describe the phenomena of experiment to be explained since they are for the most part entirely amenable to description in nonspecialized terms, and I provide the actual explanations specific to them as so described.

That a crisis of ontology exists for contemporary natural science is shown most clearly—though by no means exclusively—by the philosophical turmoil in physics, an ongoing and fundamental controversy, over the general interpretation of the "quantum" phenomena, of which the problem of nonlocality is only one aspect. An array of physical perplexities are encountered when the interactions of radiation with matter are studied in detail, and though they present no permanent obstacles to mathematical modeling and technical achievement, they are yet without a basic resolution for purposes of understanding. In the current state of science, there remains only the predictive algorithm to fill the role of "description" at this level of nature, and this facile but disastrous misapplication of "description" and "explanation" has been one of the basic characteristics of scientific nihilism. This misapplication brings the illusion that the functions of science to which these words refer are being fulfilled in a manner compatible with the emerging doctrine that the extent of the natural world in and for science is limited to the attenuated immediacies of visibly present things and measured relations. A related feature of scientific nihilism has been that even as traditional epistemological problematics are being discredited in their foundations by diverse traditions of general philosophy, philosophy of science has been generally retrograde in this respect, adopting the legacy of Hume and Kant transmitted by way of a certain methodological tradition of *physics*. The result is to bind notions of scientific knowledge to epistemological doctrines. Along with the resurrection of causalist physics, science needs to reassert its original freedom from the clutches of epistemology.

Reflections by working scientists on the overall state of affairs encompassing the breakdown of mechanistic models and the nonlocality result have typically proceeded as follows: "This must mean that causal/descriptive explanation permanently breaks off. Are not our 'narrative' physical concepts after all rigidly bound to the 'classical' models of particle and wave? Does this not mean, then, that some things are permanently beyond language and knowledge, so that the quest for scientific understanding of nature has 'reached the end of its rope' for this area, unless quantitative models and technical terminology can somehow be considered explanation? Beyond materialistic causal explanation, are we obliged to move in the direction of idealism or teleology, toward a role for mind, consciousness, and intentionality in the generation of phenomena, if we insist on sustaining the quest for explanation?" Physicists have explicitly raised and continue to raise these questions. But I would urge altering the course of such reflections in their initial stages.

To state the aim of Part Two concisely, it is to show that it is possible to produce a physical account of radiation in which its mode of activity as propagation and its cosmological status, its place in nature, is definitively understood. In other words, a question is here resuscitated that has gone into dormancy since Augustin Fresnel and James Clerk Maxwell: What is light? How does it propagate? As I said, it turns out that pursuing this question requires that a physical ontology or general theory of nature be simultaneously developed. The need is to reestablish a basis for the very expectation, at one time generally upheld, that a certain domain of physical occurrences whose effects are catalogued by science in a sophisticated quantitative knowledge might, apart from these formal and technical scientific successes, be uncovered as a coherent aspect of the interconnected workings of a comprehensible nature. To raise this question again to the light of day, and to *reopen prospects* for narrative explanation in an area where, according to a current vague consensus of scientists and philosophers, it has proven unfeasible—these are the tasks defining physical ontology for the present.

The explanatory proposals I submit here amount to a theory of radiation, covering some related questions about space and time. The strategy to be adopted for producing some first answers to ontological questions about radiation and fields as physical and causal occurrences, reestablishing prospects for the basic comprehension of these realities, can be accurately termed "Whiteheadian." But the primary point is not to do an exposition of Whitehead's physical theory and theory of natural science, whose riches I scarcely tap, but rather to explicate and appropriate his basic guiding intuition, pursuant to the aims of causal ontology. In the final two chapters, Whitehead's alternative theory of relativity is interpreted in its overall narrative aspect (excluding its algorithmic aspect). What marks the present proposals as Whiteheadian throughout is the claim that the category "events" has a specific extension beyond the features of nature from which the notion of simply given, present matter is derived. It is suggested that physical objects (together with certain aspects of space and time) are as Whitehead said "derivative" from, or explained in terms of, certain physical events and processes, and that a physical theory in the full sense will be some kind of concrete explication of the characters and relations of these explanatory events and processes. The expectation is for thorough narrative explainability, simply because the very origination of physical actuality becomes a subject of scientific investigation and narrative disclosure. To follow through on such a strategy of explanation requires developing a concrete conception of physical events and their relations on a basis other than that of local space and occupying objects, since it assigns these the status of manifest aspects of nature, things subject to explanation rather than irreducible and pregiven elements. But I completely reformulate the project of an event or process ontology as Whitehead conceived it. The difference is that I seek to produce narrative accounts of causal structure, whereas Whitehead's physical theory develops a kind of geometry of abstract "events" and their abstract structural relations (apart from a set of laws derived therefrom, which is outside the present concern).

My reason for considering the ideas of major philosophers is obviously not primarily to further academic scholarship or criticism; instead, physical ontology finds itself empowered by appropriating some of the thinking in this area which these philosophers have done. Another philosopher important in this regard is Maurice Merleau-Ponty, of whom I also attempt some textual interpretation. At the time of his death Merleau-Ponty had begun a major work which was to be in part concerned with physics and conceptions of nature, and some drafts and notes for the work are published.[2] I submit that in its own way the present book accomplishes some clarification and corroboration of this uncompleted project (although Merleau-Ponty did not intend to emphasize the working-out of a theory of nature, as the present work does[3]). The third titled section of Chapter 7 draws upon some of Merleau-Ponty's finished writings on the "concept of nature" in which are found anticipatory formulations of some of the content of the planned work.

I must take the trouble to respond to a certain objection to the kind of claims I make in the following chapters, one that will readily and habitually arise for the contemporary mind. The thought will be that any explanatory description of a process causally antecedent to the observable field of nature, and in this sense outside direct observation, can in the end amount to no more than one of a set of possible explanatory models, none of which can have any special claim to truth as against the others, since only a direct "checking" observation could decide among them. A further conclusion based on this doctrine is that the scientific value of such a proposed explanation can only be decided by its *utility* in furthering science; no criteria exist for determining its correctness or incorrectness apart from this utility. Though in important circles these conclusions have the status of well-worked-over and virtually settled truths, they are nevertheless refutable simply by considering some well-known items of scientific knowledge. These solid counterexamples to standard empiricist theory of science involve the recognition of specific physical structures that lie outside the possibility of direct observation. In these cases alternative proposals for causal structure with comparable plausibility do not exist, and there is no reason to think the explanations are subject to scientific skepticism or future revision. I will describe two such examples.

(1) Certain effects observed in experimentation with light, effects commonly known as "interference," have been known about since the time of Newton, but they were first successfully explained by Thomas Young at the beginning of the nineteenth century. The explanation led to the general acceptance of the wave theory of light during that century.

What happens in a typical "interference" phenomenon is this: A barrier with some slits in it is used to discriminate two separate transmissions of light having the same particular frequency (color) and originating from the same source, but proceeding in slightly different directions. When a screen is placed in the region of their overlap, the screen is illuminated in a pattern of alternating bright and dark regions. Young realized that if light consists of waves through space, such a pattern would be produced in the following way: at some points in the region of overlap the cycles of the waves would combine their opposite phases and at other points would combine their same phases, so that an alternating cancellation and addition of phases would result in a repeated pattern of graduated variation. The interaction of separate processes to form these result- ant patterns is known as "superposition," and its mechanism is easily demonstrated with waves on water or sound waves.

From what is known today, it is clear that the conclusion can- not be drawn from the interference effects that light actually propa- gates as vibrational disturbances in an all-pervading substance, since this model has proved unworkable, and since the ether itself vanished from the applied conceptions of physics. Nevertheless, Young's explanation using the principle of superposition remains unchallenged as the way to make sense of these effects. What this means is that a definite *causal characteristic that cannot be directly observed* has been solidly determined by science and for scientific knowledge, namely, that whatever form of physical process it is that we call "light" and "radiation," it has the cyclic characteristic which is required for superposition to operate. If a skeptical philosopher complained against the truth of the explanation that he could never actually view the superpositions of phase in *some process analogous to a wave* resulting in these effects, and so could not be certain that Young's explanation is correct, this would at best amount to an idiosyncratic philosophical view—certainly it would not be science, but rather a futile opposition to accepted and clearly correct scientific conclusions. The discovery of particlelike effects early in this century merely highlights the obscurity of this form of physical process, it does not argue against its reality or even against its potential intelligibility (and it certainly does not speak against Young's explanation apart from its mechanistic interpretation). Similarly, although the old fluid model of electricity is definitely dead, it can never be a piece of nonscience to claim that electric current is *some kind of physical flux or propagative process*. Notwithstanding the claims of empiricist/antirealist philosophers, skepticism is inappropriate to these scientific understandings, and this is because they involve fundamental recognitions that specific causal structure underlies certain natural phenomena.

(2) By the late nineteenth century the theory that matter is composed of atoms and molecules was generally accepted as a working hypothesis. A few scientists were skeptical, and thought that this theory about entities which apparently lie outside the sphere of possible observation would never be more than an ad hoc hypothesis and useful model. Eventually Jean Perrin conducted a series of experiments, involving a considerable variety of independent procedures, and each had the result that it would have if atoms and molecules existed (and none of the experiments disclosed the atomic structure of matter to visual confirmation). After these results were well digested, skepticism about atoms was no longer a supportable position for a scientist.

These cases show that specific physical structure not accessible to direct observation can nevertheless be securely disclosed in the course of science, or in other words that statements about specific causal characters based on features of phenomena they generate can have a *solid* status as scientific truth (an unpopular rubric at present). The explanatory conceptions of nature and natural processes to be given in these chapters fall into the same category as these examples of scientific explanation, so that the vague empiricist complaint that unwarranted extrapolation from the evidence is taking place—since only the confirmatory glance can provide "indubitable" evidence—is an inappropriate response which would miss entirely the kind of scientific knowledge exhibited in the above cases. These examples are incompatible with an empiricist theory of science, so that one has to reject either the theory or the examples.

More than this: I believe that the explanations of radiation phenomena that I am proposing actually combine the distinctive strengths of each of these cases. In the case of Young's explanation, this strength is its unique capacity to make sense of a particular phenomenon. My explanations propose to identify as concrete physical process (without recourse to locality) the wavelike character of light propagation that is established by Young's explanation (a character that until now had been conceived under the constraint of the concepts used in the description of familiar waves), continuing the narrative theory of light after a long hibernation, sans material ether. The strategy concerning the more recently encountered radiation phenomena is thus to entirely eschew the path toward a materialistic wave *in order to* uncover in some detail the exactly suitable causal structure of the wavelike propagation in question. The special strength of the claim for atoms and molecules is the wide array of different facts of observation of which it can make physical sense, and the following chapters cover a wide range of problems precisely in order to make this "convergence" argument for the truth of the overall proposal. The aim is to show that not only can a nonmechanistic and nonlocal concept of physical events be generated by a shift of ontology, but also a new background conception of spatial and temporal relations, which extends the power of the explanation to include questions under the heading of "relativity" (Chapters 9 and 10). Beyond making this general range-of-success argument, it is more important to stimulate and orient inquiry than to address as many physical questions as possible (a scarcely ponderable task). Speculations of physicists opposed to Copenhagenist orthodoxy tend to focus on special problems in isolation, but the procedure here is to point to the *overall* prospect and potential of a shift in the most basic terms of physical description.

A related objection arising from conditioned intellectual reflexes might be simply that since no new experimental result which can be tested is predicted by the initial and general outline of a theory of radiation presented in these chapters, nor is there the mathematical format that such a predictive model would require, the proposal cannot have any scientific worth. This complaint arises from the standard mistake of identifying scientific value with predictive value, a mistake closely allied with the assumption that present-day theories of physics are (paradigmatic) explanations. In fact, it says nothing about the correctness or incorrectness of an explanation if it does not immediately yield new predictions, though this may well affect its utility for the progress of experimental/technical knowledge. To understand what constitutes an explanation, and the kinds of grounds that actually support the correctness of explanations in natural science, one should consider explanations in biology and geology. The likelihood that an explanation is correct depends primarily on whether the proposed generative processes are in fact suitable to explain the observed facts, and on the scope and variety of facts explained. Certainly, as I pointed out in Chapter 1, the timing of discovery of the different things the explanation explains or would explain is irrelevant.

Yet another source of beforehand dismissal of all claim to truth on the part of the proposals here presented is the dictatorial and utterly hollow contemporary notion, the roots of which were traced in Chapter 1, that no investigation of nature or texts can turn up anything but the ideology of the investigator, can attain anything but the expression of a life and a time.

To summarize the philosophical standpoint of physical ontology: In contemporary physics there is a practical terminology designating entities by species and category (which has a purely surface resemblance to physical description), and there is algorithmic "theory" (often mischaracterized as physical description or explanation). Neither of these constitutes an answer to persistent and basic explanatory needs, so that in order for explanation (which in natural science, on my view, is narrative disclosure of physical events) to begin to catch up with experimental research, the traditional elemental concept of a physical process must be replaced by a different elemental concept. If this should prove feasible, then there may be an alternative to settling for the radically circumscribed and phenomenal natural world of empiricist/ nihilist doctrine. Some might want to remind me of Merleau-Ponty's remark that in relation to science "philosophers must find a way nicely between conceit and capitulation."[4] Am I not erring on the side of conceit? As I see it, the project aims to correct a hegemonic imbalance on the side of "capitulation" enshrined in the procedural doctrines of philosophy of science and reflected in general attitudes of thought.

In proposing a shift in the concept of physical being and giving this proposal shape, the ontologist does not dabble in the specialty of physicists; but neither does he proceed in accordance with that tradition of philosophy of science which, guided by the neo-Humean outlook, understands itself as confined to the role of reflecting on specialized concepts and constructions found in science as sources for elaborating a concept of science or a theory of scientific explanation. Physical ontology charts a course of investigation which is different both from traditional philosophy of science and from physics in its contemporary form. This course independent from physics is made possible by the striking fact that the pursuits of explanatory science and technological science in the same domain can under certain circumstances be quite distinct and divergent. Though physical ontology is driven by traditional and time-honored scientific impulses, to a techno-scientific intellectual culture it is an unfamiliar apparition.

6

Radiation and Causality

A USEFUL WAY of formulating the overall aim of physical ontology is to say that it seeks to develop an answer to the basic scientific question, *What is the causal structure of light propagation?*—and to do so within and for a fully naive standpoint, that is, without resorting to technical terminology or mathematics or other specialized knowledge. This brute question of natural philosophy lies in obscurity now, though at one time it was sustained by the provisional answer, "vibrations in an omnipresent substance." The established textbook answer "electromagnetic oscillations" is not an answer to *this* question because it refers to electric and magnetic fields, which are themselves subjects of the same physical-ontological question more broadly conceived. Merely to proceed on the supposition that such questions may have definitive answers goes directly against contemporary habits of thought.

This chapter examines the results of the original investigations that forced the abandonment of the "classical" mechanistic models of light propagation. As such models were abandoned, so were all attempts at narrative explanation of radiation phenomena, due to the ontological limitations whose overcoming was never envisaged. But the primary goal of this chapter is to show how progress can be made toward understanding the wavelike character and other properties of light, simply by practicing that openness to possibilities urged by Samuel Clarke and supposing that radiation and fields of force involve physical events that conform to no mechanistic models, but may be comprehensible in some other way this side of teleology, that is, as natural and "constant" occurrences. The inquiry treats particular problems such as the nature of the light "wave" in the framework of an advance orientation toward the *whole* of the physical-ontological question. It takes the position that the breakdown of the mechanistic wave theory was the first stage in (unexplained) experimental discoveries eventually leading to the disclosure of nonlocal connections, and that this series of discoveries has a single background physical explanation. Stated conversely, the neo-Whiteheadian project sees the recent face-to-face encounter with nonlocality in the course of the study of radiation as a late feature of the general collapse of mechanistic concepts for the description of a class of physical events and processes. Through an analysis of the earliest, simple experiments that signaled the *breakdown of the bond of causality with locality,* the chapter makes a start at developing a nonmechanistic narrative conception of the form of physical activity by which light accomplishes the crossing of space. In the process a background conception of the physical structure of "fields" in general is introduced which is later shown capable of dispelling the fundamental perplexity that nonlocal connection has presented for the traditional conception of physical being.

Only the results of Part Two as a whole can be regarded as an answer to the ontological question about radiation that is even approximately complete. This chapter is devoted to demonstrating the basic vitality and viability of physical explanation by showing how the essential content and structural possibilities of the general notion of a "causal process" are broadly applicable in the description of wavelike process alternative to mechanism, that is, it shows that there is already a conceptual foothold, possibly a secure one, for proceeding with a physical story of what mode or manner of "wave" this would be. This strategy sets its own course of questioning guided by the physical phenomena, and at the same time keeps in mind a general proposal of Whitehead's: that one might speak of events (by way of physical explanation) which are distinct in character from local motions undergone by matter. It is argued in the final section that a precisely appropriate narrative structure for a physical explanation of the famous source of bafflement known by the heavily mathematicized name of "wave-packet (or wave-function) collapse" is made available by this procedure. The overall point is to show that even the complete breakdown of a *particular form* of causal explanation, mechanistic models, does not after all represent a permanent impasse for causal inquiry.

Enigmatic Physical Activity

All observation relevant to understanding the transmission of light has among its conditions the interaction of transmissions with items in our perceptual field: emission from a glowing source, the illumination of objects and the casting of shadows, an efficacy in the growth of green plants, some distinctive phenomena with reflecting and refracting material—and a complex array of effects in the action of light upon matter under experimental scrutiny. We observe the effects of ubiquitous propagations, but we do not see light, notwithstanding the familiar expression "visible light." The explanation of light phenomena must be sought through conjectures about the unseen form of activity that generates them.

That there are transmissions across space interceding in optical phenomena is a primitive scientific comprehension, which is to say, not a theory, given a certain degree of familiarity with these various phenomena. Theories of light have been theories about the nature of the transmission, not about the hypothesis of its existence. The case of light thus bolsters the view about physical observation shared by Whitehead and the critical realists: it would be extremely difficult to explicitly maintain for the case of light that the only real events are a collection of well-correlated and calculable "effects" directly perceived as changes in objects. The effects have the overt physical form and character of manifestations of physical transmissions and interactions.

Before the present century the scientific presumption that the study of observable radiation effects would indeed lead to their causal explanation had found considerable encouragement. But twentieth-century experiments had results which called into question in a thoroughgoing way the prospects for such explanatory inferences. To this day there has not been any satisfactory explanation as to why and how these new experimental effects arise in their specific characteristics from physical propagations, and the remarkable upshot has been that the quest for a narrative understanding of light, of simply what it is and how it propagates, is no longer alive.

For the present purpose it is necessary and desirable to recount the story of these developments in physical science without resorting to the technical language and constructs of physics that are the standard terms in which such discussions are framed. What is needed is a carefully formulated reanalysis of the basic physical observations under a distinctly different approach. The projected goal and the guiding assumptions of this study are alien to the existing professional and popular explications of these issues.

Prior to the historic disappearance of the quest for causal comprehension, scientists supposed that light could be scientifically understood by using models drawn from familiar physical processes such as wave motion or the flight of an object through space. During the nineteenth century the theory that light is transmitted as a wave in an interspatial medium was generally adopted. An important boost to the wave theory had occurred at the beginning of the nineteenth century with Thomas Young's success at explaining the "interference" effects by recognizing them to be patterns of "superposition," as explained in the Introduction.

But at the dawn of the twentieth century, experiments were beginning to reveal major facts about the effects of radiation which were not attributable to any kind of vibrational disturbance in a material medium. In fact, the term "particles" became a more useful modeling term for forms of radiation. Even so, there can be little doubt that Young's explanation of "interference" is correct. It is not merely that the wave model continues to be broadly applicable as does the particle model in different contexts; we can also conclude as a definitive scientific result that light has a wave*like* structure, whatever the fate of the particle model as explanation. The consequence is that as long as it is to be maintained that light transmission is some actual and comprehensible physical occurrence, then any proposed explanatory description must on the one hand *incorporate the general idea of a propagation with a cyclic succession of phases*, accounting for superposition and other facts, yet exclude the more tangible specification of this as vibratory motion in a medium, ruled out by experiments since the beginning of this century. In sum, the starting point for properly understanding the wavelike activity of light is to fully appreciate the scientific demise of the ether as absolute reference space and as material electromagnetic medium, along with the general break off of mechanistic explanation. In view of the more recent findings adding up to the recognition of nonlocality or nonlocal connection, it seems intuitively clear that if a correct narrative account of this form of propagation is possible, it must escape not only the idea of vibration in a material medium, but the idea of local motion in general. It should become clear in due course that there is no concrete solution along the lines of a combination of a mechanistic particle and a mechanistic wave, and that the whole conceptual framework of mechanism has proven entirely naive and artificial for an actual physical account of radiation.

That no example or concrete description of a wavelike though nonmechanistic propagation is immediately available does not establish that such a thing is ruled out logically. The collapse of the "classical" wave conception does not produce a paradox, but a problem (given some persistence in the quest for a causal understanding) of how to understand a propagative and reiterative process which is not, apparently, of any familiar sort, and which also produces certain effects *resembling* collisions of particles with atoms, giving rise to a discourse of particles (the last section of this chapter is a suggestion toward explaining the particlelike effects).

The culmination of nineteenth-century theory of light, before the demise of "classical" conceptions, came with the work of James Clerk Maxwell. By the time of Maxwell's work the wave theory, worked out largely by A. J. Fresnel, had produced the result that light must be a purely transverse wave disturbance, meaning that all displacement in the wave and hence all elasticity in the hypothetical medium of light waves known as "ether" had to occur along directions perpendicular to the direction of propagation, without any displacement by compression occurring along the direction of propagation as in sound waves. Fresnel had simply reasoned that this transverse shearing action without a component of longitudinal compression meant that the ether somehow had this very property. Maxwell carried through these results in an unforeseen way through the determination that light is an *electromagnetic wave*. This insight was attained through carrying out the unified mathematical analysis of electric and magnetic fields in interaction. The overall result was that electric and magnetic fields and light are connected in the following way: An oscillating electric field induces a transversely oscillating magnetic field, which induces further oscillation in the electric field, and so on, producing a continuous propagation perpendicular to the plane defined by the vectors of these fields. The "oscillation" of the wave on this model is variation in the magnitude of the field vectors. A "wave action" with this abstract structure is purely transverse, as required by Fresnel's results. Maxwell's work on the electromagnetic field was a multifaceted success; it unified many phenomena, found confirmation in new experimental discoveries, and even calculated the velocity of light in agreement with experimental measurements.

But does Maxwell's electromagnetic theory succeed in describing light propagation as a sequence of definite *events*? Does this conception of a wave "displacement" amount to a concrete causal narrative, as does an account of sound propagation or a succession of collisions of billiard balls? In other words, does "oscillation in a quantity of force" really clarify a type of event sequence? The basis in observation for talking about forces of attraction and repulsion in general is that objects move toward or away from one another in the relative proximity of one another, and for Maxwell the parts of the all-pervading material ether were subjected to the forces, rather than ordinary objects. Now, forces as they are directly evidenced in observation do not occur only when there is motion, since the object can be prevented from moving when the force is operating upon it. Because of this, Maxwell's electromagnetic theory is not necessarily a conception of wave *motion*. It only tells of periodic variation in the quantities of strains or stresses in the ether due to forces, which is not to tell of consequent undulating motion corresponding to these periodic variations.

Historically, the problem of physically interpreting the electromagnetic wave theory was soon complicated, because the very object undergoing strains due to forces in Maxwell's theory, the material ether, vanished from the applied concepts of physics with the Einsteinian revolution. As a result there arose the additional problem of interpreting a force vector which is purely an "occurrence" in empty space, *without* the usual physical context of some present substance (in this case ether) responding to the forces by motion or resistance. From our point of view today a major limitation of Maxwell's theory as physical description is that it completely begs the question of the *form of activity* of the field, and hence of radiation, because it invokes forces without any story of the physical events which would causally explain the existence of forces in general. *What happens* when a force is exerted, apart from motions that it may or may not produce? An answer to this is necessary if Maxwell's construction utilizing force vectors can amount to a story of a definite sequence of events. What sort of "displacement" is due simply to the existence of a force? Already for Maxwell "vibrations in the ether" was merely a convenient way of understanding and communicating the theory. Nevertheless it has remained scientifically established that light is a propagation through the electromagnetic field whose relation to that field is one of triaxial perpendicularity. But electromagnetic/optical phenomena in general have remained without any understanding in the sense of descriptions of the physical processes that produce them, a fact that Maxwell seems to have fully appreciated given his view that true physical explanations belonged to the future.

The wave theory of light had accounted for all effects associated with reflection, refraction, polarization, and superposition ("interference"). But in the twentieth century, in addition to the downfall of the ether, the wave-mechanical explanation of how electromagnetic radiation is absorbed and emitted by matter did not withstand detailed exploration of such interactions. These experimental studies will be put in focus here, because they are a source of detailed evidence toward the ontological understanding of light.

The breakdown of the wave model began with an experiment carried out by Phillip Lenard in 1902. The experiment was set up to measure the effect of light on a photoelectric cell, with the arrangement such that the intensity of the light could be diminished to any degree desired. Light of a single uniform frequency was used. If the intensity of the light was made extremely low, the cell would register separate individual pulses of light coming one after another. Reducing the intensity further, it turned out, would slow down their rate of occurrence, but would not change the quantity of effect (in terms of the measurement parameter "energy") produced by each individual pulse. Their individual efficacy could, however, be increased or reduced by varying the *frequency* of the light, which left the rate of occurrence of the detection events unchanged. This is contrary to what would be expected if light were a continuous process of vibratory motion. If it were such a process, its *frequency* would determine the rate at which cycles of the wave arrive at detection material, not its intensity or brightness as the experiment shows; and intensity would be a function of the kinetic energy of vibration, a quantity attributed to the individual wave cycle, not to their rate of occurrence. But something like the reverse is what is actually observed. When it comes to the detailed measurement of the properties of light, which entails interactions, "frequency" turns out to be a parameter for the energy or efficacy that individual pulses of light contribute to interactions, and "intensity" is measured as the numbers of these obscure individual occurrences ("photons") arriving at the detector in a given time period. These relationships show that to conceive the incoming events antecedent to the interaction of light with matter as periods of mechanical vibration is at best overly simple and is probably fundamentally mistaken. The phases of propagation engender one another and generate the interaction event in a sense that so far simply escapes thinking.

But the broader facts insist that there is some physical propagation with a wavelike structure. Is there a way to begin to understand this structure in view of recent knowledge? We can only begin with the very beginning. In order to produce the individual measured effect in the ultra-low-intensity experiment, the individual phase-event of the propagative process arrives (in some sense) at the detection material. By specifying the character of the propagation in this deliberately vague way in terms of individual phase-events, I am trying to avoid introducing comfortably clear pictures that would prejudice an emerging description at the margin of physical understanding, and also to maintain a basis in the description for the wavelike characteristic necessary for superposition, in light of Young's solid success at explaining interference patterns. We should not allow the familiar talk of the photon or light particle to cause us to slip into picturing light as a train of objects in motion arriving at the screen, since this directly clashes with the fact of superposition, especially as illustrated by the well-known double slit experiment: there cannot be a single particle that passes through one or the other of the slits and then interacts with itself to produce the characteristic double slit interference pattern. We should attempt to steer our way through accurate, if limited, characterizations, while avoiding paradox.

Each arriving phase-event contributes a quantity of energy to its detection (interaction) event, and this quantity is a measure of what under normal circumstances is called "frequency." Einstein discovered that these measured interactions could be given the right equation, solving a major theoretical problem, if this relation of frequency to energy were given a certain constant of proportionality, an infinitesimal quantity which Max Planck had earlier discovered in a related connection and which is known as Planck's constant. The success of Einstein's solution means that the propagative pulse arriving at the detector always brings with it a definite integral number of unit contributions to the energy of the interaction, a number varying according to frequency, in a single "bundle." Here again we are poised on the edge of paradox with the suggestion of a moving object model; also, "energy" here is a simplification for the more complex parameter "action." One can surmise that the individual pulse of radiation itself has a composition of events, of causal parts, the "atoms (or quanta) of action" whose measure is Planck's constant. Through the latter we know of an irreducible "grain" in physical causality composite of radiation and fields.

To insist on sustaining talk of causally connected and composed propagation and interaction events conflicts with the conventional account of these early "quantum" developments, which speaks of an encounter with fundamentally discontinuous process; but in my view this standard conclusion, which would in the end thwart the significance of "production" in causal narrative, is not warranted. Neither Planck's constant nor the discreteness of atomic interactions demonstrates a fundamental discontinuity. Why should the "atom of action" composing the pulse of radiation be atomic *in the sense of discrete*, unless one insists on what I have called the discreteness/contiguity preconception about events? (This whole language of events composing the pulse might seem to trained minds a peculiar prosaic concoction, but this is the very idiosyncrasy to which the study is committed, that narrative understanding be the goal at all times.) And what is true of observed patterns of effect, for instance their discrete overt character or discontinuities in measured energy exchanges, is not necessarily true of propagations antecedent to these effects whose actual description is unknown. As I have been pointing out all along, to describe the antecedent processes (if possible) would in fact be a different procedure entirely from the useful modeling of their effects in terms of particles or discrete "quanta." One must keep in mind, notwithstanding the empiricism and operationalism of *practical* physics, that it is events *underlying* the observations that are the quarry of narrative explanation, not characters and quantitative relations of observed events of interaction with matter. The prevalent models and thought-pictures are merely convenient similes, because the phenomena show organized structures of individuation whose basic form so far eludes scientific thought. I submit that there is nothing logically preventing a unit event of propagation from being causally and continuously connected to its predecessors and successors and also giving rise to a discrete measured effect, and the final section of this chapter explains how this might take place. The familiar conclusion of discontinuity is assisted by the assumption that any possibility of continuity and connection is tied exclusively to the model of a mechanical wave. But the present study ultimately seeks to show that there is a distinctly different possibility for the general description of a periodic and connected process with suitable properties.

Though one may persist in the supposition that light transmission is some real, physical process and that there is a live possibility for its description and understanding, even these very basic experimental results seem to leave one at a loss as to how to proceed beyond this point in the quest for a concrete causal story. However, this chapter is making the simple point that though experimental science has advanced beyond the mechanistic wave model construed as physical explanation, this in no way leaves an inquirer without any conceptual basis on which to proceed with an inquiry into physical causes. Concepts exist, I maintain, both for successful (if tentative) application to the wavelike characteristic of light, and for developing an account of the resemblance of certain effects to the action of particulate entities. Whether these conceptual applications amount to a model or models which are of use to *practical* science, even potentially, is not a question with which this book is concerned. The aim is only to make a *first sense* of things, which entails steering clear of paradox and incoherence, careful characterization of the phenomena, and at some point, convincing causal explanations.

The rudimentary conceptuality that has so far been applied is this: light is some propagative succession or reiterative–genetic process in which the individual parts or stages have a further causal composition, the irreducible "grain" in all physical process (not to be pictured as composition by *discrete* entities) that is determined by mathematical analysis and appears in the equations as Planck's constant. Little concrete understanding has thereby been achieved, but it is demonstrated that so far, inferences from the evidence concerning explanatory physical structure encounter no intrinsic impasse or paradox. This is a small and very particular point, which should acquaint the reader with the pace to which Part Two overall is geared.

What I would like to emphasize here is that however bereft science remains of an answer to the narrative questions physics abandoned—What is light? What goes on in the physical field?—it already possesses substantial knowledge about radiation as a resource for pursuing the questions, namely that it is some wavelike, causal, propagative-regenerative process, and that it has a detailed structure whose basic manifestations in ultra-low-intensity experiments are not so very complicated in themselves, but whose discovery was the beginning of the complete collapse of mechanistic explanation in this area.

Empty Space Events?

It is my view that Alfred North Whitehead hit upon the right direction for furthering the understanding of the physical properties of light; whether and to what extent he achieved actual physical explanations is another matter. Shortly after Einstein's two-part theory of relativity took hold in physics, Whitehead strove to construct a different sort of theory for the category of observations that Einstein's theory was designed to cope with— not those of the "quantum" area, as the direct study of radiation came to be called. One difference is that while Einstein did away with the ether as a spatial reference for motions and propagations in his theory, Whitehead preserved the ether, though he proposed to transform it completely. The alteration he introduced was that his ether was to be an "ether of events" *rather than an ether of material pervading all of space. His summary claim for the ether of* events was that "something is going on everywhere and always"; but this was not to mean that a substance is present everywhere undergoing vibration or oscillating stresses. In remarks introducing the ether of events he said:

> Thus primarily we must not conceive of events as in a given Time, a given Space, and consisting of changes in given persistent material.[1]

In this statement Whitehead is clearly suggesting that there may exist actual occurrences of a type for whose description neither occupying matter nor a backdrop of reference space are presupposed. He is raising the possibility, in other words, that physical events may not necessarily be instances of matter (or point-objects) in motion or be analyzed into such motions. For him, the very structural basis of a mechanistic process is something from which natural science must become conceptually disengaged if it is to adjust intellectually to some truths disclosed by physics. In this guiding idea, related to his critique of "simple location,"[2] can be found an intuitive anticipation of the need for a "nonlocal" concept of physical causality that would be encountered (but not recognized as such) by later quantum physics.

Applying this line of thought to the results of the investigation of light which have been recounted so far, a Whiteheadian approach would argue that the breakdown of mechanistic models does not mean the end of naturalistic explanation, and that a conception of underlying events might be constructed exactly in accordance with this development. Can this general suggestion of an "ether of events" have any scientific validity? Ether-based theories have been proposed by physicists in the twentieth century, though their status in physics is always speculative and idiosyncratic, and usually the contemporary ether is little more than a name for some kind of substratum with certain mathematical properties. Philosophically, the present-day issue of the ether shapes up as a dispute between two camps: those who believe that there must be an ether as a substratum of fields and those who feel that such a thing is obviously outdated. Here the Whiteheadian ontologist interjects with a question: What in this dispute is meant by the claim for the existence of an "ether"? Does it mean there is an omnipresent substance permeating both empty space and atoms of chemical substances themselves? Or does it simply mean there are *things going on, that is, physical processes*, at all places and all times? The anti-ether forces have a strong argument in that the ether of the first claim seems clearly to have outlived its usefulness and its place in theory. But what about the second and possibly more general concept of the ether? The physicist W. H. Watson, who is in general admirably advanced philosophically, says this:

> Can we ever have a space free of physical processes? Of course not. We have to introduce noise to stand for our ignorance, so the idealist concept of the vacuum state must be recognized for what it is—an aid to mathematical theory. It does not stand for a physical vacuum, for we believe that quantum fluctuations are occurring all the time—virtual photons being created in the electromagnetic field and virtual pairs in the electron-positron field. That is our physical view of the matter.[3]

I believe most physicists would respond similarly to Watson's question (though the specifics might easily become even more technical and recondite). If I am right, this means that even though the mechanistic ether is surely dead and buried, physicists unhesitatingly believe in the physical existence of the "ether" in Whitehead's sense! But what Watson says here does not prevent him from saying at other points in his discussion that the ether is an "imagined world" as opposed to the real world of observations.[4] Overall he appears to be saying that the real physical processes of the ether can only be spoken about as a fully amorphous background "noise" and treated with some useful but misleading models. This would be an expression of what I called nihilism version (a) (see pages 9–10), that field events are beyond the resources of narrative description.

What about the *notion* of empty space events? Whitehead did not intend nonmechanistic events to be confined to instances of radiation as transmission across empty space. They were to comprise also, for example, "subatomic" events, so that an atom of matter itself would be something resulting in some way from the composition of background events. The essential idea of nonmechanistic or premechanistic events is that they do not *depend* for their actuality or their character on locally present objects, so that "subatomic" events, for instance, are not themselves motions of localized objects smaller than the atom and composing the atom, but some other form entirely of events and event-composition. Whitehead's approach would explain the various subclasses of "objects" in terms of events, as those portions of field events that are in a special sense "occupied," so that "matter" and "empty space" refer to "occupied" and "unoccupied" field events, respectively[5] (this usage of "occupation" that is intended differently from the traditional meaning as applied to matter and space is problematic and appears to have been dropped by Whitehead in favor of other technical terminology).

"Events *in* empty space" has the oxymoronic suggestion that the space is after all occupied; it is better to say that "empty space events" designates an as-yet-unknown relation of certain events to the space of their occurrence. A more complete designation might be "events related to space in a manner independent of the empty-occupied contrast," since the relation to space would be the same for occupied as for empty space insofar as radiation penetrates matter without interacting and inasmuch as matter itself is thought to be ultimately constituted of fields. The problem is to clarify the nature of an event that is supposed to be distinguished from all forms of local motion. How would its distinctive, determinate character be described? *Some* positive content fitting the general concept of an "event" has to be given, beyond simply the employment of this word, especially if the aim is to formulate causal explanations. What could this content be, if it cannot involve any kind of motions of matter or even of point-objects? Can the study of Whitehead's explanatory projects assist us in furnishing a more concrete physical meaning to the general causal concepts which we have been at pains to keep in use for understanding radiation?

Whitehead's account of his event-ontological theory of relativity is only minimally helpful in this regard. As I have noted, his overall aim was not to produce a physical explanation of radiation phenomena. Whitehead's event construction displays his mathematical gifts, and resembles systems of geometry in which the definitions of entities such as points and lines are given only by and within the system, not as independent axioms. The "events" have only the relational characteristics given by his technical terms: *extension* (by one event "over" another), whose meaning is neither spatial nor temporal extension, but an extension from which these special modes of extension are derived; and *inclusion*, meaning simply that some events form parts of others. Extension and inclusion in fact appear to be aspects of a single interrelational property of events, in that primordial extension takes place in and through the inclusion of events within other events wholly and in part.

It is fairly easy to see how events in general can be said to have extension and include one another. Changes and motions extend spatially and temporally, and any actual motion can be viewed as a participant in some broader motion in nature, which in turn can also be so viewed, and so on; for example, from the standpoint of the sun the motion of the earth is in a sense included in the motion of a cyclone on the earth. But we know only familiar examples to illustrate these relationships. What would they mean as applied to events quiet *unfamiliar* and so far rather opaque to knowledge, the events forming the "ether" of physical space (empty or occupied)? My own interpretation of these relational terms as applied to the physical field is provided in the course of these chapters; the long-range interpretive procedure is to show how genetic structure as a fundamental concept of physical causality rounds out the Whiteheadian view. For Whitehead, that natural events include one another in a primitive and irreducible sense is the key to the interconnectedness that escapes the Humean picture of causation as discrete or contiguous series. But concerning cases outside the order of the familiar the questions immediately arise, What sort of inclusion? Extension in what sense? What *goes on* in these primal natural events? What changes take place? These questions find no detailed answers within Whitehead's project, because all that goes into the theory are these relationships expressed in abstract constructions.

Whitehead's basic narrative claim is that the explanatory range of "events" extends beyond mechanism, indeed in such a way that "events" are ontologically prior to the ordinary space, time, and material that are involved in experimental observations such as the relativistic measurement phenomena that spawn his explanatory project. Such events, whatever their nature, form a background to any and all physical circumstances, and ordinary space, time, and material are with respect to these events "derivative" physical elements. The question for a causal-ontological interpretation is whether "derivation from" in Whitehead can be construed as "explanation by" in the concrete sense of a narrative of generation describing specific physical events and processes. The aim here is to interpret the event ontology from the standpoint of an explanatory project with a different orientation than Whitehead's, namely, toward a revival of causal-narrative explanation in the area of radiation and fields.

Whether or not Whitehead's event construction succeeded on its own terms and in its own aims, these relational terms of "extension" and "inclusion" in Whitehead's texts, defined internally to the theory, do not flesh out the events themselves with any content of "something happening." What would this require at a minimum? For something to be an actual event with extension in space and time there must be some mode of differentiation between its earlier and later stages. In cases of local motion this is a differentiation of spatial locality through temporal moments. In such cases, that *something happens* (apart from an empty duration of time) is only possible because of the contrast between the object and the field of space through which it moves: it becomes present where it had previously been absent, and passes through a continuum of places in between. This is the way stages are differentiated in cases of local motion: in abstract terms, purely as differences of spatial position involving the relation of occupation. But consider the possibility of "empty space events," leaving aside for simplicity those Whitehead calls " 'occupied' events." If such a thing can be concretely described, its differentiation of stages will have to be conceived on some other basis than the contrast of presence versus absence of an object or objects. If the latter were the only possible basis for physical differentiation in general, "empty space" could only mean a uniform extended region which is not occupied but could be occupied, a condition itself devoid of any features which could differentiate stages of events, features such as the moving object occupying first one place, then another. The elemental structure of a mechanistic account—an entity present in space, an entity in motion through space—could not play a role in the description of a true empty space event.

One possible objection to the idea of events lying outside the mechanistic framework of description is that of operationalism. Einstein and others have established, it will be said, that an event that is not located in reference to some material body or part thereof would *have no* "place of its occurrence," or at least none that is "meaningful" or "useful" to science; meaningful scientific discourse in the domain of physics is thus confined to frameworks of localized space in considering events. Robert T. Herbert, a Wittgensteinian philosopher, argues forcefully for the possibility of an *event-place* that is specified neither by reference to material bodies, nor by reference to an "absolute" system of spatial positions. (He also affirms, quite interestingly, that an event might even *"'bring about the existence of a place,' "* quoting from G. E. M. Anscombe while calling attention to differences in the way the two philosophers understand this phrase.[6] Though the explanations in these chapters end up following out precisely this possibility purely in the course of investigation, it is another matter to think this thought on its own terms in the context of Anscombe's or Herbert's work. Philosophers have skills for assessing pure formal possibilities if thinking calls for it, and their thought is best appreciated in its formal simplicity.) I find Herbert's critique of the "coincidence assumption" compelling, but apart from this, the present study has been antioperationalist from the outset in that it has sought to identify and delimit the function of such methodological doctrines of physics in and for knowledge of nature.

In order to have a role in a narrative physical explanation, not only will empty space or nonmechanistic events have to be at some point accepted philosophically as a general possibility, but also such events will have to be identified and understood as a specific class of physical occurrence, and this means at least that they will have to be conceived as having stages or *contour* of some definite kind. This means that a concrete description will have to apply *some* unforeseen concept of differentiating structure. In addition, this structure must meet a specific requirement: because of the present commitment to causal explanation, Whitehead's "derivation" of basic features of physical reality from events must be understood not as mathematical derivation, or as any sort of derivation other than *bringing about* and *producing* attributed to these events. But Whitehead's "event" and "process," at least in these writings on physical theory, have only such basis for causal structure as can be derived from the relational modes of extension/intercomposition formulated abstractly.

Rather than getting involved in the intricate details of Whitehead's event constructions, the most important requirement for understanding and interpreting Whitehead's theory of empty space events is to develop a sense for his radical fundamental strategy in theory of nature as outlined in Chapter 1 (pages 31–34). There I claimed that Whitehead carried forward in an original way Faraday's mature insight that the causal structure of the field might be a property of physical space as such. For Whitehead, "field" identifies the fact that transition in "passage" is the fundamental structure of natural being, surpassing the traditional presumption that a physical explanation presupposes things present in space as the subjects of motion or other localized change. From this point of view, to explain forces in terms of "ether" or special "fluids" is to invert the true relation of ontological priority between "process" and "substance." There are active and effective fields because nature as a whole *happens*, that is, *passes*, and within this irreducible transition local events such as field phenomena are traceable to antecedent events conditioning them; this is the bottom card of the account, on the basis of which differences in the detailed structure of fields must at some point be explained in the course of physical ontology. A major aim of Part Two as a whole is to understand and apply this ontological possibility—if such it is—more thoroughly and concretely. The concern at present is to grope our way toward a background theory of propagating radiation with the help of these suggestions.

Earlier it was pointed out that light propagation could still at least be talked about as genetic succession in events, as *phased process* in this abstract sense, even after the mechanical model of wave action is abandoned—and indeed, that at least this much description must be correct if light is some real propagative process that can effect superposition. A more concrete concept of *differentiating structure* is called for here just as it is in a causal-narrative interpretation of Whitehead, but from a different source: an accumulation of experimental knowledge, which, while requiring that a wave*like* characteristic be retained, also removes the possibility that this is adequately and correctly described as vibratory motion. In the attempt to interpret Whitehead's theory as narrative causal explanation, an unforeseen concept of differentiating structure is also demanded, but in this case the need is for a way of understanding in general a causal process consisting of nonmechanistic (empty space) events.

One can give arguments showing the advantages of a White-headian "event ontology" as opposed to a "substance ontology" for specific areas and problems of physics, and thereby join the ranks of a few eccentric theorists who appear to have little immediate prospect of overturning the regnant views;[7] but there is a mere basic point to be made, which is that as soon as one allows that light propagation is some actual physical process whose structure does not conform to any mechanistic causal models, one has already accepted what is called an "event (or process) ontology" as the premise of any physical understanding.

Causal Contour

I will now consider more closely the potential for successes in understanding contained in the philosophical strategy that has begun to take shape.

The need has been disclosed for some structure to fill in a conception of nonmechanistic, wavelike physical process. But this conceptual need, at least in this rough and initial form, may already have been met. For a provisional description of the wavelike characteristic of light, what may serve as a concept of structure and differentiation is *the causal relation itself, as simply the production of one stage ("event") of a process by another, or alternatively, the emergence of one stage from another.* The concept is one of causally connected differentiation, or in other words, genetic succession. *This* "model" is not refuted or outdated by experiments (despite talk of "discontinuity" by physicists, which as I indicated results from the prejudices introduced by mathematical analyses and by the use of convenient and clear models). It brings to bear in an abstract rendering a general cyclic structure as required for superposition and other wavelike effects. However, it is not yet clear whether and in what sense this application of the abstract and general concept of a causally connected and differentiated process amounts to a *description* of the concrete physical processes with which we are concerned. Perhaps no such description is yet being given, but instead, the elements of a language of physical narrative are being mobilized for the initial stage of the inquiry, a language developing in response to a set of challenging discoveries. At this stage, in any case, only a small portion of the evidence has been brought forth.

An objection even to this very general concept of causal succession comes immediately from the corner of the Humean tradition on causality: "You may be speaking correctly of a series of events, each with its own placement in space and time. But what is needed is presumably a story of *how* one event brings about another. Will not any events supplied for this story of transition be also discretely situated in space and time? Connection cannot get established."

My Humean adversary has introduced surreptitiously the idea of discreteness. I was not thinking of the events as discrete, like objects lined up in a row, but as connected, yet individuated— which has to mean connected by transitions, like colors in a natural spectrum, though involving space and time in a different way than a qualitative variation ordered as an array in space. What, ultimately, is my story of transition? This question leads beyond the present stage of discussion and into the next chapter. So far, "causal transition" has functioned as an abstraction; but a story of transition is not ruled out by discreteness in the events, because *the conception is not one of discrete difference,* but of connected difference. Why *must* succession in events be thought fully analogous to a row of bricks? Nothing, I submit, is flawed or question-begging about the general notion of genetic transition, though it may need to be philosophically flexed and exercised in the early stages of some new and extraordinary applications.

As I pointed out at the start of Chapter 3, the limitation imposed on the meaning of a causal process by the Humean analysis is in essence the same as the confinement of causality within a mechanistic understanding. If events in a series are, for example, translatory object-motions, then individuation in the series is marked by collisions. When a collision is the very basis of individuation in a succession, it is analyzed as spatio-temporal contiguity between the events. A wave process (conceived mechanistically) also individuates at space-time slices marking (arbitrarily) beginnings and endings of cycles. The effect of confinement to these possibilities is that events are pictured exactly as side-by-side objects in a space having an additional dimension of temporal extension. In the original Humean thinking this object-reduction occurs as a result of the idea of sense impressions as the elemental content of empirical knowledge ("impression," "percept," and "sense datum" all reduce the content of perception to the aggregation of the flat results of causal processes prior to any perception, the traditional cul-de-sac of theory of perception). Both the mechanistic physical analysis and the Humean procedures end up picturing events as discrete. Individuation through transitional connections is a clear and quite possibly applicable counterpossibility to this picture. Physical events which succeed one another in this general mode, rather than as contiguous or completely separate, would be analogous to succession in biological reproduction, so that one event (or phase) of a process would be in some sense *born of* its predecessor. Connected difference has a clear illustration in the biological case: there are distinct individuals—parent and offspring—and there is also connection through phases, namely, transitions from embryonic form to adult form. An ordinary wave in a medium is already broadly comparable to this, although it is a homogeneous process in that it lacks the nonarbitrary positioning of phases within the cycle that occurs with a genetic succession of individual organisms, which already suggests that the latter is a more accurate analogy with light propagation in view of the ultra-low-intensity experiments. Without relying on a metaphor drawn from biology for any concrete explanation, my suggestion is merely that the general concept of regeneration might encompass a "quantum" process whereas mechanistic conceptions cannot, and if so it would avoid the usual reduction to discontinuity. In other words, the required individualization of light into "photons" might be accomplished with the aid of the abstract idea of genetic succession, dispensing with the mental picture or the verbal suggestion of a train of discrete objects.

A possible objection to the biological simile is that one is not likely to say that in giving birth a mother *causes* a child; here there is a difference between "bringing forth" and "bringing about." But so far I am only suggesting analogies to give thinking some initial bearings. The idea is that certain *as yet unilluminated* cases of physical causality, which ultimately constitute all physical process, might possibly be intermediate in structure between biological genesis and meanings of "cause" derived from familiar examples. This is not to stretch a meaning, but to hypothesize a possible meaning: the detailed and concrete concept of physical causality to be applied in the study of radiation may retain an element of analogy with biogenesis even though the two contexts of meaning are very different. The procedure is to blaze a process of conceptual enhancement that can only ultimately find itself in the course of the physical explanations that guide and motivate it.

Now it is important to call attention to a potentially troublesome ambiguity that lies at the heart of the abstract "productionist" concept of causal/genetic process as differentiating transition. It seems to involve two curiously entangled modes, namely, *regenerative succession* in a particular series of events, with transitions *connecting different* events, and the differentiating transition that is the course of development of a single event of the series in question. There is identity to the individual event as a process of development, even though it is essentially engendered and engendering. What is peculiar is that identity and difference are mutually pervasive in the concept. In less philosophical language, "process" and "event" each apply *specifically* to aspects of the concept, though never *exclusively*. A full comprehension of these relations will unfold in due course.

My point in this section is not that an abstract and general concept of causal succession, clarified in this limited way, succeeds even partially in answering the ontological questions about radiation. Instead, simply insisting on the general intelligibility of the physical structure of radiation and going no further is a way of making certain points; for instance, that the Humean reductive analysis of the causal relation is only a distraction from the recognition that the concept of this relation as a connecting relation can be applied *intact*. The point is also to sustain the intuition that there may be no warrant for assuming that mechanistic conceptions and models are a *primitive requirement* of causal explanation in the domain of physics. This is demonstrated positively simply by *proceeding* as the inquiry has so far with the initial stages of causal narrative in cases for which mechanism can be assumed to have permanently collapsed as an explanatory approach. This is possible because talk of "production" and "propagation" has a narrative structure conceptually independent from mechanism, namely, differentiation through causal or genetic transition. The language of causality itself thus remains entirely suited to an overall characterization of enigmatic empty space wavelike transmission, and one would expect the "productionist" concept of causation to have further application in specifying its detailed structure. It is, in other words, perfectly possible *as a first understanding*, if the facts call for it, to conceive the crossing of space during an interval of time as accomplished not through any form of local motion, but nevertheless as accomplished through some form of successive engenderment of transitionally connected phases. Also (to go out on an anticipatory limb), instead of a particle (or for that matter a wave) arriving at the site of interaction with detection material, we adopt the less sharp-edged alternative understanding that the interaction is a terminus of a certain genetic transition, the propagation being its antecedent condition in a sense which at this point remains obscure.

Merely to have affirmed the conceptual possibility of serial causal contour, a vaguely conceived event topography outside the framework of mechanism, as a feature of empty space is not to have furnished a concrete narrative explanation of light propagation; certainly the problem of the "medium" is not addressed. There is a set of mysteries in the experimental effects of radiation, and one would hope these could be given explanations as solid as the explanation of interference patterns as resulting from superposition. If successful these should bring into relief some additional structure of the propagation. What I will claim for the inquiry so far is that it sustains the initial thrust of causal explanation, steering clear of pitfalls and objections. The gist of my argument so far in this chapter has been this: No amount of obscurity in the detailed structure of propagating radiation should prevent an inquirer from speaking with assurance about the causal reality of its *known* general features, nor from pursuing further the understanding of its detailed causal character based on observable effects. By itself this claim is unassuming, but it opposes some well-established contemporary beliefs. It is important to endure in full awareness for the moment the fact that we have at this point no idea whatsoever of the "medium," context, or concrete physical meaning of genetic differentiation in empty space events.

As consumers of the popular literature of physics and scientific cosmology, we are used to witnessing a procession of intriguing mysteries and dazzling proposals, where the hope is that a solution can be wrought in a formula, a graph, a word, a phrase; but physical ontology finds it necessary to proceed in minute stages, abiding in the *questions* and in the quiet movement of the possible, away from all technical ingenuity.

The Transition in Interaction

Chapter 5 included a remark by Howard Stein, cited by Salmon, making reference to a deep perplexity that was uncovered by quantum physics but finds to remedy in the success of quantum theory. It concerns the character of the interaction of the individual pulse of radiation with matter. Though to all appearances radiation propagates as a wave front of some kind spreading through space, when the individual wave front is absorbed by matter this interaction occurs at a pinpoint atomic locality. The transition between the spread-out propagation and the interaction, presuming there is such a transitional connection, is a complete mystery. This "collapse" event is certainly nothing observed, and being essentially an antecedent structure of the primitive event detecting radiation it is not itself something whose causal characteristics could be probed experimentally. Indeed the physical reality or nonreality of such a transition seems to have no practical import, and physicists can pragmatically suppose that there is no such thing, and that the interaction simply occurs as a "quantum discontinuity." (Often the problem is unnecessarily complicated by uncritically supposing that the propagation prior to the interaction is "described" by the laws of quantum theory, so that the localizing interaction is interpreted as the "collapse of a probability wave"—where this is thought of as a physical potentiality or propensity—"to" some actual determinate outcome; but this is to think from the outset in terms of purely predictive theory.) This transition (or discontinuity) is also thought to be the interval in which indeterminism arises.

What follows is a proposal for an explanation of how and why this localization of the "wave" occurs. The explanation at the same time *begins* to develop a positive conception of the causal background structure of space (empty or occupied), which in the neo-Whiteheadian theory serves as a physical context supplanting the omnipresent ether. The immediate proposal is a highly specific causal story that defers a general contextualizing account until the next chapter. It will function as a reference point for the ongoing ontological investigation, and for now it serves as encouragement concerning the prospects for overall success at comprehending radiation.

A literal and causal rendering of the Whiteheadian guiding supposition discussed above is that there is a class of natural events *genetically antecedent* to matter and three-dimensional reference space (ignoring the question about *time* that arises in connection with problems of "relativity"). On my reading, Whitehead conceived the field as essentially activity or transition of a type more fundamental than the local motions of objects by which its activity is made manifest. According to this ontology not only the effects of fields on matter are traceable to background events; the material objects themselves are "derivative" or manifest features of fundamental natural transition or "process." Whitehead's theory became more explicitly a *genetic* account of space, time, field, and matter as his thinking progressed. From the standpoint of causal ontology, such an account proposes that "field" and "radiation" phenomena are explained as resulting from particular causal structures within a permanent *background or substratum of fundamental engendering activity.* The rest of the book will follow out this interpretation of Whitehead's path of physical explanation over a broad range of applications. When the words "genesis" and "engenderment" are used here one should not think of "arising out of . . . " as if this were merely a halfway point on the way to postulating the "something" to fill in the ellipsis. One should retain only the image of *arising as such,* that is, nothing more than *transition in development from stage to stage* (for Whitehead, "passage" or "process"). This sustains for the present purpose the central Whiteheadian postulate that transition in "happening" or "passage" is fundamental and irreducible for knowledge of nature. In the present proposal, that which is engendered, or perhaps better expressed, what comes about in the arising, is "locality," which designates the actual source in nature of our simplifying scientific concepts "matter" and "space." "Engenderment" here refers to a structure of arising or emergence on its own, not the production of one event or thing by another.

The implication of this causal-narrative interpretation of Whitehead is that locality, meaning any given system of spatial relationship determined by the presence of actual atomically and "macroscopically" aggregated matter, as *physically engendered or emergent actuality,* is in each case purely a foreground or manifest aspect of a physical circumstance, as are the localized material bodies themselves. As a proposal for ontological identification of the physical field this is consistent with Einstein's suggestion, which is not at all eccentric, that matter can be reduced to particular conditions in and of the field. "Physical engenderment" here does not refer to an ultra-ancient event originating the universe, which would presumably be irrelevant to the explanation of present-day and nonastronomical experimental realities. It refers instead to an active aspect of all natural reality which, because its physical structure is that of the *processes by which localized space comes about,* has a relation to such terminal localized space which is not that of motions with respect to a given background system of space in three dimensions. The constituent events of the engendering processes will *as such* be of a category different from local motions, since their essential relation to locality is that of antecedent transition to terminal place. On their positive side, if these background processes are actually those in the course of which physical locality arises, they must have the general systematic characteristic of thorough interconnection via genetic transitions with all parts of space and aggregate matter; this is the sense in which they occur "everywhere," that is, are never absent or displaced from particular places. They must therefore constitute a "fourth dimension," an initial comprehension of the causal structure of physical space, having so far just this meaning: genetic process. This fourth dimension would extend along the genetic succession terminating in actual systems of locality. This is not the concept of a fourth dimension in Einstein spacetime as presented in Minkowski diagrams or in the popular conception of "time as a fourth dimension." For the same reason it has no immediately clear connection with nonuniform Riemannian space-time, because insofar as the latter can be forced into a mental representation, the available conception is one in which "fourth dimension" means that time as uniform linear extension in duration is affixed to space (perhaps as the "temporalization of space"), whereas genetic process, according to the strategy of physical ontology developed so far, is a continuity of difference or structured continuity, not involving any reference to undifferentiated extension in space or time. This conceptual contrast is discussed in more detail in the next chapter.

The present proposal for identifying and understanding the "medium" of light involves a physical conception that for the moment will have to remain somewhat abstract, identifiable only as a general structural possibility. Pursuing a neo-Whiteheadian approach to causal explanation has so far arrived at the supposition that this "medium" amounts to the *process by which* space and matter "localize," that is, come into being as aggregate locality; auspiciously, this is precisely not a substance present in all of space. The next problem, then, is to explain roughly how both wavelike propagation and the puzzling process of individual interaction at material sites might obtain their causal structure from a background genetic transition (process) terminating in local space, that is, in matter and physical place referent to matter. The whole array of explanations comprising Part Two is based on learning to conceive localized physical space as a product of a process of development whose unique transition is the background physical structure that needs to be explicated to provide a background explanation of radiation and fields or forces. Fully identified, this genetic trace connects localized space with an engendering totality of nature, as explained in the next chapter.

The conception I am introducing here, which will form the basis for all the explanations in the ensuing chapters, is that of a *prelocal dimension* in the sense of genetic structure antecedent to localized space and matter and as such forming a permanent background to all physical place. Extension along the "axis" of this depth-dimension of nature cannot, according to this description, be identified with extension through any actual and determinate local spatial distance, nor is it simple extension in time (duration). I am supposing that this dimension supplementary to local space in three dimensions comprises in a certain way a "medium" or physical context of activity ("ether of events") for certain propagations. The justification for introducing this difficult notion of prelocal process is, first, that a basic comprehension of fundamental natural phenomena ought to be possible, and second, that these propagations in their detailed structure have systematically and thoroughly perplexed traditional concepts of physical processes. So far in this inquiry these enigmatic processes are described only provisionally in the abstract terms of the structure of genetic differentiation in events; and now my proposal is that when general structural possibilities contained in the concept of causal or genetic transition are applied in this extraordinary postulate of an engendering dimension, possibilities are opened for explaining the more perplexing properties of light. The context of activity of light, known to physics since Maxwell as the electromagnetic field, is initially identified ontologically as an aspect or dimension of physical space that is inherently "in process" as the genesis of given locality, making the latter an emergent aspect of physical space fully conceived.

It is useful for a limited heuristic purpose to form a schematic picture of the extension into the fourth dimension formed by transition in this "process" as a series of strata, a succession in the engenderment or emergence of spatial locality, with the relation between local space and its antecedent succession made clear in the representation. The following diagram (Figure 6.1) will serve to illustrate. The three dimensions of ordinary localized space are collapsed to a single plane at the top of the diagram.

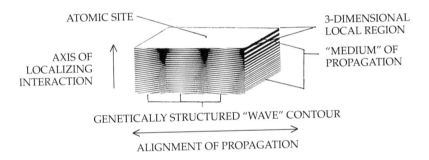

Figure 6.1   Prelocal Genetic Structure

This shows the prelocal structure of space as a physical context for two basic aspects of radiation, propagation across space and atomic interaction. Along the vertical lies a fundamental genetic transition, shown artificially analyzed into a gradation of discrete levels. A genetic substructure of (or to) space is the replacement for the defunct idea of an ether-substance *present in* space. (This is only a diagram of structural relations. The question as to what a light wave "looks like" is entirely misguided.)

The objection might be raised that this account of the medium of propagation relies on an analysis of genetic transition as a series of parallel strata, which has the Humean problem of connection between discrete particulars. But this abstract model is only the result of conforming to the idealization of localized space as a plane terminus of fourth-dimensional structure. The account can be framed in the concrete by saying simply that the propagative phase of radiation is aligned transversely to a structure of continuous (primordial) development. It should be emphasized that the graded transition in genetic structure is the *mode of differentiation* of the "parallel" strata of the diagram; apart from this graded or graduated structure, there is no differentiation. To speak of a vertical "axis" here is to pragmatically frame in abstract linear terms what is *essentially a gradient* in which difference or structure is not dependent on and does not presuppose uniform linearity, whether in the form of simple extension in space or simple extension in time or in some other form. Such a presupposition however natural would appear to be incautious in light of what has been gleaned of the demands of a nonmechanistic concept of physical events along the lines of a Whiteheadian event ontology.

This initial conception entails a *process-product relation* in which the "process" (genetic transition) constitutes a background or substructure that in no part occurs at a separate time or place from its "product," namely, *physical place,* but rather permeates space as a permanent context for forms of ongoing physical activity interacting with local objects. These structural aspects are the *emergence* and the *givenness* of space, respectively. The idea raises a multifaceted philosophical issue called "the ontological difference" in the Heideggerian tradition[8] and "becoming and being" in the Whiteheadian tradition.[9] "Analytic" philosophers and others are likely to raise this issue in their own way by pointing to a basic conceptual difficulty that seems to lie in this idea of permanent structure of genetic antecedence to physical locality. Must not causally antecedent events precede their products (in this case three-dimensional space and matter) *in time,* and hence antedate the latter's existence? In what manner could such events be ongoingly and pervasively interconnected with actual, present locality in order to comprise the context for the physical field and radiation? What is the basic status of time in the conception? Whitehead asserts as part of his explanations in the area of relativity that time itself in an ordinary or "abstract" sense is derivative from events; but how could this be? Does not any event or process presuppose time? Is "genetic process" *in time,* and how else could it be related to time?

These objections to the whole strategy adopted here have to be attended to, and the next chapter will address these objections by exploring *the sense in which prelocal genetic transition, as a process, is temporal.* Until then, the explanation of localizing interaction based on this postulated physical structure will be entirely schematic. The immediate need for a specific physical explanation does not require the more satisfying philosophical account given in the next chapter, but only requires the structural relations shown in the diagram (Figure 6.1). Let me clarify this order of the argument more closely. From the previous section, the meaning of the causal transition *in propagation* is so far only our "genetic succession" model based on the "productionist" approach to causality, which is not yet a full-fledged physical description but only a hazy application of an abstract and general conception of causal process. In the next chapter the meaning of both propagations of this class and their context of occurrence, the engendering background to local space, is rendered more satisfactory by a decision concerning the central question of physical ontology as to the elemental structure of physical being and physical entities. The reason for this procedure, which might seem inverted, is that the diagram shows only that structural aspect of prelocal background process which is useful at this time, and later some alternative and more satisfactory ways of conceiving it have to be mastered. With the help of the diagram the materials are already available for a narrative explanation of the famously inscrutable event of localization of the "wave" upon absorption by matter, in "bare bones" or schematic form. I present this causal story at this time primarily because it serves to round out the demonstration of the general causal intelligibility of radiation. For the philosophical qualms concerning the background conception the prescription is patience and involvement with the process.

Figure 6.1 serves to provide a rough and schematic idea of the relation that would exist between a propagation across space and the genetic "substructure" to space, when and if such a "substructure" can indeed be conceived as the context of activity of radiation. The *engenderment* of a certain three-dimensional part of space, supposing such a thing can be defended and explained as a physically concrete occurrence, would involve extension along a fourth-dimensional "axis" which, in a specific but so far abstract sense of process terminating in product, is *perpendicular to this spatial region as such,* that is, *to all three of its linear dimensions equivalently.* This is merely analogous to the perpendicularity of each of the three dimensions to one another, an analogy which tends to cover up the unique character of the engendering dimension. One might reserve the alternative term *orthogonal* for this relation, to be distinguished from perpendicular in the ordinary sense. The primary distinguishing mark of the orthogonal extension is that it is inherently structured in the mode of genetic development, not as a uniform linearity.

Keeping in mind the limits of particular representations, one can for the schematic account collapse the three dimensions of physical locality to a plane terminus of the fourth dimension conceived as a genetic succession, as in Figure 6.1. Physical space in this conception has a background dimension structured in stages of development such that three-dimensional space, coinciding with the presence and absence of matter, is a terminal or cumulative aspect. This is consistent with Whitehead's analysis of events as composing into four dimensions, with present matter and localized space (and time as simple duration) purely foreground (terminal or "derivative") features. All genetic strata are represented in the genetic constitution (referring to a physical trace along the prelocal dimension) of any actual place in nature whether occupied or unoccupied, and the strata themselves differentiate as phases of a unitary development. Levels of genetic development antecedent to fully constituted locality, that is, to material-spatial aggregates, form a gradient shading off as the structure of emergence of place—thus this antecedence does not extend away from certain sites in space and toward others, but graduates into a depth dimension in a certain hazy sense. This would be one application of the general concept of causality as active differentiating transition.

The diagram represents the genetic stratification in the abstract as a series of parallels which lie across a fourth-dimensional extension. If the event-succession in light propagation were aligned along this parallelism of strata, proceeding—that is, genetically reiterating—*transversely* to this "axis" of genetic transition in the coming about of local space, it would necessarily be a transmission *across* space (though not *through* space, as I explain further in Chapter 8, pages 299–302), since (a) all strata occur in the background of any actual place along a fourth dimension and are not "somewhere else," and (b) local spatial order, marked by the positions of actual matter at sites of source and detection, forms a particular (terminal) stratum participating in this parallelism; therefore the transmission would *lie parallel to* (in the peculiar sense mediated by antecedent genetic strata) this simple spatial extensiveness. The fourth-dimensional extension also contains a possible framework for describing the composition of the propagative phase-event: it could be formed of a superposition effective as a particular alignment through the sequence of genetic strata, as suggested by the vertical alignment of phase in the moiré pattern of the diagram (this is worked out somewhat further in the penultimate section of Chapter 8). Such a superposition across a stratified substructure might be the fourth-dimensional analogue to a transverse wave action.

The context of the "process" dimension or background structure of physical space is so far proposed as a possible source of causal structure for propagation. What is wanted at the moment is a story of what happens when this propagation is absorbed at an atomic locality, of *how* it makes this transition to a microcosmic interaction. Now, a general Whiteheadian physical ontology would claim that this context of events also explains the atom itself as the "derivation" or outcome of a particular lineage along the genetic sequence, or in other words that this "occupied" feature of space has its own depth aspect, its route or course of emergence in terms of "substratum" events, so that the atom itself is actually the terminus of a specific formative trace extending "orthogonally" along the prelocal "axis" of genetic structure. For the radiation to be absorbed at an atomic locality there must then be a shift from its propagative alignment, which lies *across* the fourth-dimensional graded extension and is thus confined as propagative activity to prelocalized event-structure, to an orientation *along* this original genetic gradient terminating in the atomic locality within aggregate matter. The interaction, then, happens when a propagative series (occurring, as a succession, transversely to the depth-"axis" of space) shifts its alignment in such a way that it engenders a particular phase of its activity "vertically" as represented in Figure 6.1, so that the genetic structure of this phase conforms to or retraces the genetic transition that is the route of formation specific to an atom or local group of atoms. ("Alignment" and "orientation" clearly cannot here mean "in space" in the sense of having reference to a given localized system of space. They refer directly to prelocalized structure, and are related to local space only mediately, *by way of* this structure and its relation to locality.)

The "collapse" from wavelike propagation to localized interaction is thus given a context with the continuous and interconnected features suitable for a transitional and mediating event-succession: first, a process oriented "horizontally" as represented in Figure 6.1, that is, parallel (in the sense given by fourth-dimensional graded structure) to local ordering in space and propagating through the primal causal background of local space; and then a shift to a "vertical" orientation by which there is emergence from out of this background prelocal dimension, retracing a genetic route to a definite atomic site. As argued in Chapter 8, second and third sections, the indeterminism about this site would be simultaneously explained; the key to this explanation is that the transition to the site of interaction is not a localized transition. A narrative with causal connectedness is maintained, even though the shift or transition from propagation to interaction is fundamentally untraceable *through the space of measurement* because it can be identified with no localized trajectory or set of trajectories. This explanation would supersede the prevailing doctrine of a wave-particle duality embraced by the "principle of complementarity" by claiming unequivocally that the interaction effect appearing to result from the arrival of a particle at a site does not in fact result from the arrival of a particle at a site. Without a traveling particle the two slit paradox, for instance, does not arise.

Conclusion

In retrospect it is clear that certain experimental developments at the origin of quantum physics signaled the collapse of mechanistic explanation for radiation phenomena, but this collapse does not appear to entail the end of my possibility of narrative causal explanation. Causal concepts remain applicable, though so far their application remains to a certain degree abstract and has the character of a foothold toward an uncertain end. Light propagation is some wavelike and cyclic physical process, and to the extent that the Humean and mechanistic pictures can be laid aside, the causal relation itself provides at least an initial working comprehension of this cyclic structure through the general concept of genetic succession. A neo-Whiteheadian option for the continued physical interpretation of radiation phenomena remains open (providing certain philosophical objections can be met) which not only confronts no apparent impasse in known phenomena but can even attend to a major perplexity in quantum physics already by furnishing a schematic causal story.

The official doctrine of a wave-particle duality is hasty and superficial. Light propagation is clearly wave*like* in structure, but one is not forced to conclude from any character of the interaction that a localized object collides with an atom. There are two basic sources of the "particle" terminology: the pinpoint localizing interaction mimicking an atomic collision, and the fact that mathematical analysis shows the "photon" (e.g.) to be composed of an integral number of irreducible quanta of causal efficacy. Neither compels the conclusion that the propagation actually has a particulate character, especially since it appears that alternative accounts are possible that do not lead to paradoxical interpretations of experiments.

The initial stage of the theory of radiation and "ether of events" is the projection of a strategy, which is to suppose that nature is fundamentally an "upsurge,"[10] rather than being reducible to given space and aggregate matter as in the Cartesian-mechanistic ontology. This upsurge is a coherent transition and therefore contains the possibilities of "horizontal" and "vertical" alignments suitable for the causal structures of propagation and interaction.

7

Time, Space,
and Genetic Structure

I HAVE UTILIZED in explanation the schematic idea of prelocal genetic structure, a context of events and processes permeating all actuality—"permeating" not in the sense of "present everywhere in space," "filling" pregiven spatial extendedness, but rather as a structure extending into a fourth dimension along an axis formed by the "process" of engenderment of locality or place. The idea is that instead of the traditional "ether" and its universal localized space of occupation there is fundamental engendering transition— whose nature has not yet been explained concretely—terminating in actual occupied and unoccupied elements of locality. This is pro- posed as the common background hypothesis for the preceding account of the localizing interaction as well as for the explanations of the other basic properties of radiation that are given in Part Two. Once fully explicated, prelocal genetic structure affords a basis for understanding physical space as inherently *causal*, that is, as itself active and potentially effective in a variety of ways. The idea is Whiteheadian, in that Whitehead proposed a theory in which any given local space and matter (and in a sense time as well) are "de- rived from events"; I have interpreted this "derivation" in terms of causal antecedence, or process-product differentiation, "process" being a physical background dimension to any actual material– spatial aggregates. It is the job of this chapter to bring the structural analysis of prelocal causality (or derivation of localized reality from events) into coherence with our concept of time, and generally to make its physical-ontological specification more concrete. These goals require clarifying the extraordinary relation between time and physical causality (or between time and "process") that is involved in the idea.

In the three chapter sections to follow, the proposal for an ontology of prelocal process is developed along three different routes, presenting the background conception from three different angles. First, the way to incorporate time into the conception comes to light by way of an exposition and interpretation of Whitehead's theory of the primary and derivative meanings of "time." The resulting account of the primordial interconnection between space and time is then shown to render fully intelligible for the first time a certain essential fact about any perception or observation, making a first genuine physical sense of this relatively familiar fact. In the third and final section the need for a shift in physical ontology, and the direction in which this need points, are found to be articulated with profundity in Merleau-Ponty's writings on the requirements of a new conception of nature.

Whitehead on Time and "Process"

The guiding insights for physical ontology do not appear on the surface in Whitehead's writings, nor can they be drawn out without a struggle. They emerge as a transformed emphasis moving away from Whitehead's textual personality with its special usages and technical constructs.

Whitehead's philosophical discussion of time in *The Concept of Nature*[1] occurs in the context of his elaboration of a *general theory of nature and of our knowledge of nature*. This theory is primarily an answer to the transformation in physical science that was marked by Einstein's theory of relativity. Einstein's work unquestionably showed that the concepts of physical space and time had to be enriched or elaborated in such a way that they are assimilated to one another in a common conception; for Einstein, the now-familiar "spacetime" (or "space-time"). But Whitehead felt that Einstein's theory, whatever its value, was "faulty" as explanation[2] due to some general philosophical problems. His alternative approach was to suppose that the new developments in physics, notably the problems they raised about simultaneity, required scientific thought to seek the *origin in nature* of those aspects of time and of space that are the referents of these terms in our ordinary and scientific practical usage. Despite the eccentricity of this procedure then and now, it is actually in keeping with an ancient and primitive assumption according to which it is of the essence of natural science to seek the origins and causes of things in nature.

But to speak of "the origin of time in nature" has a problematic ring to it, which is due to a quite natural and commonsense belief that time must be among the preconditions of any narrative explanation; for example, it seems that time must be pregiven for any account of the evolution of the universe and its beginning (widely conceived as the "Big Bang") in time. If this is so, there can be no explanation of time in terms of its genesis in nature. Whitehead's project of thought indeed does provoke and challenge this quite natural conclusion of reflection. But the objection can be answered by showing it to be based on an unnoticed and inessential constraint confining the conceptual imagination, one that is connected to mechanistic assumptions. In these late stages of science, failing to examine the limitations in ordinary practical concepts for physical description renders silent any basic physical cosmology.

As I said, Whitehead's overall physical theory, which includes a theory of the origin of time (or to put it another way, of the primordial meaning of "time"), was developed as a response to the need for adjustment in the essential concepts applied to understanding certain features of the domain of physics. (Primarily it was a response to Einsteinian physics, not to the emerging "quantum" physics of radiation, which had not yet developed its full consequences.) The philosophical background content of the theory is a general critique of what he calls "materialist science"; this is his way of characterizing the previous epoch of physics and its traditional set of elemental and irreducible explanatory concepts. *The Concept of Nature* and the other books of this period seek to fundamentally revise this conceptual basis for physical description and for the basic comprehension of nature. This involves a new vision of the "natural elements"—or as he also calls them, "natural entities"—among which he includes time along with space, objects, and events, in their primary interrelationships as required by his physical theory. In view of this, it seems to me that one can only describe this period of Whitehead's work as *philosophical* inquiry *in and as physical science,* though it is not at all clear that Whitehead would have termed it such.

Whitehead informs the reader at the outset of his chapter "Time" that he is going to discuss in sequence the whole set of basic "natural elements," beginning with "Time." But he does not proceed immediately to talk about time, at least not on the face of it. The topic that he does take up initially is one that confronts the reader directly with Whitehead's challenging central conceptions and presents difficulties for exposition. For provisional purposes I will identify this subject of discussion as that of which there is a single preeminent example for every discussion or process of thought, namely, *the whole of natural events comprising the "present" of perceptual awareness of nature, conceived as a single total occurrence.*

Such a contemporaneously occurrent totality, according to Whitehead, has two basic structural features, and at first he calls these the "discerned" and the "discernible." These terms are soon left behind, and to bear down on them with too much weight of conceptual criticism would be a waste of time. We should simply ask what the distinction is to which he attaches these labels. The discerned is that part of the physical world we see, touch, and hear around us. The discernible includes the discerned, but extends beyond it to encompass all of contemporaneous nature that is not part of this spatially and temporally immediate perceived environment. This "beyond" is known to us only as an "inexhaustible" reach of determinate places and times, which are known *as standing in such-an-such relations* to the discerned, but not *as* having the specific characters or qualities that belong to the discerned. Events at these places and times are known purely as those "relata" at the other end of spatial and temporal relations, at this end of which are things accessible to the senses and having, of course, their own spatial and temporal organization within the scope of "discerned" nature. Whitehead thus conceives a structured totality, disclosed in perception, or more precisely in an aspect of perception that he calls "sense-awareness." This structure is comprised, first, of everything in one's field of vision, and things one hears, encounters by touch, etc., and second, of all those places and times that have only entered one's knowledge through a kind of spatial and temporal extrapolation (a function of perception itself) from the spatial and temporal relations in this immediate environment, yielding all together an awareness or experience of nature presently occurrent as a spatially and temporally organized and indefinitely extended whole. It is important to see that Whitehead does not thematize perception in the manner traditional in philosophy, but seeks to convey as a starting point for knowledge of nature a meditative, contextual apprehension, a "feeling of connectedness with nature," that is likely to be most accessible in a mountain wilderness setting in fair weather: *"Here we are* under this Oregon sky, on this knurled continent, and the cosmic 'now' reaches across a scarcely ponderable totality"; the feeling/apprehension takes on a special intensity when the blue veil dissipates and the celestial depth is stunningly manifested. This is not a case of detached observation or contemplation, but of captivation in awe of immediate factuality. Whitehead's account of natural perception presumes in the reader some recognition of such "experience," which is not at all a merely subjective reaction, but (at least for Whitehead) a founding event of natural science, the traditional paradigm of "objective" inquiry.

Whitehead first seeks to illustrate the extrapolation from the discerned to the extended whole of the discernible that takes place in meditative-perceptual awareness, through an example focusing on spatial relations and leaving out temporal relations "for simplicity." His presentation here is confusing, but I will try to render it more simply as follows: If one sees, hears, or touches something, there occurs along with this perception a determination of spatial location, the "where" of the thing seen, touched, or heard. Because of this the instance of perception "discloses," according to Whitehead, a system of relations including this place of the thing and including also places other than that occupied by the thing. Suppose one has *seen* something; call this something "percept A." At about the same time one might, for example, *touch* something else without also seeing it, thereby disclosing a *definite thing-character* (specific shape, texture, etc.) at a different place than the thing seen—"percept B." A and B are perceived through different senses, though they are perceived at the same time and within a common system of spatial relations. *Seeing* in this case does not "disclose" percept B in its particular thing-character, but does "disclose" *its place* in relation to the place of the thing seen. It does not belong to seeing in this case that there is anything of any particular character located at the place of percept B *in* the system of spatial relations "disclosed" in seeing. *For seeing* it is simply a place. Whitehead calls this general fact about perception, the disclosure of bare places and times beyond itself, "significance":

> An entity merely known as spatially related to some discerned entity is what we mean by the bare idea of 'place.' The concept of place marks the disclosure in sense-awareness of entities in nature known merely by their spatial relations to discerned entities. It is the disclosure of the discernible by means of its relations to the discerned.
>
> This disclosure of an entity as a relatum without further specific discrimination of quality is the basis of our [Whitehead's] concept of significance. In the above example the thing seen was significant, in that it disclosed its spatial relations to other entities not necessarily otherwise entering into consciousness. Thus significance is relatedness, but it is relatedness with the emphasis on one end only of the relation.[3]

Discerned entities are "significant" not in the sense that they are worth considering, but in the sense that through their spatial and temporal determinateness they function to "signify" a system of places and times "undiscerned" in specific character—such as "occupied" or "empty"—but disclosed as relata, a system extending throughout a totality of times and places. The discerned is the environment around us insofar as it is perceived in its specific characters and qualities. But the discernible includes also all of nature extending (in a certain sense) beyond this discerned region, that which in the process of perception is disclosed *purely as spatial and temporal relata*, encompassing a totality of definite "wheres" and "whens" about which nothing is actually discerned, though of themselves, presumably, they have distinctive features and qualities potentially discernible. The function or role of the discerned in disclosing spatially and temporally remote *relata*, which Whitehead calls "significance," is also the feature of discerned entities that allows their "cognizance by relatedness," as opposed to "cognizance by adjective," a distinction Whitehead explains at length in *The Principle of Relativity*. "Relatedness" means that every event is *marked* by unique systematic relationships toward a background of "factuality" with a definite spatial and temporal structure that Whitehead finds difficult to formulate to his satisfaction; he suggests the term "totality" with the reservation that he does not mean an "aggregate" which is "the sum of all subordinate aggregates," but rather, "the concreteness of an inexhaustible relatedness among inexhaustible relata."[4] The present work will proceed on the hope that any suggestion in the terms "totality" and "whole" of a collection or aggregate of things can be left behind, in part by applying a conception of event-structure as opposed to object-structure, so that these terms can be retained as expressions for ultimate natural finitude.

What is this adventure of thought leading to about time? With this account of a two-part structural totality given in perception, Whitehead aims to specify for purposes of his theory of nature *the senses in which nature in the aspect of totality is extended and bounded spatially and temporally,* and what he is struggling toward is not at all a simple or familiar conception. The analysis of nature as a totality of spatial and temporal relations given in meditative-perceptual awareness is intended to supply *primordial meanings of space and time* as the foundations of a transformed physics. He is leading toward a theory of the derivation of ordinary time and space from "process," the ongoingness of nature in totality, which in *The Concept of Nature* is being developed as a primordial extensiveness underlying and constituting space and time. Though he had basic disagreements with Einstein's procedures, Whitehead recognized in Einstein's results the truth that any "theory of relativity" must in some way assimilate space and time into an encompassing conception, and his theory of the common origin of space and time in primordial "events" (or "process") served this purpose.

This whole problem of conceiving the spatial and temporal boundedness of nature in the aspect of totality and determining the applicable sense of "finitude" might seem artificial, a construct of an idiosyncratic theoretical reflection. But it happens to identify a profound conceptual difficulty in current science arising from the limitations of ordinary physical concepts. Consider the discussions about the ultimate origin of the universe in a "Big Bang": Matter and space come into being at a certain time (to put it simply and roughly); but space and time are fused in physical theory as "spacetime," which changes the way time fits into the account. A fundamental ambivalence arises between "the beginning in time of . . ." and "the beginning of time."[5] There is also a lack of basic coherence connecting the event of "creation" or origination of space and matter at a point in time with the subsequent, continued "expansion of the whole" understood solely in terms of the general outward motion of the galaxies and other aggregate material of the universe. "The expansion of the whole of space" cannot mean that at one moment the universe is a sphere of a certain volume and at the next moment a sphere of a larger volume; so what then does it mean? Is it identical to the creation of space or somehow subsequent to it? To face the situation squarely, a coherent positive concept of "the universe" as an evolving whole cannot apply the spatial and temporal framework of an aggregate totality of dispersed matter, though discussion of the origin and dimensions of the whole proceeds (and apparently must proceed) as if this framework were applicable. Far from being a focus of official research, these problems in cosmological concepts, if raised at all, are regarded as conversation pieces in contexts of more or less playful philosophical discourse, not as potentially productive questions for science. There is deep acceptance of, and adjustment to, the principle that remote domains of the physical both "microcosmic" and "macrocosmic" are by nature inaccessible to such demands of "ordinary" understanding. On the other side of this coin, the widespread impression that science has in hand in its basics the ultimate cosmic story results merely from the expository and discursive need for *some* interpretative framework.

Whitehead struggled toward an original conception of totality within his approach to theory of nature. An early concern in this chapter on time is to specify how space and time are differentiable in respect to their ways of being derived from a *perspectival totality of occurrence*. Having described the role of "significance" in the spatiality of ordinary perception, he then begins to distinguish the temporal and the spatial aspects of the structural whole comprised of discerned and discernible as follows:

> The disclosure in sense-awareness of the structure of events classifies events into those which are discerned in respect to some further individual character and those which are not otherwise disclosed except as elements of the structure. These signified events must include events in the remote past as well as events in the future. We are aware of these as the far off periods of unbounded time. But there is another classification of events which is also inherent in sense-awareness. These are the events

which share the immediacy of the immediately present discerned events. These are the events whose characters together
with those of the discerned events comprise all nature present
for discernment. They form the complete general fact which is all
nature now present as disclosed in that sense-awareness.[6]

Some things need to be said about Whitehead's use of "present"
here. We would be positively misled if we understood this term, in
keeping with spatial examples given "for simplicity," to have the
sense of "present (somewhere) in space." Not only is this ill-suited
to the pervasive language of "events," but more specifically,
Whitehead attacks as belonging to materialist science the idea of
"nature at an instant,"[7] that is, of nature as simple presence-in-
space, as an aggregate totality of aggregates composed of "simple
locations," some occupied by matter and some devoid of matter,
independent of any reference to temporal passage or any sort of
activity (though time as pure duration, and changes of local matter
in time, may be tacked onto this materialist conception). Accordingly, to think that by "all nature now present" he means all nature
"present in space" as a totality (a problematic conception for independent reasons) must be to completely mistake his meaning. And
"immediately present . . . events," classified together as comprising
the spatial aspect of totality, cannot be referring to a set of *present
things.*

One can avoid conflict with Whitehead's important anti-
materialist conviction by understanding "present" and "now
present" in this passage to mean *occurring simultaneously with the
present time,* that is, *referent to events,* without any additional implication of things present in space. A "present" whole of nature thus is
not something defined as occupying a common space with the perceiver, as two people co-occupy a room; there is nothing identifying
the occurrences comprising this whole as being located either
"here" or "there" *in a pregiven space;* they are only delimited as a set
from other events that are not part of this particular whole (N.B.) by
being identified with a particular *transitional totality,* namely "the
present" with its essential structure of "passage." Natural relations
for Whitehead are fundamentally *event* relations, which ensures
the essentiality of temporal passage to the actuality and
relationality of the parts forming a temporally unified whole. One

result of this interpretation whose significance is still unclear is that times and places "discernible" but undiscerned are *event-times* and *event-places.* If the spatial relations of these parts with respect to a natural whole were reducible to simple differences of location, there could be a corresponding reduction to instantaneous time, to time as a series of idealized instants in the uniform subsistence of space and matter, a conception of time that Whitehead repudiated. An obvious feature of Whitehead's work on physical science is that he is generally inclined to employ the language of "event," "process," and "occurrence" rather than that of "thing," "entity," or "object." With this, on my reading, he seeks to conceive the referents of these terms as entities more primary in nature than matter and local motion; and on my interpretive extrapolation, this means that events and processes are referred to by way of *causally explaining* these manifest material facts and their detailed observable properties, though this is not developed in Whitehead. I would argue, therefore, that in defining the discernible (which he comes to call a "duration") as a "present" whole of nature, he means by this *nature as a single event, in the sense of a totality of natural events, which is ESSENTIALLY AN ONGOINGNESS contemporaneous with the present time.* In this same discussion Whitehead formulates a sentence in a way that avoids "present" and its ambiguity: "The general fact [the discernible] is the whole simultaneous occurrence of nature which is now for sense-awareness."[8] It is not simply that temporal passage is fundamental, which by itself would allow reduction of time to instantaneous series, but that a *differentiated or structured transition* in "passage," referring as much to the actual happening of the whole of nature as it does to "time," is fundamental: "There is time because there are happenings, and apart from happenings there is nothing."[9]

Another thing to note about this usage of "present" with reference to the discernible is that the discernible as Whitehead describes it is nature as a whole taking place "in the present of" a perceiving being, that is, simultaneously with this perceiver's "process" of perceiving. The structure of significance, discerned-discernible, must therefore consist of *extension, in some sense,* outward into the rest of the cosmos from this perceiver at this time and place (this extension will trace through events, not through a system or systems of localized space). But this essentiality of cosmic standpoint is no reason to attribute a radical subjective idealism to Whitehead, since it only means that what is under consideration is the relation to a perspectival whole on the part of one class of natural events, "percipient events." The relation of events in other classes to a contemporaneous totality may be supposed to have an analogous structure, about which these chapters aim to make some progress. Whitehead does not say or imply, at least in these middle period works, that durations exist only with and for perceiving beings.[10] Occasions of perception, thought of in the context of general physical theory as themselves events participating in the process of nature, may just be conspicuous examples of how the irreducibly transitional structure termed the "passage of the whole" has its imprint on all events.

We should keep in mind, again, that because of Whitehead's commitment to an "event" ontology, the structure of significance cannot be an extension simply outward in space, that is, toward outlying regions of an encompassing system of localized space including the physical environs of the perceiver. In understanding Whitehead, for whom matter and space have a derivative status in nature, we must explicitly renounce our handy conception of the universe as a spatial whole of aggregate material, dispersed throughout a single system of locality, a conception in which the most remote galaxy lies "on the other side" of the universe just as the next city lies down the road, these differing only in magnitude. The point is not to renounce this conception only to throw up our hands and institutionalize the renunciation in the temple of science, but to move on toward a conception that succeeds. The outward extension into a totality which is now for primal awareness of nature must mean something with an entirely different structure, and what has been gathered so far is that it would be extension through differentiation *of events* as spatial and temporal relata, a meaning of "extension" intrinsically involving temporal passage of nature in totality.

Whitehead continues by explaining that spatial relations throughout nature originate from relations among those events that belong to a particular presently occurrent whole, while time comes from relations of this temporally bounded whole to "other events":

> It is in this second classification of events [all events simultaneous with the present] that the differentiation of space from time takes its origin. The germ of space is to be found in the mutual relations of events within the immediate general fact

which is all nature now discernible, namely within the one event which is the totality of present nature. The relations of other events to this totality of nature form the texture of time."[11]

At this point Whitehead phases out the terminology of "discerned" and "discernible" and talks about what he calls "durations," and this remains a central term for the rest of *The Concept of Nature*. (A useful supplementary text on this same topic is the earlier work *An Enquiry Concerning the Principles of Natural Knowledge*, especially part 2.) A duration is a particular temporally bounded whole, encompassing all events "occurring simultaneously with" the present, that is, comprising the natural present. One should not think of the temporal boundaries as sharp, as if they correspond to particular times in the past and in the future, because Whitehead makes it clear that he does not mean by "duration" "a mere abstract stretch of time,"[12] an interval of time in the sense of pure duration. He intends rather a simultaneous whole of actual events, thus a structural unit of the physical "passage of nature," involving fundamental transition essentially; this is what he means by "a concrete slab of nature limited by simultaneity [as simultaneous occurrence]."[13] The "texture of time" is formed by relations of the presently occurring duration to "other events." The guiding principle of his thought, which is perhaps more fruitful for our understanding than analysis of formal definitions, is that "nature is a process." The connection between "durations" and nature thought of as a process is explained as follows:

> It is an exhibition of the process of nature that each duration happens and passes. The process of nature can also be termed the passage of nature. I definitely refrain at this stage from using the word 'time,' since the measurable time of science and of civilized life generally merely exhibits some aspects of the more fundamental fact of the passage of nature. . . . Also the passage of nature is exhibited equally in spatial transition as well as in temporal transition. It is in virtue of its passage that nature is always moving on. It is involved in the meaning of this property of 'moving on' that not only is any act of sense-awareness just that act and no other, but the terminus of each act is also unique and is the terminus of no other act. Sense-awareness seizes its only chance and presents for knowledge something which is for it alone.[14]

Whitehead speaks of the happening and passing of "durations," each of which is a particular *whole of nature* specified as comprising the present, that is, including within it the occasion of perception as it discloses nature for knowledge, a disclosure that is in each case *now*. But one should be careful about understanding "passage of nature" to mean a continuous succession of like entities, "durations." What is it that each duration arises out of and passes to? Other durations, it would seem. These other durations must comprise structural parts of nature with its temporal boundaries expanded, *before* the seemingly arbitrary limitation to simultaneity with "the present" is introduced, that is, it must consist of those durations and their constituent events which are *not* simultaneous or contemporaneous with the present time ("duration") but lie in "its" future and "its" past. But then it would seem that one had referred the explanation of time, as a derivation from the arising and passing of durations, back to a process *in* time, which is unacceptable. The arising and passing of durations cannot therefore be understood as a sequence of events in time. The solution, as I will presently explain, is to understand this primordial passage of nature as *in a certain sense time itself;* this solution requires a critical reformulation of the matter. We will continue to see how Whitehead's physical theory in the formulations he gave it has a problem that recurs at certain points about the relation between derived elements and that from which they are derived. As to the present instance, we should note that the definition of a duration in *The Concept of Nature* cannot be considered final; the passage cited above (page 230–31) from the later work *The Principle of Relativity* describes how Whitehead finds that the terms he resorts to are persistently inadequate to express what is intended by the term "duration."

Whitehead says that for him to call the passage of nature "time" at this stage of his discussion would tend to mislead because of a certain prejudice he thinks comes from common usage. But he does not say definitely that the passage of nature is something other than time. Rather, he wishes precisely to connect the arising and passing of durations with "time," though not carelessly, by imposing the prejudice of an *ordinary and practical concept of time*. In fact, he says that the relations of each duration to events of other durations "form the texture of time." Whitehead thus seeks to mobilize a distinction between, on the one hand, time in an ordinary and practical sense, and on the other hand, time in a fuller and more primary sense, the (temporal) passage of nature. For Whitehead, time in its authentic meaning is a unitary becomingness constituting, *as transition,* a whole of nature. It can only be regarded as irresolvable whether this transition is in fact physical *time* as opposed to physical *process.* On traditional practical and technical assumptions, by contrast, time (even in cosmological applications) is a universal and uniform succession in simple duration, a *precondition* for all events, processes, and histories insofar as they are temporally extended and temporally situated.

The distinction between the primary and derivative meanings of "time" is based partly on the stipulation that a "duration" is a special type of *physical occurrence*—namely simultaneous totality—rather than being an "abstract" temporal period, whereas the latter concept embodies our ordinary and practical concept of time. This distinction works only because "serial and measurable time" are aspects of a quite ordinary sense of "time" arising from our practical dealings with time: pure duration *within which* events occur, something that is presupposed in and for those events, whereas time for Whitehead is a certain *relation within* complexes or totalities of physical events that is not independent of and prior to events. "There is time because there are happenings, and apart from happenings there is nothing." By saying that the passage of nature consists of a sequence of physical "durations," in each case a "concrete slab of nature [as a whole]," Whitehead marks the distinction between on the one hand an ordinary, simple, and practical concept of time, which is abstract and devoid of structure, that is, which gives one the idea that reduction to uniform series of ideal instants is possible, and on the other hand the ongoing fundamental actual occurrence, the passage of durations, that he claims is the physical basis for ordinary "derivative" time. The everyday and practical/scientific concept of time tends to suggest that time is something independent from, and necessarily pregiven for, actual events. Whitehead thus exploits an ambiguity in the meaning of "temporal passage"—is it pure duration, independent of actual events, or is it ongoingness *of* actual events in totality?—to frame his distinction between derivative and primary senses of "time."

It is important to remark that, in contrast to time conceived as simple extension in duration or uniform temporal passage, a "duration" has the character of *structured or differentiating transition:* it is marked by arising and passing, keeping in mind the problem of regressivity that resulted from understanding this apparently cyclic succession in an ordinary way as a process "in time." It has a structure of development, in contrast to pure duration (in the ordinary sense of the word) or simple extension in space. A particular "duration" is identified in temporal terms: each is composed of events that are contemporaneous (within a perspective) in the sense that they form a total structure of occurrence, a unitary transition or "passage." Nevertheless, the "process" or transitional character of durations is not accurately understood as a process *in time.* Rather, the meaning of this differentiation through transition is that of a primal sense of "process" and "extension" that is the common origin of time and space in the ordinary senses. It is nature "always moving on" as an extended whole, conceived rather fuzzily as fundamental transition in "arising" and "passing." That particular time-series are derivative facts means for Whitehead that the transition comprising a duration, the "creative advance of nature," is "not properly serial at all."[15] It is not obvious how to understand this distinction between serial and "not properly serial" process.

Causal concepts might facilitate the distinction. Salmon's conceptual strategies under our critical revision brought to light a possible distinction between an aspect of causal process that may be labeled "production" and another aspect that may be labeled "propagation"; for deceptively simple examples, the action of a light beam on the wall in producing a spot, and the transmission leading up to this interaction. What the distinction exemplified here ultimately comes down to, I think, is this: There are two forms of what may be called "differentiating transition through the stages of a process." This can refer to the stages leading into and through a single event such as the "production" of an individual physical effect (the interaction); insofar as this event has some extension, each stage differentiates from the previous only in the sense that genetic *development* occurs. In such a case "process" is applicable only in reference to an *individual graded transition,* not in the sense of a series of events. Or it can mean a succession of like or identical events, forming in some sense a cyclic reiteration. One might argue that the first case can be called "succession" only as an imposition of analysis, so that preanalytically this is continuous (structured) transition. This points to a possible sharpening of the distinction: joining Whitehead's voice with the resistance to analytic reduction one can say that an individual genetic transition, as opposed to the other mode of structured process, is "not properly serial at all." This is a way to understand Whitehead's general insistence on the essentiality of unitary transition as against reduction to linear seriality.

If a process is "not properly serial," then, it incorporates regions of antecedence and consequence into an individual structure of development, without being a series of successive reiterating structures. This transitional structure appears in Whitehead's thought as the authentic concept of natural totality, and its unitarity identifies the applicable sense of finitude implicit in the concept. How to understand the relation of one whole to others that "precede" and "succeed" it remains a problem. The problem might at some point be resolved on the basis of the tension between "identity" and "difference" that was mentioned in the previous chapter in connection with causality. If unitary and "not properly serial" transition is fundamental or primary and reiterative structure is derivative or secondary, then differentiation in the sense of reiteration has reference not primarily to linear order in space and time, which is itself derivative from physical "passage" and arises along with the reiteration, but to an encompassing *unitary* differentiating transition. There is no serial structure, hence no simple linear order in space and time according to Whitehead, without the concrete event of transition in "passage." These relations are explored further below.

For the purpose of this inquiry natural totality as "process" is best thought of with reference to the essentially interrelated character of particular physical events, and vice versa: the natural event as such has a unique physical situation with reference to a single total perspectival transition or "happening." Whitehead's own primary focus was on reconstructing the derivation of ordinary time through his central physical hypothesis that the relation of any event to other events in totality can be stratified into a diversity of exact time "series" in the sense of successive slices of simultaneity each with a purely spatial extension, as discussed in Chapters 9 and

10. From the standpoint of physical ontology this formal construction of multiple "time-systems" proves to be a plausible framework of applicable physical theory with a tenuous status as explanation. The present aim is to avoid these abstract constructions and conceive the relation between time and fundamental physical "passage" along different lines.

One feature of the difference between primary and derivative senses of "extension" is becoming clear. Simple extension in space and simple extension in time (the latter of which I have called pure duration) are in themselves uniform and without structure, featureless apart from objects or events that may or may not lie within or take place within them. Each sense of simple extension respectively means (a) an expanse *within which* objects can be located and can move from one position to the next, and (b) a span *along which* and *during which* events can *occur*, can run their course and pass. "Transition in arising and passing," on the other hand, even in the "not properly serial" mode, corresponds in its essential content to *genetic differentiation*; it has inherent *structure*. If extension as transition (in this sense) is "more fundamental" than spatial and temporal extension, this just means that it is a form of extension identified only *as* this transitional differentiation, making no reference to simple extension in space or in time against which or within which it takes place. As ontologically prior to the two forms of simple extension, there *is no* uniform linearity to this extension, only graded structure. Transition in the "passage of nature" is a particular character of extension with inherent structure or contour, while the simpler and more familiar forms of extension are featureless or what I will call "flat."

This distinction leads into a landscape of concepts and cases that is treacherous for philosophical thinking if it is not carefully scouted out: "different," "discrete," "separate," "distinct," "connected/disconnected," "continuous/discontinuous." For example, it is easy for the assumption to take hold that for genuine continuity and/or connectedness to obtain in a physical process, a concept of *uniform extension, devoid of structure* has to be applied, at least in the spatial and temporal framework of the process. This is because to speak of events "in time" and "in space" imposes a picture of contiguous relations. But given that a process may have a structured continuity, for example as a genetic sequence, might not the continuity of its spatial and temporal context also be so structured? For a rudimentary image, consider the example of the streak of a meteor which enters the atmosphere as a faint light, brightens, and suddenly burns up and vanishes; here is a unitary occurrence with structured continuity from stage to stage. Is this continuity dependent on the *uniform* continuity of a background space? Yes, insofar as the example is conceived as a trace simply lying before us momentarily with all stages mutually present. But as a process of development, the image and its continuity are potentially independent of any pregiven setting of space. Conceptually at least, the space of its occurrence may be developing along with it and thus possess a structured continuity of itself. In the theory of radiation, according to my arguments, this is in fact what provides the basis for understanding genuine empty space events. In a related context, the problem of causation is really the problem of conceiving structured continuity with the additional feature of "connection" in the sense of dependence or derivation.

In our interpretation of Whitehead so far, "flat" or uniform extension in space and in time are derivative forms or secondary manifestations of primordial extension in (structured) transition belonging to the passage of nature. Whitehead's basic procedure, then, for surmounting problems of physics is to postulate that the time and space of measurement, as simple forms of extension quite distinct from one another, are themselves products (or "derivatives") of the natural processes ultimately being investigated by means of measurement, so that explanations must finally turn to the structure of these primordial processes. So far in our interpretation this structure is merely the element of transition contained in the general ideas of genetic development and genetic series.

It is valuable to understand how "serial time" is derived (or arises) from relations among durations.

As I discussed in the previous chapter, Whitehead's "extension" is intimately related to "inclusion," the composition of events by other events. Extension means "extending over," nothing other than a whole-part relation *in a sense specific to events and durations:*

> Durations can have the two-termed relational property of extending one over the other. Thus the duration which is all nature during a certain minute extends over the duration which is all nature during the 30th second of that minute.

[Let us keep in mind here the dangers of a false understanding of the spatial and temporal outlines of this "all" that arise from habitual representations.]

This relation of 'extending over'—'extension' as I shall call it—is a fundamental natural relation whose field comprises more than durations. It is a relation which two limited events can have to each other. Furthermore as holding between durations the relation appears to refer to the purely temporal extension. I shall however maintain that the same relation of extension lies at the base both of temporal and spatial extension. This discussion can be postponed; and for the present we are simply concerned with the relation of extension as it occurs in its temporal aspect for the limited field of durations. . . .

The continuity of nature arises from extension [in a special sense!]. Every event extends over other events, and every event is extended over by other events. Thus in the special case of durations which are now the only events directly under consideration, every duration is part of other durations; and every duration has other durations which are parts of it. Accordingly there are no maximum durations and no minimum durations. Thus there is no atomic structure of durations. . . .

The absence of maximum and minimum durations does not exhaust the properties of nature which make up its continuity. The passage of nature involves the existence of a family of durations. When two durations belong to the same family either one contains the other, or they overlap each other in a subordinate duration without either containing the other; or they are completely separate. The excluded case is that of durations overlapping in finite events but not containing a third duration as a common part.[16]

(The term "finite events" referring to events other than durations is a source of confusion. Durations are finite in some sense; their finitude was conceived initially by Whitehead as temporal boundedness. What distinguishes events other than durations is simply that they are not *totalities*-in-transition, whatever sense of boundedness/unboundedness in space and time this involves.) The "family of durations" is intended to form the connection between the unitary transition characterizing a particular duration and derivative "serial time." The concern to distinguish different families of durations within the "passage of nature" is an outgrowth of his theory of relativity, the heart of which is the claim that multiple "time-series" or "time-systems" always exist in nature.

To bring Whitehead's detailed discussion of relations among durations—belonging to the same family or to different families—into the range of tractability I suggest getting a firmer grasp on the term "duration." To this end a different terminology is needed, one which avoids the confusion that is almost inevitably generated by the fact that by this term Whitehead precisely does not mean a part of time in the sense of an interval of simple extension in time. Since what he does mean is the "whole simultaneous occurrence of nature which is now for sense-awareness," it may not be unsuitable to employ a terminology developed by Martin Heidegger in a discussion of time: "the now."[17] For Whitehead's theory of nature this is the encompassing "now" of natural totality. As with Whitehead's "duration," the now for Heidegger is "intrinsically transitional" and as such has a specific mode of extension, so that no grouping of nows is merely a succession of ideal instants in a linear time; nor does this "transition" have the sense of mere extendedness through a period of time, but instead is an asymmetrical and structured dimensionality:

> In the now as such there is already present a reference to the no-longer and the not-yet. It has dimension within itself; it stretches out toward a not-yet and a no-longer.... Correlation of the manifold of the nows—where the now is taken as transition—with a point-manifold (line) has only a certain validity, if we take the points of the line themselves as forming beginning and end, as constituting the transition of the continuum, and not as pieces present alongside one another each for itself. A consequence of the impossibility of correlating the nows with isolated point-pieces is that the *now*, on its part, is *a continuum of the flux of time*—not a piece.... This transition belongs to the point [in time] and is itself, as now, not a *part* of time, in the sense that this time would be composed of now-parts; instead, each part has transitional character, that is, it is not strictly speaking a part.[18]

Heidegger's discussion using this simpler term helps to clarify what Whitehead means by the composition of "durations," by making it clear that Whitehead's illustration of the "extending over" of durations—"the duration which is all nature during a certain minute extends over the duration which is all nature during the

30th second of that minute"—is as effectively misleading as the term "duration" itself. Part of what Whitehead really means to point out is that the characteristic dimensionality or transitional structure of the now is variable in its scope. Heidegger says:

> Because it [the now] has this peculiar stretching out within itself, we can conceive of the stretch as being greater or less. The scope of the dimension of a now varies; now in this hour, now in this second. This diversity of scope of dimension is possible only because the now is intrinsically dimensional.[19]

A now of greater scope may "include" a now of lesser scope *in a sense specific to this peculiar structured dimensionality,* and not according to the picture of a uniformly divisible temporal extension; nor is it the case that a particular period of time is composed of a succession of nows arising out of one another in a cyclic regeneration, as Whitehead's "arising and passing" might superficially seem to suggest. "Time is not thrust together and summed up out of nows . . . " (Heidegger).[20] Whitehead's point is that transitional nows include one another *in a peculiar and irreducible sense* to form what we count and measure as uniformly extended periods of time.

Whitehead's aim in talking about families of durations is partly to specify how particular periods of time, say, from eleven to twelve o'clock today, or a time interval in scientific measurement, derive their structure and their identity as instances of temporal transition from the perspectival and transitional character of "nows," each of which encompasses nature as total occurrence. It is the relation of "inclusion" or "extending over" in the sense peculiar to the dimensionality of the now, which is perhaps better stated as "passing over,"[21] that underlies a particular countable sequence of time. The very identity and unity of a period of time—an hour, a week, a year—is formed of a "family" of nows, a series "extended over" by a unitary dimensionality of variable scope; we only view it simply as a certain segment of a universal and natural time-stream because that sums up its ordinary and practical aspect.

So far this analysis is independent of the question of alternative time-systems raised by the new physics. If there were only one universal passage of time, then the potential scope of any now could include all nows of lesser scope, that is, there would be only one family of durations. The crucial part of Whitehead's definition of a family of durations that establishes the possibility of different families is the "excluded case," the specific relation that distinguishes durations not belonging to the same family. These have in common events other than durations (temporally perspectival totalities), but do not have durations in common (by "inclusion"). Durations from different families intersect in subordinate events, but not in whole durations; they are analogous to planes which intersect and therefore have particular lines in common, rather than being related as a succession of parallels. Different durations in the same "family" compose, either by direct overlap as durations or through some mediating durations, into a single "not properly serial" (though structured) transition, a more inclusive duration. It is important to keep in mind that this "extension" relating durations "lies at the base both of temporal and spatial extension," so that the series formed by a "family" of durations is not a series *in space* or *in time*. A family of durations is a group of "nows" that identifies a *particular overall unity of temporal passage,* and as such its members have an aspect of serial order in their relations *due to* the peculiar relation of "inclusion" made possible by the variability in scope of the "now." According to Whitehead transition in perspectival temporality is the source of physical time *and space* as ordinary modes of extensiveness and seriality. The immediate interpretive point concerns time: for Whitehead, simple succession in time and the abstract linear representation of time have their source in the fact that "durations," through their variable extensiveness, are organized in "families," different members of which form a succession of "parallels" in the sense given by these structured "slabs" of totality. A family of durations is a particular natural succession that can in some way be derived from the inclusion of nows commonly within a single now, a unitary transition in "passage." *Unitary structure of transition in natural totality* is the fundamental explanatory fact, with all serial structure secondary or derivative fact.

Time conceived as inclusive-extensive transitionality in a perspectival whole of nature, a unified structure with a serial derivative or derivatives, shows an overall similarity with *causal or genetic process* as it has been explicated so far, but there is a sharp difference as well. In each case there is a peculiar interrelation between individual graded transition and serial reiteration, but to complicate matters the interrelation is different in the two cases. Temporal passage identified with a family of nows is not genetic sequence; the mode of *serial* ordering is different. On the other hand, any inclusive-extensive "duration," extending from the no-longer to the not-yet, or as Whitehead said, "the vivid fringe of memory tinged with anticipation,"[22] *has the very structure* that I have called genetic (or causal) transition in the nonserial mode; and for whatever it is worth, to characterize the structure of a particular "duration" as genetic is one point of agreement with the later metaphysical Whitehead. We interpret "genetic" here not metaphysically, but as physical engenderment of what Whitehead calls the "derived" natural elements: ordinary matter, space, and serial and measurable time. That a "duration" is divisible in the sense given by inclusion in the family relation bears some correspondence to the reciprocity of identity and difference that was discussed in connection with causal or genetic process (page 212). However, the inclusion relation of durations and the serial aspect of physical causality are not the same, since a family of durations is not a genetic succession. But consider the encompassing transitionality of the durational now: First, it is not a transition that *takes time*, but is so to speak a suspended structure at once and at one with itself, while at the same it is essentially "passage," in some sense *processive*. Its encompassingness involves not only the boundedness of an individual extended element of transition with dimensions of past and future, but also a spatial-temporal depth into the perceived universe disclosed in the relation of "significance," shading off into the indefinite dimensions of a unitary whole. Consider also the premise that physical time as structured transition is not prior to and ontologically independent of nature as a total happening. Whitehead spoke of "becomingness" and "creative advance of nature." The full-fledged conception here is really that nature as a transitional and unitary gradient whole of physical-genetic process and time in its fundamental, essentially perspectival meaning coincide in one and the same actual structure. Of course, this structural correspondence does not mean that temporal passage and genetic process are the same; rather, the unfolding of time is an aspect of fundamental engenderment and vice versa. Engenderment terminates in locality and in the serial derivative of time.

The overall result so far of this interpretation of Whitehead on time is as follows. Time and space in the ordinary, practical applications of these terms are derivative or manifest aspects of physical nature as "process," where this is conceived as transitional structure comprising a given perspectivally concurrent whole, and their respective senses of "extension" are both derivatives of a primal meaning of "extension" intrinsic to this "process." The concept of time as simple duration comes from an element of seriality derived from the relation of "inclusion" characterizing the transitional structure of primordial time, and "in the course of" this derivation the purely spatial extension and the purely temporal extension mutually unfold. That from which the practical concepts of science are derived is of course nature itself, while the derivative concepts are useful "abstractions" or simplifications of natural event relations with their aspects of specifically spatial and specifically temporal passage, more or less differentiable. ("Abstract" does not mean space, time, and matter are constructs or scientific fictions.) So far there has been no discussion of the possibility of "multiple time-systems" derivable from transition in nature-as-process, the centerpiece of Whitehead's physical theory. For the present purposes discussion can be limited to identifying the natural basis of "the ordinary and practical concept of time" as a linear element of "passage" that is pure natural *product*. I have all along been interpreting "derivation" in terms of physical causality, and if such derivation is fundamentally "process" it seems to me straightforward to construe "process of derivation" as "physical engenderment," which, as process-to-product transition, identifies both the "source" of ordinary space, time, and material, and the "process of derivation" of these elements from that "source." Whitehead himself came to formulate "process" explicitly as genetic process in later works.

Heidegger, in his lifelong investigation of the relation between "time and being," reached determinations that parallel Whitehead's and corroborate the four-dimensional and genetic analysis of time and space. Besides the discussion in *Basic Problems of Phenomenology* utilized above, the lecture "Time and Being" is noteworthy.[23] Stanley Rosen has interpreted Heidegger's "Being," insightfully it seems to me, as "process" and "genesis without a generator,"[24] and Heidegger does struggle to express a meditative disclosure of motionless and suspended transition, particularly in the late termination of the question of Being in the "event of Appropriation." The concern in these works of Heidegger is with general ontology and "phenomenology," not with physical science and not directly with nature. (The spate of recent books attempting to show that Heidegger's philosophical ideas are a primrose path to evil political outlooks may well be enough to ensure a period in which Heidegger is not seriously read. It should not be necessary to point out that there is not a hint of politics in my remarks here, and I do not discern any politics in the works of Heidegger cited; but just to reassure some readers, my own convictions are firmly behind the social ideal of democracy, whose popularity merely hides the fact that the general level of thought given to it stays at the level of half-baked slogans and formulas.)

A response can now be given to the basic philosophical objection to a Whiteheadian physics that I voiced in the previous chapter. The objection claims that no explanation of time in terms of its genesis in nature will be conceptually coherent, because time must be presupposed in any account of a process of engenderment. Whitehead's structural analysis of primordial time/process in terms of "durations" appeared at first to succumb to this very objection. But his intent apart from provisional formulations was uncovered in its basic content and can be articulated as follows. The above objection is valid insofar as "the genesis of time" is understood to say that time *first* arises at and as the terminus of an antecedent process, or in other words, that it is reducible to a product, as are the manifest and simple aspects of matter and space and, on the Whiteheadian view, linear and serial time. But the Whiteheadian insight on my interpretation is precisely not an attempt to say that at some point in a process time originates, but rather the idea that the *structure of time* corresponds to the "process" of natural origination, that is, time in its primary physical meaning is none other than a primordial structured transition such that along with the emergence of space there comes about an organization of perspectival totalities into one or more "family" groupings. In its unitary and "not properly serial" structural aspect physical time is coextensive with nature as a total event of origination and is therefore not a pregiven and universal substratum or vessel of duration for any and all events. This does not deny that time is also measurable and countable duration, but proposes a twofold meaning of time, primary and derivative, corresponding to process and product. In my terms, time encompasses and is encompassed by primordial physical engenderment, it does not *preexist* any and all evolving physical actuality.

The significance of the twofold meaning of time for the concept of prelocal process is this: though the engenderment of locality, as a process, is temporal, it is also the engenderment of *particular time-space systems* such as enter into measurement and form essential given contexts for any local motion, and such a system involves the product or derivative meaning of time. This entails underlying and pervasive "process" having a form of differentiation, and specific causal properties, independent of present material or localized space. The answer to the philosophical objections that arose from applying the ordinary concept of time for understanding the concept of prelocal causal structure is to point out that there can be an underlying genetic structure *of* time, specifically, the structure of origination belonging to natural totality accurately conceived, which does not suffer from the incoherence of a process outside of time. Whether one feels that this is a complete resolution of the objections, or whether the sense remains that there is something essentially engimatic about the whole idea of a process of origination of space and time, at some point the physical arguments for its correctness have to be taken into account.

But what about this enigma of transition in "durations," or passage of nature, a "process" that is not *in* time, but rather encompasses time and *is* time in the sense of the essential temporality of the occurrent emergence or unfolding of time and space? Is it not fanciful or intellectually extravagant to stretch ordinary concepts in this way? For Whitehead this fundamental becomingness or productive ongoingness is "stubborn fact" and something directly in evidence if we can only direct our thinking aright. It is the basic character of the "givenness of the actual world," and as such is the fact to which the meditations of philosophers ought to be attuned, but which is completely missed by traditional doctrines of philosophy (particularly philosophy of science). To illustrate productive transition as the basic structure of perceptual life he gives the example of starting a sentence and being "carried on" to finish it:

> The sentence may embody a new thought, never phrased before, or an old one rephrased with verbal novelty. There need be no well-worn association between the sounds of the earlier and the later words. But it remains remorselessly true, that we finish a sentence *because* we have begun it. We are governed by stubborn fact.

> It is in respect to this 'stubborn fact' that the theories of modern philosophy are weakest. Philosophers have worried themselves about remote consequences, and the inductive formulations of science. They should confine attention to the rush of immediate transition. Their explanations would then be seen in their native absurdity.[25]

This "immediate" fact of transition is not a "moving knife-edge" of the present proceeding through a uniform series of identical instants, but rather the differentiated structure of time, with past, present, and future conjointly inherent to this "stubborn fact."[26] If it seems to elude our intellectual grasp it is because our developed philosophical procedures interpose themselves against a concrete apprehension of organic perceptual life, the starting point of all science.

There is a strong correspondence between this primordial transitionality of which Whitehead speaks and what Merleau-Ponty in his lifelong work on perception sought to "rediscover," in the manner of an archaeologist, as the "structure of the perceived world [that] is buried under the sedimentations of later knowledge."[27] These overlaying sediments are not only philosopher's theories, but also the detritus of specialized sciences in our ordinary ideas. Merleau-Ponty sought to bring to light perception in its original and buried content, before the standard and automatic "scientific" explanations of perception come into play and impose inappropriate meanings, leading to the quandaries of empiricism and rationalism in philosophy.[28] The convergences between the concerns of philosophy of perception and theory of nature are material for a separate study, though these already find a common ground in the argument of the section to follow.

How does primordial transitionality uncovered in Whitehead's thought as a fundamental concept of nature serve to further the conceptual foundations of physical ontology? A basic claim that has been emerging and will continue to develop is that when the causal relation is applied to the genesis of physical space and physical time, in each case process underlies and permeates product within and through their differentiating relation, in such a way that there are (connected) process and product aspects of both time and space; within this common structure space and time are interconnected one with the other. This ensures a conception in which the transitional differentiation of antecedent and consequent in the genetic structure of the physical is not conceived on the basis of simple differences of place or of time: the conception requires a form of extension primordially structuring all physical space and time. To talk about prelocal causal structure as primordial time/process is to employ a somewhat uncanny notion of causal antecedence. But the neo-Whiteheadian claim is that for an advanced conception of physical being, time (as fundamental "process" or "passage") and causality are *cofundamental;* it is not that causality is somehow more fundamental than time. The key to this whole ontological prospect is to see that if causal transition in the form of "process of becoming" is truly inherent to physical being, then there is *genetic-transitional physical process* that eludes the representational models ordinarily involved in speaking of occurrences "in space" and "in time," the models, that is, of succession in spatial position and of uniform succession in time, but which can nevertheless be fully and concretely conceptualized in terms of genetic structure. This class of process can only pervade all spaces at all times.

I do not seek to promote the "atomic" or "epochal" theory of time of Whitehead's later period, which is his own way of carrying through the earlier work on time in the theory of nature studied here, but which seems to me problematic. It belongs to the metaphysical outlook of the later period, which is not especially compatible with physical ontology, as I explain briefly in Chapter 10.

Whether or not Whitehead would have agreed with this interpretation, the neo-Whiteheadian ontological proposal so far based on the middle period work is to think of physical "time" as unitary transition in and as a totality of ongoingness. Time is emergent in the sense that it has primary and derivative aspects, the former coinciding with the transitional structure of fundamental engenderment; this avoids the illogic of a pretemporal process. Physical space, on the other hand, is emergent in the sense that actual locality systems in three dimensions are termini of originative process, and as such are conditioned throughout their extensiveness by transition in emergence as a structurally independent, prespatial background. But insofar as the meaning of "space" is released from the confines of flat three-dimensionality, it is correct to speak of the "process" aspect or dimension of space: the prespatial background is a primordial spatiality in the sense of a dynamic "time-space" formed by primordial transition; this prespace is always a condition of and for space in the more limited sense of given locality. The nineteenth-century ether was material everywhere *present*, a naive conception belonging to a stage of science for which space was simply and irreducibly given presence, void or occupied. But the explanatory lacuna left by the demise of the ether is now to be filled by a structure of activity forming an antecedent context for locality or place, hence for "presence" as such. Physical being in its primary structure is transition in presencing. "Derivative" time in the concrete is the durational aspect of particular material-spatial systems, and from this comes the concept of a "mere abstract stretch of time." Time and space are thus differentiated in their process and product aspects in different ways. *Temporal* transition is antecedent to space as the temporality of its engenderment, but the measurable, countable aspect of time is coterminal in the genetic process with localized, three-dimensional space (more on this in Chapter 10).

"But must not the transition in 'presencing' proceed *from* some *place*, precluding its definition as physically prior to place?" This thought reinserts the identity of space as a backdrop to the transition, and so misses the essential description. Fourth-dimensional extension has identity only in the coming about, that is, in a structure of development; it is a case of identity as and through differentiation. However one represents this transition in a spatialized model for one purpose or another, the essential thing about it is that no part of it, no stage of development, can be *absent* anywhere.

The next section demonstrates how the conception of time and space in terms of genetic structure can be used to lend a basic coherence to a fact that is well known but is also disastrously misdescribed by the simple, practical uses of physical concepts.

A Sense-Making Application

I have been talking about physical engenderment, by which I mean the origination of the manifest elements of nature themselves, rather than that of particular natural objects or phenomena. This notion will be given some more empowering formulations in the discussion of Merleau-Ponty in the section following this one. Here I will demonstrate, through an example that is difficult to think about but is nevertheless in a way quite commonplace, how engenderment or genetic structure as I have conceived it is inherent in our perceptual and observational encounter with nature. This will help dispose of possible objections that such a thing is either a physical or a logical/conceptual impossibility.

First, consider that the familiar scientific story of the beginning and overall evolution of the universe as portrayed by cosmologists in physics seems to be essentially at odds with the present notion of primordial engenderment. In the cosmological conceptions normally employed today, the origin of the universe itself is situated along a single time line which is formed by numerical extrapolation backward beyond the periods marked out for geologic and solar history. In this conception (at least in its popular presentation) time is not portrayed as co-originating with nature, that is, there is nothing extradimensional about time in this universal history; it has the same basic sense as "in years past," "yesterday," and so on. (Relativity theory has already clued physicists not to settle for such a simple and straightforward understanding of "time" when it comes to the Big Bang theory.) But my cosmological story is a tracing of events into a "process" substructure of space and time, and cannot be framed in terms of simple temporal pastness. How are these two quite different basic stories for the ultimate natural explanation to be reconciled?

This question and my answer to it call attention to what was discussed above about present-day cosmological science (really an extension of physics), namely, that the concepts still being applied in the scientific discussion of *nature as a totality and in its ultimate origins* are simply not adequate for understanding nature in these aspects. Actually, the standard accounts of the event of "creation," such as they are, and the concept of physical engenderment developing in the present context are not exactly contradictory stories; rather, they serve quite different purposes: the conceptions of the whole and of the "beginning" in use today belong in the category of useful working models serving scientific progress (as are conceptions in other departments of physics), whereas a coherent narrative explanation is something altogether different. The conceptual underpinning of current cosmology remains a useful basis for interpretation of data in the course of progressing research. That underpinning is the conception of nature in its wholeness and finitude as a *total aggregate object:* the universe began to exist at a certain point in time and as a localized whole of matter and space, and has since undergone a series of transformations of character and a continuing expansion. This basic conception persists in its utility even as it becomes clear to scientists that "the universe" is not at all correctly represented as a total volume of space in the ordinary sense; that it does not have an "edge," and that there is no actual location, no matter how approximate, among objects which is privileged as a "center." And "where" did the Big Bang occur? What is an expansion that is also the "creation" of space? As in recent microphysics, when such conceptual difficulties arise in cosmological science, they have little bearing on the practical and so give rise at most to a discourse of speculative play. It is not hard to see after some examination that the basic "story" of cosmic origination popularized by scientists lacks narrative coherence in spatial and temporal terms.

But the general inadequacy of the object model of nature under the aspect of totality is indicated by a consideration more basic than these conceptual problems stemming from scientific models. This philosophical consideration is even further removed from the actual specialized inquiry of scientists, though I believe it is of fundamental importance for philosophy of nature. As I will presently show, the total cosmos (or even a substantial portion thereof) as it is known to general astronomical science apart from any special investigation has a spatial and temporal characteristic incompatible with representation as an object or configuration of objects present in space.

The characteristic of the field of observation of which I speak is one that is often called the time lag in observation, or as Whitehead put it the "belatedness" of all perception;[29] but it is only misleadingly conveyed with these expressions, and their use can quickly lead to confusion. I would prefer to describe it in a fresh way, deliberately evading sources of perplexity in the terms of description, as follows. Because light is a transmission, with a finite and measurable velocity, the outward distance from ourselves of astronomical objects seen or detected is known to involve a *kind of extension* distinct from that of simple spatial distance, though this is not simply extension backward in time in the ordinary sense, but rather extension along a span of universal transition in cosmic unfolding necessary for the course of the connecting transmission in the perceptual relation. In other words, the exact character and position exhibited by the astronomical object represents a stage of developing totality at least somewhat prior to that of the standpoint of observation. In the terms developed in the previous section, spatial depth in all perception and observation has an element of temporality *in the dimension* of genetic antecedence.

In scientific discourse and literature this fact is not, of course, described in any such terms, but is treated as the easily acknowledged and perhaps curious fact of a "time lag," certainly not as a source of genuine scientific questions or problems; indeed, it is not treated as an issue of natural science, but as an issue of epistemology. But it nevertheless has the intriguing implication that to probe *outward* with telescopes into cosmic space is to probe "backward" into cosmic evolution (in a sense not easily conceived correctly, despite the usual facile treatments). The most distant objects are known to belong to cosmic periods remote from our own and perhaps differing in the kinds of objects featured; for example, there is a theory that quasars, extremely remote objects, are protogalaxies. The fact that remote celestial objects are also primordial in the sense of cosmogenetically early does not, I think, force the standard conclusion that a radio telescope is a time telescope detecting objects that are "no longer there." The way I would describe this general fact is that the field of astronomical observation displays a genetic structure; in Whitehead's terms, it discloses a "duration" as a transitional whole. This structure contrasts essentially with the representation of the universe on the model of a spatially extended object. The object model induces a breakdown of narrative understanding by attempting to conceive the universe as an aggregate presence in a single system of space both total and "local," somehow a colocated whole. On this model, the finitude of the velocity of light means that at least part of what we see in the sky today is a remnant of the ancient past, and certainly the stars are not seen "as they really are today."

"But is it not an allowable extrapolation in science to simply *correct for* this irremovable time lag in observation, calculating with a certain degree of precision and at least *conceiving* the characteristics and the total extent that the universe has *as a spatial aggregate simultaneous with the present moment* and as differentiated internally by simple spatial distances, with, for example, protogalaxies evolved into galaxies (or exploded or imploded), and thus conceiving of nature as forming one co-present whole of matter and space having the general characteristics of a single stage of cosmic evolution, the one to which we belong? This would be to coherently adapt the object model to the observed cosmos in spite of the fact you describe."

But *what universe would it be* that is conceived in this way? Not the one observed with telescopes, in which there are, for example, quasars, and a general regression through cosmic stages. According to this reconstruction the genetic regress and primordial objects have only an apparent existence at the time of observation. The universe thus conceived would always remain an exercise in the reconstruction of the observed, approximate at best, but more importantly, neither yielding any particular data nor explaining anything; though perhaps to proclaim its possibility is useful heuristically as part of a background picture—which means, as a concocted description that conforms to the object model, which itself is quite useful for guiding investigations. In fact, to thus scientifically reconstruct the field of observation is to tacitly believe that the contents of this field *as science comes across them* are a kind of illusion as against what is "really there" or "really happening" today, that these contents are a mere residue of past cosmic events. The neo-Whiteheadian approach to background explanation would propose a wholly different way of understanding this fact about distant observation, an understanding that stays with what is presented by nature for scientific study. This interpretation, which I have already utilized in the exposition of the problem, would learn to avoid the bewildering conclusion that arises from employing an ordinary notion of "time" in thinking about a "time lag," namely, that the thing observed as we see it is really a phantom of the past, whose true condition *now* is open only to observation in the future. The star we see might "really" have vanished!

There is no contradiction in the claim that a star we see might have gone extinct.[30] But to understand this as a genuine reality/appearance fracture by saying that it may be a "thing of the past" and *really not there* is to insist on the preconception that a perceived object and the perceiver occupy a single, universal coordinate system of spatial positions (to which time as simple duration is affixed). Though this spatio-temporal picture may suffice for many practical purposes, cases of remote cosmic objects expose its artificiality: here the concept of a purely spatial differentiation as among aggregated present things does not describe the actual natural relation to the observed, because of the inseparable element of temporality. What is troublesome is that this concept imposes itself with the language of "there for us" and "really not there." But it is perfectly natural for Whiteheadian physical explanation to assume that nature as a finite whole, just as in any other fundamental aspect, will present itself to science as a gradient structure of genetic differentiation, a structure hidden from the expectation that present things and localized motions and propagations must populate physical explanations. Any perspective upon totality has for its standpoint a terminus of "nature as a process," so that the perspective is conditioned throughout by genetic transitionality. *There now* intrinsically involves genetic/temporal structure. The transmissions converging on our locality from the stars are constituents of the fundamental causality of the universe, and the temporality essential to transmission only points to the fact that the natural relation of perception is always "in the making and because of the making," as Whitehead said.[31] Cosmic perspective is thus *temporal* perspective in the primal sense developed in the previous section, that it is permeated by a total process or history in a systematic way, with a series of features along a process of development that are unique to that perspective.

There is no such *possibility* as a perspective or viewpoint on an emergent whole for which that whole is an aggregate object/space system, because genetic structure is essential to the concept of such a totality. The instantaneous moment in the existence of the universe as a spatial aggregate with all parts reducing to localities mutually present in a total locality is only a construct arising from a useful simple schematic of description; it corresponds to no actual temporal unity of the whole. Concrete physical simultaneity, on the Whiteheadian view, is simultaneous *occurrence* with all the dimensions of physical occurrence, a "slab of nature" with causal thickness; it is not mere coexistence "at an instant," as of material components arrayed in space.

When a scientist says about the supernova which was observed in February 1987 that it "actually occurred 170,000 years ago," it sounds as if this event were simply being given a date of its occurrence, a date far in the past, without any further clarification being necessary. But this "dating" of the event cannot be understood in the same way as, say, the dating of a solar eclipse at that time. The supernova is not an event embedded in the process of evolution of the earth and solar system, an event of local history, as if in February 1987 scientists actually got a glimpse into the ancient past. The fact behind this assertion by the scientist is the finitude and specific measure of the velocity of light; but here the notion of "velocity" tends to introduce the grid space of the object model in a way that escapes examination. (Contributing to bewilderment over this matter is the careless talk of "seeing the light from stars.") The basis of this "assigning a date" is precisely the impossibility of any temporal correspondence or simultaneity between the explosion and earthly events prior to the observation in the ordinary sense of "in the past"; it did not "really happen at that time." The supernova event lacks a shared framework of temporal passage, the basis for contemporaneity, with events of the ancient geological past; in Whitehead's terms, these systems of events belong to different and obscurely related families of durations, they are not encompassed by an inclusive temporal transition. All temporal commonality of the supernova with events of earth history is in fact constituted in the "now" of the observation of the occurrence. The cosmic "now" is inherently perspectival. At large distances, the genetic regression of the whole presencing transition constituting the observational present becomes a significant temporal depth in the field of observation. This is commonly interpreted in terms of the automatic practical conceptions of quantitative science, in this case the extrapolation of a system of space and time applied within our local object-complex, resulting in the idea of the inherent slowness and delay in the messages of perception traveling along light beams.[33]

(This issue is related to that of the traditional "causal theory of perception," in which a muddled philosophical situation arises from thinking that when one sees a green tree, for example, there is an act or process of interpreting a terminal "datum" that presents itself to a rational consciousness, having been generated by an antecedent physical process mediated by the sense organs. Though repudiated by philosophers of diverse schools—Whitehead, Wittgenstein and his followers, and Merleau-Ponty[34]—this traditional "scientific" account of perception now seems more firmly institutionalized than ever in its long history; this irony results from the subordinate status of philosophy [excepting philosophy of science] in relation to the official sciences. The traditional basic story is that some external transmissions lead, through the mediation of the senses, to some internal processing, which terminates in the "event" of perception. But whether or not there is such an "event," the content of this "event" is in no sense internal but rather extends outwardly from us to the tree and toward a world. In perception per se there is no essential role either for a "perceptual datum" or for an interpretation. For instance, "I see a green tree" does not mean that the tree is an actual and objective set of primary qualities which I then see _as_ green through a mental reaction; this very "as" is out of place. Though the presence (at whatever distance) of the perceived object and the actuality of pervasive radiation belong to the "necessary" physical conditions for perception, the notion that a particular physical story "explains" the subject-object relation in perception is wrongheaded. I argued in Part One that there is no cause of intentional action, and there is also no cause of perception in the sense of a process outside of and prior to perception somehow generating world and things "as" perceived. The philosophical impasses resulting from this doctrine are well known enough that it is time to give attention to Whitehead and Merleau-Ponty, who both seem to be saying that perception is grasped only in its immediacy.)

Consider an ancient geological event, such as the superscale flood that occurred in the western United States when a giant body of water broke through an ice dam left by receding glaciers of an ice age. To fix the date of this event is to make a scientific determination through procedures of measurement and calculation, analyzing its effect on land formations and so on, producing a piece of knowledge about earth history. By contrast, to say of a supernova now under observation that it "actually took place 170,000 years ago" is really a picturesque way of noting its great distance in light-years. Beyond this, the assertion is a scientific curiosity on exhibit (along with so much that is enigmatic in contemporary physical science). But the original brute fact disclosed by science is that of genetic/ temporal depth in the field of observation, though in speaking about it a scientist may merely resort to the convenient and beguiling notion that "we see into the past." To comprehend the fact positively calls for an application of the results so far concerning the relation between time and causality. According to these results, tele-

scopes do not probe into the past; instead, they simply probe outward into cosmic space, which means, through primordial temporal
structure that is due to the *process character of the cosmos as a totality.*
Depth in the field of cosmic observation extends into *genetic
primordiality,* along a structure in the genesis of place (and thus a
structure of perspective) that is already temporal to whatever extent
it is spatial, a structure that natural science will eventually come
across wherever it looks. Seeing stars on a clear night is not receiv-
ing a set of perceptual messages, belated or not; instead, the perceived universe itself is a heritage of perceiving. Perception is
literally the appropriation of the perceived world with its temporal-
spatial horizon.

Outwardness in cosmic space, just as we encounter this dimension and its remote terminal features, is not simply spatial extension; but neither is it extension backward along a local sequence of
transmission events leading away to another part of a localized
spatial whole at a previous moment of its existence. (The standard
space-time model based on special relativity pictures a spatialized
manifold of events with causation consisting of transmissions along
causal lines that "act on," rather than bring about, particular events;
this procedure inserts locality conceptually by sleight of hand and
binds it to a purely serial-successive mode of causal structure.)
What is shown by physical science is really that cosmic depth in
perception extends prelocally, that is, extends from terminal locality
along a differentiating transition structuring a totality of genetic
process. I submit this Whiteheadian or neo-Whiteheadian account
as the narrative-explanatory replacement—or explanatory supplement, if this is preferred—for the simple extrapolation of local historical time employed for empirical purposes in cosmological
science. The advanced instruments of astronomical detection and
observation reveal features of nature at different stages of a "history" or "process," reveal a *structural totality,* as do those instruments of refined detection employed by the physics of the
"microcosm." We observe stars and galaxies in their "past" states
only in the sense that these observed stages of development of remote circumstances identify stages in a fundamental genetic structure that are antecedent to the standpoint of observation.

The observed supernova and our observation of it are not
two events separated in time; this is not because they occur at simultaneous instants in a uniform passage of time, but because this

ordinary framework of temporal description is too simple for their relation. Instead the causal transition underlying this relation involves specifically the dimension of time coinciding with primordial genetic structure, *not* "derivative," practical, local time, nor for that matter a single universal temporal passage. In Whitehead's terms, observer and observed belong to the same "duration" and are thus concurrent with a unitary event-totality, despite remoteness along the spatial aspect of this transitional whole. The *difference*-structure of perception and observation in general, that which for instance makes it possible to point to something, is constituted by the *identity* of concrete primordial transition (it is certainly not constituted by extension along a process of transmission *in* space and time). The thought-picture of an encompassing linear temporal passage for all events represents the supernova as simultaneous with ancient earthly events; but on our view this seemingly inescapable simultaneity (as though the observation is like the reading of a message long delayed) would appear to be a construct filling in for the absence of any simple and direct temporal relation whether of simultaneity *or* sequentiality. The usual assumption is that an undifferentiated temporal passage, a uniform series of identical instants, underlies all events, so that direct cross sections of simultaneity through events can always be established. This picture leads to mere befuddlement over the fact that though we are seeing the supernova happening *now*, the perceived and perceiver represent separate perspectives on a total originative process, so that it may already be a mistake to assign their two local histories a temporal correspondence of their successive parts in a shared framework of time (though of course they have *some* systematic mediating relations within overall cosmic evolution). There is a powerful pull toward a picture of these processes as taking place on a common cosmic stage, so to speak, despite their disparate situations, and even this virtually irresistible picture has to be resisted if identity of temporal transition does not belong to this direct comparison. It is helpful to think of transition in temporal passage as essential to simultaneity and not only to temporal difference.

   Observation at the cosmic horizon of perception is conditioned by the structure of emergence of the spatial field as a totality. What the physics of light shows is not that we see into the past, but that the natural perceptual horizon of the present is extended spatially as prelocal or presencing transition and temporally in the genetic dimension of time.

These considerations buttress through a concrete illustration the claim that there can be cases of causality in which the antecedent process or causal trace does not extend linearly into the simple temporal past, but has a different temporal structure. There can be a "causal past" conditioning the present ongoingly, so far in two possible modes: in the case of empty space events, transitional-genetic structure in the background of local space forms a substratum of physical processes concurrent with that space and integrated with it; and second, genetic transition temporally structures the field of observation within the present of observation. Conceptual qualms may linger concerning emergent time and space, but these cannot be based on the necessity for a spatio-temporal separation or dividing line of contiguity between process and product. Even if this differentiation is thought simply in terms of different events, nothing prevents an event A from being *in process* contemporaneously with an event B which it directly engenders. *Though this would not be a case of simple seriality,* such a relation of contemporaneity involves temporal transition inherently, since an analysis into instantaneous slices of time would leave out causality and extension in the events, that is, leave out their whole character as differentiated occurrence. Talk of the emergence of fundamental natural elements does not disrupt the relation between time and causality, as happens for example with talk of "reverse causality," but shapes the concepts to the needs of an improved understanding.

The temporality of outwardness in space is not accurately conceived in terms of a time line extending into the past, but only in terms of that structure to which genetic process and temporal passage are equally essential. The corresponding spatiality has the structure of transition terminating in the standpoint of perception. Cosmic observation thus explores toward the horizon of a *concurrent* event-totality. It will gradually become clear in the remaining chapters that this is the correct way to finally surpass the clearly inadequate concept of the whole as an aggregate of dispersed material and hence as a single, total system of locality differences, as if one could in thought lay a single coordinate grid over the whole.

The Ontological Shift

Merleau-Ponty gives a number of clarifying and constructive suggestions on the topic of the "ontology of nature," its history and prospects, including comments about cosmology and about contemporary physics.[35] He describes how the traditional ontology which had served as a framework for centuries of natural science has proven inadequate both philosophically and scientifically, which calls upon the philosopher to show the way toward a more adequate concept of physical being and physical causality. This material lends itself to exposition and interpretation as a resource for physical ontology.

The traditional concept of nature is also termed the "Cartesian," because it is definitively stated in the metaphysics of Descartes. It holds that nature (insofar as this is ontologically distinct from the human "soul-substance") consists ultimately of visible and invisible *matter*, either filling all space in different configurations, as Descartes himself thought, or discretely occupying space as a void. Matter as present being is fully explained as the pure product and remnant of the act of Creation, itself causal only derivatively and strictly in accordance with the natural laws it embodies. This is the ontological basis of the Laplacean universe of perfect predictability in principle. In Merleau-Ponty's discussion the fundamental decisions and assumptions of this traditional conception of natural being are given focus and made explicit. I will give this material an exposition with my own interpretive emphasis.

To *create* is to bring into being, not merely to actualize or activate, but to make present, to bring to presence; in this way Creation has been correlated with the fundamental status of matter in ontology of nature. In this traditional ontology the physical thing is forced to conform to a fully localized object or object-aggregate undergoing various modes of activity. Nature is thus thought as pure product, referring outside itself for all fundamental explanation, with a purely inertial and dependent effectuality. The usual notion is that the teleology of Creation is the appropriate affirmation of the spirit to compensate for a purely mechanistic or "dead" universe while affording to science the absolute dominion of natural law; yet it is precisely the *created* natural universe in the traditional ontotheological sense that is in the first place "lifeless" and "devoid of meaning."

Merleau-Ponty's exposition of this concept of nature brings out some puzzles that it presents for reflection. For example, at what "point" does the hand of the Creator cease to determine things and the rigid mechanistic laws "begin" to take over? How is there any transition or mediation between these two absolute determinants? How could the act of Creation itself escape determination by natural law at the very moment that mechanistic nature is brought into being?

These problems traditionally have little if any significance for science; historically, the Cartesian concept of nature as matter and mechanism guided all explanation in natural science for a considerable period. But eventually, given this framework for explanation, fundamental perplexities arose not from philosophical thinking, with its historicality and wayward tendencies, but from experimental investigation. Though this poignantly signals the need for an ontological shift, scientists concern themselves first and foremost with practical progress and are under no compulsion to respond to this need (even though they may find the situation intriguing and venture much speculative musing). For philosophy of nature, however, the situation that has come to light cannot be evaded and is at the heart of its concerns:

> At the end of the experience produced by this [Cartesian] ontology, European philosophy again confronts nature as an oriented and blind productivity.[36]

The very successes of science at explanation and at enhancing human perceptual and technological means, successes stimulated by the traditional working conception of natural being, have led inevitably to an encounter with realities for which this conception is inadequate. The newly encountered realities resist all application of the simple concept of a material object, that of a "product and a dead result"[37] of a Creative act or event, with empirically discoverable properties of mechanical interaction. This is a way of expressing Whitehead's lifelong critique of material substance as a foundational concept in physical science. Natural being must now be understood as *productivity*, that is, as process and product inseparably. Further, this nature-as-productivity is "oriented," that is, it is not *merely* the sum total of a diversity of processes and products, but instead, there is a single orientation to the process of nature as a whole, an orientation toward a terminal stage consisting of the manifest natural world science explores, whose foreground context is material being in its richness of aggregate forms. The present study has described this directedness of natural productivity in terms of the one "axis" of engenderment, represented structurally as the succession in depth-strata of genetic transition terminating in the aspect of simple presence that is the material of the traditional concept.

By saying that nature-as-productivity is "blind," Merleau-Ponty means that despite the wondrous and delicate natural forms that have emerged, it would be an anthropomorphic error to apply to natural production concepts associated with intelligence and purposeful action. This would be to close off the original character of the perceived world, that of "an unmotivated upsurge of brute being"[38] (or "wild being"), a character that is for the most part hidden from everyday consciousness due to an overlayering with cultivated concepts, and to which the philosopher in the phenomenological tradition strives to return our attention. Though the "again" of the above quotation indicates that in the new encounter with nature there is an echo of pre-Cartesian conceptions,

> This does not represent a return to teleology. Properly speaking, theology understood as the conformity of the event to a concept, shares the same fate as mechanism—these are both concepts of *artificialism*. Natural production has to be understood in some other way.[39]

(This last sentence would be the perfect motto for the present book.) The same thought is also found in the *Working Notes* for the planned but uncompleted work:

> It is this wild or brute being that intervenes at all levels to overcome the problems of the classical ontology (mechanism, finalism, in every case: artificialism). . . . [40]

Merleau-Ponty's persistent point is that an adequate understanding of physical causality must somehow move beyond these traditional alternatives, a suggestion that was embryonic in the remarks of Sameul Clarke quoted in Chapter 1 (page 20). Comparably to Whitehead, "wild being" is the primitive, dynamic givenness of the world. The currently popular idea that nature is a living organism is in the right spirit as an alternative to mechanism, but it is a form of teleology or quasi-teleology to which the basic philosophical objection is that it pushes the distinction between living and nonliving nature into limbo. Resorting to such convenient possibilities of metaphoric expression only substitutes for ontological labor.

According to Merleau-Ponty's examination of the matter, the shift to an understanding of natural production "in some other way" than mechanism or teleology will bring with it a fundamental rethinking of the ultimate referents of cosmological explanation— traditionally, Creation and a personified Creator. As mentioned, the relation of this personage and this event to the world as addressed by science has long been a source of philosophical problems. A peculiar situation subsists in the Cartesian conception of nature as mechanistic process and of a Creator whose intentions and actions have no determinants:

> If nature exists only through the decision—and the continuing decision—of God, then it does not "hold" in time (nor for that matter in space) by necessity of its own fundamental laws. Nature as event or an ensemble of events remains different from nature as object, or an ensemble of objects, as does God conceived as the free creator of the world and God as the source of a causality from which derives an eminently finitized world.[41]

There is a basic discordance and ambivalence in the traditional ontology between the ultimate explanatory factor—Creation, or the will of God—and the overall causal process of nature itself under a mechanistic conception. How can a free creator be the generative ground of a necessitous physical universe? This problematic result of the Cartesian ontology directly prefigures the required transition to a new conception. It is nature as event— specifically, as production conceived outside of mechanism—that must transcend the reduction and limitation of the concept of nature to pure object and product, natural being as simple presence. The Cartesian concept

> obliges every being, under pain of being nothing, to exist completely without hiatus, and with no hidden possibilities. [Under this concept] there is to be nothing occult or enveloped in nature anymore. Nature must be a mechanism and one should be able in principle to derive the form of this world from laws which are themselves the expression of the internal force of infinite

production. . . . Everything internal is handed over to God's side. . . . Both historically and philosophically our idea of natural being *qua* object in itself, which is what it is because it cannot be another thing, derives from the idea of an unlimited being, infinite or *causa sui,* and this in turn comes from the alternation between being and nothing.[42]

It is not immediately clear what a new concept of God as the "source of a causality" would be that would go along with the new concept of nature as productivity, to take the place of (or transform fundamentally) this traditional ultimate explanatory referent. Must there be such an ultimate, distinctive source? If so, what could supplant the simple anthropomorphic concept of a Creator and his works? What "naturalistic" concept could have this kind of cosmological and spiritual significance? In any case, does not a fundamental gradient of being shade off into *nothing?* To say that it shades off into nonphysical being might only be an understandable refusal to think this "nothing." We can avoid getting in over our heads by staying with what is clear: that with the ontological shift the relation of ultimate causality and ultimate explanation comes to be internal to physical being, becomes precisely that aspect left out by the mechanistic conceptual reduction to locality and present things, which is what in the first place required an explanation outside the physical. As a consequence an ultimate "source" of all things can no longer be conceived as standing apart from that of which it is the "source," like an artist completing a work. In my own terms, this is because production or productive transition must be recognized as intrinsic to the physicality of a thing, so that any actual matter located in space is not merely a given, irreducible existent, but rather, *as* product identifies a specific route of engenderment that *constitutes the object* as emergent, as a trace through the genetic structure of space and time. In its unreduced and native reality the physical object is conditioned by "natural production"; as concrete locality it is identified with a specific formative trace in a context of engendering-presencing process. The concept of nature as object, as a total aggregation of matter, with the Creator Being inhabiting or comprising a separate category of existence (and the resulting inexplicable mystery of Creation), must give way to nature as event, with engenderment process and product together in a transitionally differentiating relation. (This noncommittal "transition from out of . . . " will be a cause of complaints, but my claim is that a whole sphere of physical explanation can resort solely to the structure that belongs to the transition, saying nothing about what might ultimately "complete" this formulation.)

"Eminently finitized world" means this: By eliminating the permanent mystery of where and how in tracing the explanation of things the laws of nature break off and the acts of a free Creator begin, the new concept of nature incorporates all fundamental explanation into a unitary causal transition marked by finitude; that is, it is *as* traceable into originative production that nature forms one whole; this would be *the* structure of totality. It is not clear how an ultimate "source" could be *outside* of the structure of engendering transition. The new concept of physical being thought in ultimate terms as a reciprocal relation to nonbeing has a gradient structure, rather than the structure given in the simple contrast between presence and absence. Anywhere along this fundamental transition are stages whose activity is manifested in the characteristics of observable events, since the very origin of natural elements eventually figures in the background content of physical explanations. The "source" of natural causality would have to lie at the furthest reaches of the depth dimension of engendering physical activity, at the heart of the world, not outside of it as an absolute ground of being. But this also cannot be a form of pantheism, claiming a simple omnipresence of the source, since this would require collapsing the crucial distinction between presence-in-space (object-being) and active (structured) engenderment.

Also, with the new concept there is no longer a creation event (whether it is thought of within a theological or a physical–scientific perspective) that might be assigned a date in the ancient or ultra-ancient past. Unless "create" is used as a synonym for "produce," with no implication of intentional action, the very notion of creation is an artificialism in the mode of teleology. What is needed in its place is a nonmechanistic concept of natural production which retains intact the structure of the physical, of "wild being" outside of all our attempts to impose a familiar and/or practical concept, which conforms only to "the gradient: not linear being, but structured being."[43] (The rejection of teleology obviously does not fit with Whitehead's mature metaphysical scheme, nor would I attempt to force any results into this scheme. Whether Whitehead's *theory of nature* involves teleology is another matter. In any case, I tend to think that for a theory of physical process, teleology should be a last resort at best.) If not only space and matter, but time as well, are in different ways emergent—that is, *physically* engendered—physical facts, then the idea of an intentional creation of the natural world seems both extraneous and incongruous.

Merleau-Ponty ends the first of these talks on the concept of nature with a point about the effects that the shift underway in natural ontology has or will have on historical perspective in philosophical scholarship. One might have thought that if what is needed is a redressing of ontological balance between "process" and "matter," then those approaches to science and philosophy of nature known as evolutionary cosmologies or process philosophies[44] would be consulted. It has always been the business of these philosophies to concern themselves specifically with the element of process in natural history in general; they are not inductive sciences that "fix themselves upon constant relations."[45] But considering the succession of these cosmologies in a history of thought during which the concept of a simple object present in space, a perfectly self-contained parcel of identity, has been made to suffice as a concept of natural being

> would lead one to establish a regression of eternal ideologies in which nature is an object identical with itself[46] and finally the emergence of a history—or, as Whitehead said, of a "process"— of nature.[47]

The new concept which arises unforseen in response to changing conditions of knowledge, as an option for sustaining the viability of naturalistic cosmological explanation (while shaping the concept to nature and not the reverse), is *physical being as essentially event, process, or history.* This supersedes the foundations for previous cosmological histories which, considering their special concern, could only "regress" historically under the Cartesian concept. Within this concept these cosmologies have been forced to juxtapose on the one hand the being of Deity as first cause and as outside of time and space, and on the other hand pure object-being, mechanistically determined and determining. Even leaving behind the mechanistic prejudice, if the ultimate cosmic explanation is God as imaginative and active originator, this does not actually give any answer to the question as to *how* the natural world of productive process, open to science as a totality of historical and causal explanations, itself came about. This calls for a story of emergent transition that is not to be found in our concept of a creative act, any more than it is found in traditional scientific concepts of physical events.

On the purely philosophical side, the "concept of nature" as irreducible engenderment process resolves in an unforseen way the philosophical predicament about nature that Merleau-Ponty expresses as follows:

> We can neither conceive of primordial being engendering itself, which would make it infinite, nor think of it being engendered by another, which would reduce it to the condition of a product and a dead result.[48]

It can now be seen that this predicament points to its own solution: physical being is essentially genetic transition. Whatever may be enigmatic about this result does not render nature inscrutable or ineffable; on the contrary, the claim is that physical being is correctly and fruitfully disclosed to our understanding as just this "enigma," resistant though it is to all reduction to something that is more familiar to traditional emphases and prejudices.

For the present purpose, these are the essential results of Merleau-Ponty's writings on the "concept of nature." Their exposition and interpretation reinforces the results of physical ontology so far in the following way. On the new concept, required for contemporary knowledge, physical being is structured as a *gradient* totality of productive process, whose stages terminate in particular three-dimensional localities and their unique linear times derived from a concurrent totality of "passage." *This* genetic transition is the background physical context for fields and radiation ontologically identified. The "axis of engenderment" cannot be specified by reference to differences of spatial orientation supplied by particular systems of local space, since this "axis" is unitary throughout a finite totality, and since as process differentiated from product it has a character of genetic antecedence in relation to elements of given locality (of whatever scale). However, in the previous chapter some use was made of a representation of three-dimensional spatiality as a flat terminus of physical emergence, yielding an image of the gradient structure of physical being. The "vertical" genetic axis can be the basis for an account of the causal characteristics of known physical events conceived in their structural alignment with respect to this axis, for instance, in the theory of radiation being developed, propagation follows that parallelism which is defined by the succession of strata that analyze (artificially or not) prelocal genetic transition, while interaction with local matter follows the axis of emergence into particular termini of the gradient. The particular physical "entity" as well as the whole-within-a-perspective are examples of this elemental structure.

For Merleau-Ponty, ontology of nature is called upon to begin to formulate for the first time an adequate concept of physical engenderment; for us, this concept must be in addition provide a context of and for physical processes that is not illuminated by current science.

Conclusions and Ramifications

This chapter has explored cosmological issues surrounding the proposed theory of radiation and has provided background conceptual and explanatory accounts supporting that theory, including general ontology of nature. In particular, Whitehead's strategy of reconceptualizing the ether as "ether of events" has been given an ontological foundation in "productionistic" terms. In the previous chapter the concept of the unit event of differentiating transition— for example the individual pulse of light, or its composite structures—was specified simply in terms of the causal relation, rescued as *production* from philosophical attack and also fully conceived as genetic transition, but nevertheless a completely abstract usage. Now the concrete physical context for the required applications of this concept has been better identified and clarified. It is a pervasive background structure of engendering activity in the sense of the "presencing" or emergence of physical place. Local environments are in this way pervaded by the actual and active influence of a total structure of physical process, where the finitude of the whole has the meaning of an encompassing genetic/temporal gradient. The way to fit propagative pulsation into this context is to say that the linear succession of pulses in particular natural environments is aligned horizontally along a genetic substructure of local space. The vertical structure of Figure 6.1 (page 218) is now to be understood as a transition connecting locality with totality via presencing process, or a portion thereof.

It is possible to speak of "the genesis of time" if this does not confine time somehow to a terminal stage of engenderment, but rather understands "temporal passage" in a sense that incorporates a genetic structure with process and product aspects. The "field" aspect of space, occurring as "presencing" independently of the presence or absence of particular matter (through specific causal characteristics depend upon this), has a causal-transitional structure that is none other than that of primordial time: it is essentially "process," that is, transitional differentiation in particular event-successions contextualized by the one genetically structured event-totality. Though all actual occurrences involve time in some way, not all are simply "in time," because simple duration is emergent and the process of emergence of time and space identifies a class of physical occurrence. The interconnection between space and time as required by modern physics is also to be sought in these relations (Chapters 9 and 10).

The objection that "nothing comes from nothing" is taken care of, because the notion of a preexisting condition of nothingness "from which" everything spontaneously arises applies the simple concept of time, "first nothing—then being," whereas the present conception structurally identifies time with being in such a way that "nothing," along with "becoming," is integrated into the gradient structure of being.

Continuing the interpretation of Whitehead's basic physical-ontological strategy begun in Chapter 1, "transition in the passage of nature" is more fully specified as the enigma of the genetic structure and the "process" character of physical totality, identifying at once the "event" of total becoming or upsurgence, the primary structure of time, and the genetic constitution of matter and physical place. How can time, which we ordinarily think of as a universal framework of events, be identified with a *particular* occurrence, namely this total physical event? The answer lies in the way "time" is integrated into the true concept of natural finitude, together with the universality of genetic structure in the broadest sense, the reciprocal actualization or unfolding of identity and difference. Primordial time/process has the "not properly serial" structure of an

individual "event" of differentiating transition, but as encompassing a totality of occurrence it is closer to the meaning of "process."
It is clearly not tailor-made for our ordinary physical concepts.
Whitehead urges that we apprehend it in a concrete and immediate
meditative perception.

The investigation has gained a sense for the *specific
interconnectedness* of physical events in four dimensions for
Whiteheadian theory, namely, their interstructuring into a contemporaneous genetic-transitional totality. The origin*ation* of a totality
of relations, as transition, is necessarily an *absolute structure* (in essential contrast to, e.g., Newton's vacant conception of absolute
space). A physical event is inherently contextual, participating in a
unitary (because primordial) event (or process) of origination. The
context of empty space events is the underlying structure of "passage" pervading emergent locality.

The irreducible element of explanation continues to be the concept of causal structure as differentiating physical transition. This
structure is not analyzable into components with a different structure, but only into further gradient event-structures; this is the
meaning of "parts" and "wholes" in this context. This descriptive
element fills all narrative functions: it is the concept of natural totality as a field of events; of four-dimensional temporality; of the individual propagative event as derived from the "passage of nature";
and of the physical field terminating in "centers" of aggregate locality (the latter is clarified further in the next chapter). Since in all
cases "entities" identified by this concept intercompose with one
another into contexts of transitional-gradient occurrence (ultimately
into a whole), there is no "further" analysis of these elements into
some different elements. The impulse to seek physical clarification
of "causal transition" per se is misguided; it would seek concreteness in the wrong direction. It takes some practice to get over the
habit of pressing to resolve all physical events into local motions on
one scale or another. "Wild being" allows no conceptual reduction;
but it does give itself in full clarity as a source of explanatory narrative rich in possibilities—in fact, just the possibilities called for by
recent fundamental discoveries. Physical causality is ultimately disclosed as complex composition among elements of genetic transition, essentially involving a gradation of levels of integration and
superposition, with aspects of reiteration in series and equally es-

sential aspects of individual and unitary structure. The Humean and mechanistic traditions on causality were not apprised of the context of unitary genetic structure—the inherently active in the sense of upsurgent character of physical being—underlying all reiteration in events. On the new conception (which the next chapter will continue to develop), this primary aspect of physical causality coincides with the fundamental physical fact of the emergence of time and space, which identifies ultimate natural finitude as gradient structure of "passage."

The enigmatic character of these results is partly only the default of picturability in the sense that physical events as disclosed by the ontological shift are not to be described as changing configurations among variables in space or space-time, that is, they radically elude representational models; by the same token there is no complete description of field elements as *isolable* facts. The atom is the limit of such picturability in nature, and even it is only accurately conceived as a physically unresolved "fuzzy blob" shading into infinity. In the case of "field" events the demand for immediate representability of local particulars without reference to a contextual totality is frustrated; but the pervasive application of a single effective concept provides a unifying intelligibility. One upshot of the chapter is a clarification of the claim that the question about the causal structure of radiation is not answered by a merely "ontic" scientific hypothesis describing local events in space and time.

Throughout the modern period science has carried out its tasks within a metaphysical framework in which the relation between present things and their originative "presencing" or engenderment is conceived in terms of creation and createdness. This has radically confined natural science to an ontology of localized givens, excluding from its proper concerns what the German idiom conveys as the *"es gibt"* ["there is"], "it gives," pertaining to natural givenness ("it" signifies neither substance nor agent).[49] Prior to the ontological shift that allows phenomena resulting from a permanent structure of fundamental genetic transition to be identified as such, science could only construct fragmentary explanations using familiar physical models.

On a somewhat peripheral topic, these results overall have some implications for a certain standard historical doctrine about the West, namely, that a religious worldview gave way over recent centuries to a scientific worldview. A reevaluation of this doctrine needs to take into account the cosmotheological framework of modern physical science. What I mean is this: When the background explanation of fields and radiation is finally found, it involves a shift in the methodological and ontological assumptions of science. With this shift matter and space cease to be essential given elements of any and all physical understanding, and become themselves *subject to physical explanation*. From the viewpoint subsequent to the shift, the traditional (modern) framework of science is seen to be "metaphysical" in the sense that it combines explanation reducing to mechanism (even, e.g., the "Big Bang") with a resort to the idea of "creation" as an ultimate mystery (if not an outright reference to a Creator) when it comes to the origination of natural entities as a whole. This cosmotheology in the background of science serves to silently support the view that nature is ultimately inscrutable (a view that today can cite in its defense hard science and cold logic), and at the same time functions to keep an essential (if unmentioned) role for a Creator personage. Thus ontology of nature is ensnared in the options of mechanism and teleology. The modern "scientific worldview" has never fully displaced the religious "worldview," simply because so far the epochally developing scientific picture of the natural world has been *blank* in its most general aspects—namely, a coherent account of finite totality/ultimate origination, and fundamental causal explanation of local phenomena—and thus has not amounted to a positive cosmological picture or story having the generality to rival religious cosmology (though to be sure these radical limitations in the account of the natural world have not been generally recognized as such). Moreover this lacunate condition of understanding naturally allies itself with traditional theistic metaphysics. Atheism does not follow from the results of physical ontology so far, but the indication is that if theology would preserve an internal bond with cosmology, it needs to move beyond the traditional naive conception of a Creator personage.

To the extent that this inquiry succeeds in disclosing the profundity of physical being, I venture to suggest that spiritual affirmation and scientific understanding have begun to coalesce in a single meditation. The bifurcation of spirit and nature that had been a consequence of materialist science[50] is being bridged in a mysticism of comprehension. By this I mean a developing epiphany in which primitive human wonder and awe at existence, which for many is sheltered in religion, finds vital inspiration and sustenance in a process of discovery and disclosure, placing no stock in the popular doctrine (whatever its merits) that certain things are essentially ineffable. One possible source of the popularity of nihilism is the thought that if comprehension were to reach into the ultimate aspects of physical processes, the universe would in a sense cease to inspire; would not a crucial spur to questioning be at an end? My contention is that this is not at all the case, and that a true comprehension precisely cannot "deprive itself of the stimulus of the enigmatic," as Nietzsche said.[51] But naturalistic mysticism has no use for a blind spiritual optimism pointing to the mysteries of physics.

Cosmological science, for its part, might occasionally release itself from the charms of algorithmic procedure and think about the need to transform and unite into a coherent understanding the interpretations of the natural universe as a spatial and concurrent whole and as a finite history of origination. A suggestion that is especially strong in the discussion of the middle section is that one can no longer assume the utility of the simple concept of time that operates when, for example, the "beginning" of the universe is understood as an ultra-*ancient* event. The key to this enigma is that the history of nature as a whole is not the history of a total evolving locality, as argued further in Chapters 9 and 10.

8

Interatomic Reality

Realism and Quantum Interpretation

THERE HAVE RECENTLY BEEN organized efforts[1] to produce and defend a "causal interpretation of quantum theory." I believe that an important intention is embodied in this phrase—the quest for a "realist" physics—but that so rendered it is at best inauspicious and at worst misconceived and wrongly oriented. What is wanted is a causal understanding of certain observations made in the course of experiments with the "microphysical" domain. Quantum theory is highly successful and broadly applicable, and Einstein's contention that it is "incomplete" seems to have been defeated; nevertheless, if the quest for a "causal interpretation" has any validity, this means of course that quantum theory "uninterpreted" constitutes something less than, or something else than, an explanation of how the experimental phenomena are brought about. It is in fact a formal apparatus of prediction functioning independently of any ontology for this domain or lack thereof. Does this not mean that the role of quantum theory in the quest for causal understanding is at best uncertain? Why, on the face of it, should quantum theory necessarily enter into this quest, let alone be its whole focus? Naively, might one not leave aside quantum theory, since it is not itself a part of the explanation sought, and simply ask what sort of entities or processes in nature might generate the unexplained aspects of these observations?

"But we need quantum theory to know what it is we are look-ing at," is a possible response. Upon examination, this claim has little going for it. Though the detailed phenomena are quite peculiar from the standpoint of causal understanding, their context of natu-ral occurrences is nothing alien to ordinary knowledge. It is that of empty space transmission, or radiation, a feature of the world that is familiar both in ordinary life and in scientific research. There is even an element of positive knowledge about radiation as a kind of causal process: "interference" effects unmistakably show the inter-action of processes with a wavelike characteristic. As Salmon points out, it is one species or genus among the multifarious causal trans-missions that pervade the natural world. Even in their perplexing aspects, radiation phenomena are something encountered in the world which form a subject matter of science independently of whether or not quantum theory was ever invented. In contrast to the situation with forces of attraction and repulsion when taken as problems of explanation on their own, with radiation the beginning and outline of a causal description is already established: it is some process analogous to a wave propagating across space. To sustain these basic physical intuitions is to ignore and/or critically resist the conclusion traditionally advanced by philosophy, that connection in causation is a fully opaque meaning.

Quantum theory is not a first disclosure of a domain of natural entities; rather, it consists of some mathematical models that orga-nize quantitative features of certain experimental phenomena, while these phenomena themselves remain without explanations even as the most ingenious and successful formulas are invented—this coexistence of extant and powerful "theory" and the default of explanation is the source of the call for a "causal interpretation." But one should above all be clear about the *type of knowledge* that is represented by quantum theory: mathematical formulas for predict-ing the outcomes of measurements. Since the call for a causal inter-pretation cannot be pointing to some deficiency in this thoroughly successful and eminently useful mathematical invention, it can only be correctly oriented as a call for a narrative causal hypothesis de-scribing the detailed structure of radiation so as to explain the regu-larities formalized in quantum theory—quite a different mode of "theory." Owing to this fundamental difference, it is not the case that to explicate the crucial unexplained effects of radiation for the purpose of a causal inquiry, or even to proceed with that inquiry, requires immersion in technical terminology, much less equations.

To characterize the approach of physical ontology toward microphenomena of radiation in terms of current controversy, it is, first, to take a strictly instrumentalist view of quantum theory, and to uphold as a starting point a "realism about entities" a la Hacking and Cartwright. As it proceeds from here to a constructive realism it of course neither expects nor aspires to produce any new formalism to compete with or supersede the existing formalisms, nor does it seek to "interpret" some existing formalism, but seeks instead, through the ontological shift, to provide a *first* basis for narrative physical explanation in this domain and to initiate explanation on that basis. Furthermore, in contrast to the bulk of projects toward a constructive realist microphysics proposed so far, "realism" here positively does not mean arguing that what answers to the name "particle" actually in any way corresponds to or resembles a "classical" physical object (or a "classical" mechanistic wave), nor does it advocate restoring either locality or determinism to a fundamental status in explanations.

The best-known "realist" quantum interpretation is that of David Bohm. Bohm is an important ally of the present point of view in that he upholds the value and validity of seeking improved physical understanding apart from any empirical consequences the new understanding may or may not have. He would interpret the "particle" realistically by postulating that something like a particle trajectory precedes the particlelike interaction, not a "classical" motion of a particle, but formed as a sort of average result of underlying processes.[2] The "quantum" behaviors of this "particle," thought on the standard view to defy all realist interpretations, are explained as responses on the part of the "particle" to a background force called the "quantum potential," which connects the "particle trajectory" nonlocally, that is, instantaneously, with its broader environment in the experimental setup. This is a laudable attempt to sustain some kind of causal explanation in the quantum domain without abandoning the *conceptual* bond of causality with locality. But from the standpoint of concrete narrative explanation it seems regressive: it cites some new undescribed generative processes and operative forces that explain the observations by way of an intermediary entity, one that is "guided" by its environment through a fully

occult mediation and is at the very least a quite unordinary "particle." From Chapter 6, final section, it appears that such an intermediary entity, following a localized trajectory whether continuous or
discrete and successive, whether linear or "stochastic," is not necessary for an explanation of localized interactions. If Bohm's "particle
trajectory" is in the last analysis a series of discrete localizations,
just what an observed "particle track" is known to be at the atomic
level, then it introduces a hypothetical series of localizations leading
up to the localization whose occurrence and properties are the
things to be explained, namely the perceptible interaction. Perhaps
some such antecedent localizing events (understood in the right
way) do occur in some cases, as suggested later in this chapter; but
the more basic explanatory need is for an account of *how atomic
localization takes place* in the context of wavelike propagations, and
in the view taken here, this is not a demand for a "particle trajectory
in some sense." Also, what is the physical origin of the nonlocal
force, the quantum potential? ("But one does not *explain* the
forces.") How does it operate, that is, what sort of connection takes
place? Bohm makes use of the fact of "quantum" nonlocality, and
makes philosophically commendable attempts to comprehend this
fact by talking about "wholeness." But there is little basis in these
suggestions for an actual physical comprehension, which would
involve an answer to the question, What sort of active physical
connection would take place instantaneously across space? From
the viewpoint of Bohm's thinking, these questions arise from nostalgia for a causal realism that is no longer possible, despite his
philosophical affinity with physical ontology on the question of the
aims and prospects of science.

    Bohm uses the "quantum potential" to account for both the
"trajectory" of the "particle" and nonlocality in general. In his
philosophical explication of the quantum potential the whole-part
relation is made use of in a highly problematic way.[3] A basic problem with the "undivided (or unbroken) wholeness" idea to which
Bohm resorts is that it seems to assign whole and part to separate
spheres of being, the "implicate and explicate orders." Apart from
the question whether there can be a whole ontologically independent of parts in the sense that it is itself devoid of all divisibility or
difference, to refer to a prior absence of difference cannot explain
difference. Narrative understanding would demand straightaway

that the *connection between* the "orders" be disclosed, but this seems to lie outside the account. As I understand him, Bohm would effectively throw out the causal-narrative format of explanation at this level by rejecting not only *connected* difference in underlying structure, but differentiation as such, as if whole and part formed a true metaphysical dualism with the "whole" being outside of the order of differences. The talk of "parts," "wholes," and "sub-totalities" leaves out of account the mode of composition that is meant. One gets the sense that the discreteness/contiguity picture reigns throughout, with connection depending on the possibility of the absence of differentiation, as in Salmon. The centrality of the notion of connected difference as the structure of causal narrative in the present study is sharply in contrast with these features of Bohm's ideas. But the explanations submitted here accord with Bohm's intuition that nonlocality is a manifestation of "wholeness" in nature.

A problem with trying to sharply differentiate a force from a particle, as in pilot wave theories, is that in contemporary physics the terms "force" and "particle" overlap in the entities to which they refer, as is clear in the case of the quasi-hypothetical gravity radiation. It seems doubtful that a clear and useful ontological distinction is made simply by using these different words, which helps to show that one of these cannot explain the other until there is a better knowledge of what these forms of physical activity *are* in general—in other words, to explain one in terms of the other does not constitute an ontological advance, but only a new construction out of these thoroughly opaque elements.

Another category of "interpretations of quantum theory" can be given the title "informationist ontology." This can range from specific comments in quantum interpretation to the general scientific metaphysics claiming that information is the fundamental content of the universe, and it is related to logico-empiricism. To invoke the encoding and transmission of information cannot *substitute* for explanation by means of physical stories, as if this were an independent variety of physical explanation (or of explanation in general). As in the paradigm case of the biogenetic information contained in DNA, the encoding and transmission of information must *have* a physical explanation. Information "itself" cannot be the "basic stuff of the universe." This general school of thought covers certain ways in which terms of discussion such as "probability waves" and "superposition of probability amplitudes" are often understood. Narrative explanation requires actual, determinate occurrences, not probabilities or "propensities" for some effect.

Closely related to informationist ontology is the "statistical ensembles" mode of interpretation of quantum mechanics, one version of which holds that "the particle takes all possible paths." These are plainly unappealing from a general philosophical point of view. A widespread form of informationism is the supposition that mathematical models such as the Schroedinger equation are the uniquely formatted "descriptions" of the physical propagations in question; but this is far from an auspicious approach, since it seems likely that despite all the wondrous successes of this and other artifices for prediction, this whole domain awaits a first, background physical explanation. Nevertheless, nearly all of the various quantum philosophies are attempts to treat quantum formalisms as if they were descriptions of nature requiring interpretation. Such approaches no longer have the aim of narrative theory of radiation anywhere in their field of view, and so, as I see it, have little prospect as ontologies. They represent the narrowing of the focus of thought to specialized problems and to technical inventions that are uncritically assumed to be examples of explanation or physical description. These are intellectual habits trained by a history and tradition of disciplined virtuosity in technological science, for which physicists cannot be faulted.

"Realist" interpretations of quantum mechanics are so called because they are alternatives to the Copenhagenist renunciation of ontology. But at least the more famous examples of these alternatives are not well fitted for the rubric "realist" apart from this specially defined meaning, since as narrative suggestions they amount to a collection of absurd constructions: a "particle" with a fully dispersed trajectory; a single particle path in multiple universes (Everett's hypothesis); or a "non-classical particle trajectory" introduced merely to restore the sense of an actual causally antecedent existent. These are testimony to the consequences of attempting to frame explanations in this area under the dominion of the causality-locality bond reinforced by the particle picture.

In the usual treatments of the mysteries of quantum physics, a variety of puzzling items are consistently brought up. In my view, some of these are real and intellectually challenging discoveries for which explanations should be sought, and others are artificial puzzles that arise from the technical and mathematical language in which the problems are framed, and from a general confusion between explanations and empirical models. The immediate task is therefore to sort out the real from the artificial problems under some workable headings to prepare for the rest of the chapter.

Some genuine facts to be explained, I think, are *indeterminism*, *localization upon interaction*, the *matter wave*, and *nonlocality*. An explanation of the second by itself has already been presented in schematic form. Outlining a physical context for the first three together on the basis of the preceding results in Part Two is one project of this chapter. Leaving behind the assumptions that make nonlocality an intractable mystery is part of the basic strategy of Part Two as a whole.

One artificial source of general puzzlement is the expression "particle-wave duality," describing radiation as having two aspects seemingly unresolved by any physical interconnection. This is done away with in the present ontological procedure by eliminating the idea of a particle from the description of the physical events in question, and by developing a conception of nonmechanistic *wave-like* process. Another source of perplexity is the talk of physical indefiniteness in measurable attributes; this talk only causes trouble if one insists at the same time on a particle ontology (more on this to follow). Another is the conclusion via thought-experiment (e.g., the cat-in-the-box illustration) that the same entity must be in contrary states at the same time. This again arises from imposing a particle ontology as an interpretation of the mathematics. The thinking seems to be that the variables combined into a "quantum superposition" (a mathematical procedure without a clear physical meaning such as ordinary "classical" superposition of waves) must somehow signify attributes of particles such as momentum or position; one ends up with a "particle" in a set of contrary states, or in no determinate (or definite) state. In the procedure of the present work it is maintained that the specific kind of connecting causal process to which talk of the various species of "particles" refers is not in any of its contexts or aspects particulate; this language is in effect viewed as *merely technical jargon*. The ubiquitous tendency in responding intellectually to the discourse that emanates from physics is to understand expressions of the facts that are merely useful practical modeling and conceptual equipment as if they amounted to explanatory knowledge. For example, to take the popular and seemingly sensible position that light is both a particle and a wave (why not?), though of course our "familiar" concepts of particle and wave cannot actually be applied, is to completely beg the question of *what it is* while confounding useful models with explanation in the standard scientific nihilist fashion. The truth is that it is neither a particle nor a wave (as commonly understood) and that what it actually is instead is to date unknown.

The rest of this chapter examines what I regard as the genuine problems of explanation and ontology. Indeterminism and localization are really co-occurrent features of the same observed events, and in the next section the focus is on indeterminism. The discovery that can be referred to as the "matter wave" will also receive a general, ontological explanation. I begin with atomic indeterminism, but only after some conceptual ground is tilled for making *causal sense* of a genuine indeterminism (as opposed to an ordinary deterministic unpredictability due to limits on the available data and means of prediction). This involves shifting for a while to a completely different philosophical context.

Richness of the Concept of Causality

Ludwig Wittgenstein criticized the belief that our thoughts and actions in all their determinate detail must have specific correlates in the detailed electrochemical events of the brain.[4] His remarks can be read either as a criticism of mind-brain identity theory or as a way of demonstrating the possibility of indeterminism or "free will" in the behavior of animal and human organisms. Rather than quoting the whole somewhat obscure and certainly controversial passage, I will try to formulate the insight in my own terms.

Consider the physiological processes in one's nervous system, processes which presumably occur as a precondition and physical substratum of one's actions and thoughts, for example, if one is writing a sentence. What sort of causal relationship might there be in general between these organized physiological conditions and the ordered thoughts and actions which presumably emerge (in some manner) out of them? Wittgenstein urges consideration of one general aspect of this question: Does the fact that there must be some kind of causal transition connecting the two apparently quite disparate categories or orders of events force us to conclude that the relationships among electrical and chemical events in the brain bear thorough and systematic analogy to the order in actions and thoughts? That is, are we forced to think that there is an exact matching, a one-to-one correspondence, between, on the one hand, the individual determinate detail in our actions and thoughts and their interrelations, and on the other hand, some individual events of neurophysiological process—chemical and electrical exchanges and transmissions—and their relationships with other events in this context? Such a thorough correspondence is the minimal claim of those who hold either that thinking and the organic functioning of the brain are one and the same, or that the relation between them is one of deterministic causation (in keeping with the metascientific doctrine that "our actions" are merely products of physical conditions and physical laws).

But Wittgenstein notes that it is possible instead to think of the particular, determinate order—one might even say the particular kind and context of order—in our actions and thoughts as *emerging*, that is, first in formation, and then at some point, formed, so that some stages antecedent to the order that emerges are, *with respect to this order, chaotic*, "quite amorphous." He says,

> If I talk or write there is, I assume, a system of impulses going out from my brain and correlated with my spoken or written thoughts. But why should the *system* continue further in the direction of the center? Why should this order not proceed, so to speak, out of chaos?

A general point of this remark, as I see it, is to show that it is conceivable, within a description of the relation of phenomena to their causal background, to account for indeterminism—which, of course, means that at least some features of the phenomena come about "causelessly." Wittgenstein apparently also thought it was mere dogma to assume that the genetic process leading from a seed to a plant is deterministic, though this has not been one of his more influential positions. "If this upsets our concept of causality, then it is high time it was upset," he remarks. But whatever the effect on the concept of causality in philosophers' ideas, from the point of view of science its possibilities are stretched and its meaning enriched. For the purpose I have in mind for Wittgenstein's insight, the implication for our concept of causality is as follows.

The usual, simple concept of causal difference or genetic succession is dominated by one-to-one repetition or reiteration, as if
every specific event in the consequent condition is correlated with a
specific event in the antecedent condition, down to the fully atomic
composition of events. On this concept there is no detailed fact
about a particular context of effects or their relationships which is
not brought about along a specific causal route or trace extending
into whatever is being called the antecedent conditions. But
Wittgenstein suggests a possibility for an actual alternative *mode* or
form of causal transition, one in which at a certain level of detail
terminal stages do not have one-to-one correlates in all antecedent
stages, or in other words, a possible causal narrative context for
indeterminism, namely this: a determinate order, a system of definite occurrences, emerging from an antecedent system or order *of a
different kind*. ("Emerging out of" cannot mean that anything about
the physiological order *explains* the *particular* order in thoughts and
actions, which can only be a function of such things as the nature of
the project at hand and the skills, influences, and sheer happenstance of the individual life.) My reading of Wittgenstein's claim is
that determinate features of an *emergent system* of events may have
no causal traces which extend indefinitely into the given antecedent
conditions. He conceives for us a certain structure of causal narrative: the graduated transition to a determinate, individuated order
of events from out of another determinate system of events against
which the former is differentiated *as a system*. "Transition" is not
structurally distinct from "antecedent system," since the conception
simply combines the determinate order in emergent events with the
process of formation of this order as a transitional structure. On the
new concept of causality, "antecedence" refers neither to a temporally pre-existing condition nor to a temporally preceding event, but
to the differentiating transition that is the process of emergence of
the generated order. Causation is not essentially reiteration in series.

For my purposes Wittgenstein's suggestion is that attempts at
narrative understanding in general should not rule out the possibility that the narrative structure of causal or genetic transition between stages of a circumstance can be maintained in a specific form,
even though the conception of causality as perfect linear correspondence between events across these stages, the causality of mechanism and determinism, is left behind. The enhanced conception of

causality would not cling to the event-event mode of causal structure, which involves one-to-one correspondence, but would propose a concurrent dual structure of antecedent and emergent
contexts of events. As developed in Part Two so far, physical causality is comprehended ontologically as integrating these two modes
of causal structure; the difference of modes corresponds to
Whitehead's distinction between serial and "not properly serial"
process.

Quantum Mysteries

Now let's take a look at the first, twofold causal problem on the
agenda. The localization aspect of the problem was described in
Chapter 6. To reiterate: Individual pulses or "wave packets" of radiation interact with detection material at discrete, pointlike sites
among atoms of the material, even though the propagative activity
of the wavelike pulse itself extends over a vastly wider region of
space, as can be seen in the effects produced by interactions with
other transmissions. The "wave" seems to drastically "collapse,"
though there is no trace of this event or any ready-to-hand way to
understand it. The observable character of the interaction taken by
itself is that of the collision effect of a particle, but considering the
phenomenon in context it needs to be described differently—not as
the collapse of anything, which would mean that a set of local
trajectories converge to a site, but as an arrival at the site "from out
of nowhere," meaning, in my account, that the particlelike effect is
without an antecedent trace along any sequence of positions in the
local space.

The second aspect of the problem is that these individual interactions appear to be fundamentally unpredictable in certain features; for instance, in a single experiment an ensemble of
interactions will scatter probabilistically over a region of the detection material, and *that this distribution will occur* for an ensemble of
events is the full analysis of the physicist's advance knowledge
concerning sites of interaction. It is widely believed that this is a
genuine indeterminism, not a randomness in the outcomes of a
deterministic process such as coin tossing where the limits on predictability are due to the impossibility of complete knowledge of all
causal factors. Genuine indeterminism means, I think, that there *is*

no physical cause for a pulse of radiation to interact, for example, at one particular site rather than another. My arguments assume that this is the case, and they claim that if it is the case, it is something physically comprehensible and not a permanent lacuna in understanding (though of course I would not claim that the indeterministic physical process is further analyzable into some deterministic process).

Wittgenstein, we saw, would account for an indeterminism of a different sort, that associated with intentionality, by suggesting that the terminal stage of a causal process might emerge from an antecedent stage in such a way that the form of order or structure, the *system* of relationships, belonging to the emergent stage is simply lacking in the antecedent stage. To apply this to indeterminism about sites of interaction would be to say that the propagation antecedent to the interaction in its spatial and temporal structure is amorphous with respect to the *form of order* represented by the locality relations established by the atoms of the detection material. This does not mean that the propagation is devoid of all determinate structure, a "pure potentiality," but only that it belongs to background events whose relational context is distinctly *other* than anything that can be described as a system of mutually present spatial positions.

This fits intuitively with my explanation of localization in Chapter 6, which said that radiation interacts with matter by coming across the formative trace of the individual instance of determinate atomic locality and shifting into coalignment with this trace. This introduces the peculiarity that while the atom itself is a traceable existent, that is, it has its own genetic gradient into nature as a totality of productive process, a trace which is available for the propagation to engage and in a strange way retrace toward the site of interaction, nevertheless the radiation activity neither in its propagative alignment nor in the transition to interaction bears any structural correspondence to the simple spatial differentiation among the atoms, as would a wave motion that could be graphed on a grid space. The event of atomic interaction is untraceable in terms of locality or trajectory, because the genetic trace of the atomic locality, hence the transition to interaction, also belongs to the background events and their special relational constitution.

Nancy Cartwright voices the idea that atomic indeterminism suggests antecedent amorphousness or chaos from the viewpoint of the representational tools of scientific thinking. As I discussed in Chapter 4, her approach to framing a causal realist philosophy of physics is to pay attention to practical experience in technological applications of quantum physics; she and Hacking are outstanding for having found a solid approach to a "realism about entities." She said, "Nature is a wild profusion, which our thinking cannot wholly confine"[5] (compare Watson's "noise"). This apparently seems to her a natural way of sustaining "realism" about underlying physical processes while accounting in a general way for indeterministic effects. Her potent formulation evoking an amorphous multiplicity of blind activity comes from attention to the relevant phenomena coupled with a rare philosophical resistance to epistemological reflexes. That thinking seeks to "confine" its subject matter through the use of representational models is precisely the crux of the ontological crisis.

Philosophers have trouble with some uses of the notion of "chaos,"[6] and it seems to me that the trouble comes from thinking that it must have a fully generalized sense, the absence of *any* structure or order, which is in the final analysis *nothing*. But the idea of chaos *with respect to some particular order or system of relations* is not unintelligible or vacuous. This might be, for instance, a spatially interconnected complex of fully differentiated (or differentiating) activity having a structure *other than* that of local motions. (Associations with the "chaos theory" that is a current topic of popularized science will only produce confusion.)

Applying these suggestions, some positive progress toward a narrative explanation of "microphysical" indeterminism might be made as follows. The propagative process antecedent to an atomic interaction involves a system of interrelationships and differentiating structure among its constituent events that is of such a character that this form of activity cannot be described as motions against a grid of spatial position, no matter how complicated the description. That is, part of our answer to the renewed question of "natural philosophy" as to how light is transmitted is that its wavelike phases are structured in an event-context of space whose relational constitution contains *nothing corresponding in one-to-one fashion* with the determinate spatial relationships given by the locality-context of the atoms of substances (or by the presence of aggregate objects), because this event-context is engendering of locality as such. The propagating process, in other words, does not occupy, spread through, or transit local space. The relationships internal to the propagative activity itself, those which structure and constitute it, might thus be radically "amorphous" with respect to the systems of detailed locality established by atomic matter. This is exactly the situation represented in the diagram in Chapter 6 (Figure 6.1), in which locality is pervaded (in a prelocal sense) by the structure of its own engenderment. What the diagram intends is that there is both a local and a prelocal aspect of any physical circumstance, contrasting with one another as systems of relationship despite their interconnection via a stratified transition, the genetic structure of the locality. Prelocal causal structure does not trace along linear correspondences of the parts of space, such as occur between one local region of space and a neighboring region. To trace the genesis of a particular locality is in some sense to depart from the engendered locality; but this cannot mean to proceed from this particular locality and toward another, because the spatial system of differences that must be given for such a local transition does not apply to the trace prior to its terminus. The crucial structure of the antecedent trace is that of enigmatic transition which in its broadest identity is unitary throughout nature as a total occurrence, and at the same time is composed of prelocal traces specific to individual elements of locality and prelocal propagations.

To bring in a philosopher of yet another stripe, Merleau-Ponty saw with lucidity in the direction of such possibilities, though he did not elaborate his suggestion in detail.[7] Considering the "signs of a new conception of nature" discernible in contemporary physics, and among these signs indeterminism in particular, he spoke of the probing by experiment of an "amorphous core of being," in which things are not "constrained to an absolute and fixed location, to an absolute density of being." This expresses the limits of the Cartesian-Gassendian concept of the physical, constrained to the simple binary contrast of occupied versus unoccupied locality, which we must now recognize is inadequate for understanding the causal structure of space. Physicist's useful models of "microphysical" events (in lieu of explanations), which taken overall convey a gradation of multifarious elementary particles, confuse the situation by continuing to express the expectation that systems of locality can be refined indefinitely, that locality occurs in nature with unlimited "density." Earlier in the same work Merleau-Ponty characterized the "amorphous core of being" as an "anterior stratum."[8] The positive suggestion here is that the event-context explored by physics is a *stratum or dimension of nature* in which ordinary space and time relations are, may we say, *preformed.*

The basic kind of order belonging to locality relations among atoms and aggregate material objects is linear order of spatial position. The present proposal is to describe the localizing interaction as an emergence, into a stratum of the physical in which such order "first" takes effect, from out of a context in which there are not relationships systematically corresponding to those of a local grid system, so that no such grid system can serve in the description of the propagative events or the transition to interaction. Causal traces of atoms and of the interactions extend fourth-dimensionally. By saying that physical being is fundamentally activity and inherently genetic-transitional, a context of description is supplied that has greater flexibility than one in which time and space are reducible to flat continuities. Mathematical physics proceeds by nevertheless inserting localized continuities in its models, resorting to many dimensions or "curved" continuities in order to expand the reach of laws; but I would argue that this entirely sidesteps narrative explanation for purposes of utility.

Wittgenstein's suggestion is sharpened and developed for our purposes into an account of the relation between deterministic and indeterministic physical causality (taking "physical" in the narrow sense excluding biological as well as intentional phenomena): there is an overall emergent "mechanistic" order of thorough determinism or linear traceability throughout events, in which the linear-serial mode of causality prevails and the spatial and temporal properties of events involve systematic linear correlation of times and places; and there is an order of events at the core of this immediately discoverable sphere of the physical, yet remote from it as an aspect or region of nature, which is to say, remote in its natural properties. Events of this core dimension may *terminate* in emergent places and times, which means, in particular systems of spatial position and particular transitions in linear passage ("systems") of time, but do so *by way of* the structure of emergence of the time and space "systems," hence not by way of "other places and times" in a simple sense, that is, tracing along linear correlations. Connection between the two "orders" calls for no special hypothesis, since it is nothing other than their inherent differentiating relation.

A by-product of this way of progressing in our understanding of indeterminism is (again) to rule out the particle picture of the antecedent process as a false attribution to it of *localized* actuality. The localization event is accounted for in a way that does not require a particle arriving at the interaction, and is precisely not this. The idea I am trying to frame is that radiation or the "field" is differentiated from that other major aspect of nature—matter—in that it is essentially engendering activity, not structured in the mode of aggregate presence, and thus different in the character of its relationships from the terminal stratum of localized being to which motion of objects in general belongs. If there are these strata differences, then one can speak of a transition between categories of natural events: the Brownian notion of suspended microscopic particles would clearly fall in the category of true motions or fully localized events, and chemical interactions would perhaps be a species of activity occupying a boundary region between local and prelocal contexts of natural activity. The individual atom would itself be a transitional entity, a specific composite of prelocal process, constituting fully localized matter only in aggregation.

This conception is in opposition to a picture of the "microphysical" that holds sway in the thinking of scientists for understandable reasons of utility, a picture that inevitably slips in the concept of spatial occupation. This standard picture, whose use is about as widespread as the recognition that it cannot be taken "literally," that is, at face value as physical description, is that the investigation of atoms, fields, and "subatomic" interactions has simply proceeded to an ultimate degree in the direction of the small, so that atoms are composed of particles (smaller objects) and particles transit the intervening space. From the standpoint developing here the wholly instrumental function of this usage is clarified: if construed "literally" it would impose a characteristic of the localized aspect of the physical upon an aspect which does not have this characteristic. The present conception would give the word "subatomic" an altered sense, one which does not bring to mind indefinitely refinable grid systems of space. I suggest the substitute, "interatomic," understood as having extension into a dimension of fundamental engendering-presencing activity. In the same vein the present explanatory accounts seem intuitively to conform to the much-discussed nonlocality requirement upon any general interpretation of the "quantum" domain, in that once the narrative has shifted away from local spatial systems at the atomic level by a strategy escaping the confines of locality as such, independence from "macroscopic" localities regardless of scale is only a matter of extrapolation. "Locality as such" refers neutrally to all levels of aggregation, whether galaxies or grains of dust, since it is synonymous with "aggregate being." Fully engendered locality—the terminus of prelocal genetic structure—is what science conceptually simplifies as aggregate space and matter with its characteristic simple time and space relations.

The term "interatomic" identifies a preatomic aspect of the emergence of matter, with characteristics determined specifically by ultimately atomic aggregate structure. This aspect or region of presencing is considered further in the penultimate section.

It lends support to this developing proposal to consider the *peculiar overall character* of the encounter with nature experienced by the quantum physicists, and their basic responses to it. For example, it was found that a definite value for the measured variable representing the *position* of a "particle" could not coexist with a definite value for the measured variable representing its *momentum*, and that the kind of measurement performed would determine which had a definite value and which did not. Thus these attributes (presumed to belong to particles) did not appear to exist prior to the measurement, which assaults the "classical" notion of a particle; also, it is difficult to conceive how a particle trajectory can fail to be deterministic. The whole "Copenhagen interpretation" is the lucid recognition that there is no "real particle" (as a localized object) antecedently to interaction with the measuring instrument; and the orthodox Copenhagenist proceeds from this clarity to the conclusion that any "thing measured" as an antecedent or generative activity lies entirely outside the limits of science. A more muddled response to the situation occurs with the frequent talk of the particle "having no definite properties"; but in the last analysis the particle (as a localized event) "is" only where and when it "arrives." W. H. Watson clearly expresses the physicist's recognition, which is prior to the epistemological theories that rationalize Copenhagenist nihilism, that the refinement of physical locality breaks off at the "quantum" level:

> When we try, as we do classically, to imagine the particle as something travelling through space like a bullet, as if it transported substance by continuous motion, we are putting the substance in the wrong place. The particle is localised only where it is created, where it is annihilated, or where it interacts with a localised field of force, and in each of these instances the localisation is crude. It is not punctual as it would be in classical physics.[9]

Closely examined, the transmissions antecedent to measurement have turned out to be nothing like particle trajectories; in Merleau-Ponty's words, the assumption that a physical entity has "absolute density of being" conspicuously fails. In spite of these physical disclosures, the utility of the "particle" terminology has sustained it, and furthermore the natural concept of a particle has guided reflection, as if something that is accurately characterized as a particle could nevertheless have this seeming indefiniteness or indeterminateness of attributes that is talked about so much, and moreover as if it could somehow not be a thing that *moves* at all but merely *occurs* as an "interaction" at a locality that is essentially fuzzy. Not taking into account what Watson is pointing to and treating the term "particle" as though it accomplished some element of causal explanation has been a main source of famous mystifications. Another kind of intellectual adjustment was to speak in a metaphysical way of the realities prior to their measurement as *potentia*, or in some other way as only latently actual;[10] nearly every investigator has been influenced by this approach in some way. But the quintessentially Copenhagenist response was that of Bohr, who saw no point in seeking to describe a reality other than the experimental apparatus itself as a locus of calculable interactions, thus opting out of ontology altogether.

I submit that the whole experience which is the source of these different philosophical reactions is explained at a stroke by supposing that the subject of experimental study (radiation) has a causal structure given by the physical context which identifies it as a form of activity, namely, a permanent structure of engenderment process conditioning any physical system, such as an experimental apparatus. Rather than being preactual or prephysical (which can only mean it is a "pure potential" or hails from a separate realm of being), the causal antecedence contextualizing the atomic interaction can be described more naturalistically as consisting of physical events having a prelocal and prematerial structure, which would mean the field is a contextual totality. This highlights one side of the argument for the truth of the theory of radiation here under development: Assume that the basic arguments of Part Two of this book can withstand any objections on philosophical or conceptual grounds to the possibility of describing physical events outside the constraints of mechanistic concepts, that is, objections to this ontological shift itself in advance of its successes or failures in specific applications. In other words, first assume that the whole idea is *thinkable*. Then, instead of looking at particular perplexities either in phenomena or in theories, consider the overall shape of the enigma posed by the past century of discoveries in physics, and the adjustment in the procedures of the science that it forced. Result: The Whiteheadian ontological shift is the obvious and unrivaled general solution. The shift merely follows out into positive explanatory strategies a supposition that can now scarcely be denied, that beyond a certain point in the microanalysis of physical events and objects, "locality" ceases to apply directly to the description of known entities; thus, as many or most physicists would acknowledge, the "subatomic particle" is not at all an infinitesimal version of an "ordinary object." Radically clarified, in this domain physical structure does not occur in the mode of the simple presence and absence of things at all, but can only be understood in terms of events forming an interconnected whole. The discreteness/contiguity picture and the mechanistic format of description simultaneously prove artificial and inapplicable at a certain level of physical structure.

(The ontological shift is acknowledged in a way by the standard formula that physics since Einstein has superceded the fundamental status of matter by reducing matter to "energy." But this is another example of misplaced ontological significance: "energy" in physics is a parameter *attributed to* entities and processes, and does not *itself* designate a physical entity or process. Confusingly enough, the convenient reification of "energy" is encountered everywhere.)

A consideration of the matter wave, which I designated a fact to be explained, will help to clarify the proposal and will add another item to its successes at contributing some causal sense to the phenomena. The recognition of the matter wave in physics results from two considerations. First, the wave equations of quantum theory (which are correlation functions and do not "describe" physical processes) bear no intrinsic limitation to the interatomic realm, though if applied to matter of the dimensions of ordinary objects they become inconceivable to calculate. In other words, the same predictive models developed for application to interatomic process, quantum mechanics, can in principle apply to the behavior of gross matter. Second, entities having nonzero values for the measurement parameter "mass," such as electrons, and even relatively "massy" atomic objects such as helium nuclei (alpha particles), can be demonstrated experimentally to produce interference patterns, which means that under these experimental conditions they exhibit a wavelike characteristic. These facts are generally considered enough for a firm expectation that all matter has a fundamental wavelike aspect, although in ordinary cases unimaginably complex, so that insofar as it is a practical possibility to arrange the right kind of experiment a material object will behave like radiation.

The best way to identify this discovery is to say that matter and radiation are found not to be *sharply* distinguished as natural categories, but rather, to be regions of a connected ontological differentiation. Our broadest inventory of fundamental natural entities, divisible into matter and fields, corresponds to the structure of engendering-presencing transition, a gradient structure of activity conjoining antecedent process and product events and things. That "matter" and "fields" are categories connected by a relation of identity was already voiced, for instance, in Einstein's remark cited in Chapter 1 (page 36) that matter should be reducible to fields. He expressed the difference within this identity as an "enormous" variation in the concentration of the energy of the field, where "energy" is as usual understood on the unmistakable model of "stuff," so that in the final analysis this is a materialistic metaphor or analogy for something with no direct description. If the material object is described and explained fundamentally as a complex superposition of localizing transitions composing to engender aggregate locality, then not only does it have a wavelike elemental structure, it is also merely the dynamic terminus or manifestation of an extended physical background of systematic activity. As discussed below, Faraday was the first to catch sight of such a relation between fields and matter.

The field is nonlocal in the sense of prelocal, but precisely as such it thoroughly pervades local matter and space through a graduated connecting transition comprising a common physical background. The relation between matter and fields is their ontological identity in and through a structure of difference in presencing transition. As a "process" dimension of space and time, this transition is a permanent and pervasive event-context for the impetus of forces and for propagating radiation.

Propagation and Spatial Representation

The alignment of a causal trace along the axis of engenderment of atomic locality nicely supplies the connecting transition for the localization phenomenon, coping effectively with a source of perplexity. But how exactly can one conceive propagation as taking place transversely to this axis, *across* the structure of genetic development? A negative component of the answer can be given here in the form of a critique of standard modes of analysis applied to causal transmissions. We will proceed here in the schematic mode initiated in Chapter 6 (Figure 6.1).

If the initial foundation for an answer to the resuscitated question, How does light propagate? is that propagation is a function of prelocal genetic strata inherent to physical being, this means that radiation involves a dimension of events antecedent to any systematic, three-dimensional spatial locality which might form a reference space of observation suitable for purposes of measurement. The actual event-structure of propagation must be independent of *any* particular line or plane in a chosen measurement "grid," so that it propagates not through this grid, but peripherally to its being, hence traversing the system of localized elements without privileging any elements. The usual vector-in-space representation of the route of propagation, in which the end points of the vector are the observable interactions, must then have the status of a simplifying picture which should not lead one to believe that the actual activity of propagation transits through the consecutive points of local

space represented by this intervening segment. The relation between (a) the physical route or course of the propagation and (b) its
representation as marked out in local space needs to be clarified.

Propagating radiation has been represented with reference to
three-dimensional space in two basic ways: as a line in either Euclidean or non-Euclidean space, or as an expanding spherical
"shell." In either standard mode of representation the transmission
proceeds through a uniform background of spatial points in a coordinate frame. The immediate comparison and contrast between this
type of account and the present conception is as follows. If the
activity of propagation follows the parallelism of prelocal strata,
then there is, on the one hand, a sense in which its actual propagative procession is *aligned with*, that is, lies "alongside," a particular
linear ordering of spatial points representing its vector of transmission, since any such actual line, as lying in a terminal stratum, will
necessarily be a participant in the parallelism of strata, and propagation is perpendicular to transition in the genesis of locality, hence
indeed *parallel* (in this sense) to local spatial lines (and planes). This
is the sense in which it does in fact *cross* space. However, in this
account the physical propagation is not an activity which *privileges*
particular local linear or volumetric trajectories, as would a moving
object or an expanding bubble. The context and "medium" of
propagation graduates off from locality as such and is thus physically independent of any such linear locus of positions. As prelocal
activity, no part or aspect of the propagation spreads *through* or
transits *through* a "flat" continuum of points, whether of zero dimensions or of one, two, or three dimensions. Propagative succession, on this conception, does not proceed *through* any "grid"
system of locality but is connected *tangentially or peripherally* to the
space of detection and measurement via the transitional structure of
prespatial process. If the four-dimensional account were to be combined with the representation of transmission as a vector (for a
limited heuristic purpose), then the fourth dimension, across which
propagation is aligned transversely, would be represented as extending away from the vector, as the set of radii of an infinite
cyclinder with the vector running along its core. Propagation would
align with the vector, as transversal to the fourth axis, and its path
would occupy this infinite cylinder without localized transition
along the central line (keeping in mind the limitations of any repre-

sentation collapsed to three dimensions). In other words, if one thinks of a linear trajectory through space, the transmission does not follow this line, but nevertheless may be coordinated with it in a certain way.

A somewhat more complicated case is the representation of transmission as an expanding sphere; I will try to show what makes this picture applicable and how it is limited. The most awe-inspiring form of the localization problem arises when, for example in the case of a photon emitted from an atom, the scientist forms a picture which is not only that of an advancing wave front whose probabilities for interaction are distributed across a detection screen, as in an interference apparatus. He may form a picture of the photon's course as a sphere expanding at velocity c. Especially if the "wave" is thought of as "collapsing" to its interaction site, at some point in the expansion it becomes a mind-boggling feat for the "wave" to have probabilities for interaction distributed around successively more vast spherical surfaces in space. And what then actually happens when it interacts? Can we rely on words like "collapse" or even "localization" here? When it interacts at a site around the sphere, what does the rest of the expanding shell do?[11]

The present conception eliminates this problem by bringing to light the precise limits of the expanding shell representation. We have been following out the idea that radiant propagation is independent in its activity-structure from localized ordering in space, due to having as its context transition in the emergence or engenderment of locality. A series of concentric spherical surfaces is a kind of ordering in local space, and so is a system of multiple spatial axes in three dimensions intersecting at a point. Consider the latter first. The outward transmission on the present view can in no part transit along some one axis or several axes among this set, as it would if it were moving material occupying a succession of positions. It is independent of any and all particulr line-axes extending outward from a point, and consequently it does not actually occupy coordinated successions of points on an expanding spherical surface.

The outward burst of transmission in its relations of succession as well as its relation of centeredness and unity has an "absolute" physical context of transition in fundamental "passage." This means that these event-relationships bear intrinsic reference only to nature as a temporal structure of totality, and are not conditioned in their meaning by any local spatial order, as are ordinary motions. For the same reason the events in its sequence are not discrete and separate in the sense of bound apart in the "occupation" of separate spatio-temporal localities, despite the outward spreading in the representation. It advances outward through a system of relationships constituting absolute transitional structure. "Nonlocality" and "the measurement (localization) problem" both find their resolutions in prelocal causality.

I hope in this section to have communicated an essential wariness about the handy and habitual representations that one automatically lays hold of when one starts to think about causal transmissions. Such representations give rise for instance to this objection: "Radiation must be local in some sense of the word as transmission proceeding from *this* source to *this* interaction. How can it also be 'pre-local'?" The answer is that it is indeed spatially related uniquely to the locality-system established by these material sites, but only by way of the mediation of the unique genetic trace of this system.

Explanation and Causal Composition

The task of Part Two is only to propose and defend a background, ontological explanation of radiation and fields in general, a compensation at long last for the demise of the notion of a quasi-material ether. A project that physical ontology must confront at some point is to explain why and how propagations in the causal background structure of space differentiate into the multifarious "particles" known to physics, with all their specific causal properties and interrelationships. (Contemporary physics, by expanding the ledger of the "particles" through ongoing technical advance, gives the appearance of producing more and more complete explanations, even though neither the proliferation of known species nor their orderly and symmetrical tabulation can illuminate this physical category ontologically.) The anticipation is that the specific genetic structure of the electron as distinct from the photon, for instance, and the causal structure of variables such as "charge" and "mass," can eventually be explained in detail. The outline of a procedure for determining these detailed explanations can already be made out. I will now try to briefly scan this horizon for future thought. By now science should know better than to expect nature to conform to the tidiness and ready comprehensibility one might prefer.

One of the basic thrusts of physical ontology so far is the gradual working-out of a new physical basis for Thomas Young's explanation of certain optical effects through the principle of superposition, leaving behind the framework of the "classical" mechanistic physical models. Neo-Whiteheadian physical explanation claims that crossing the boundary from fundamental explanations in chemistry to those in physics requires a shift in the structure of explanatory narrative from object-composition to event-composition, which in turn requires that the latter be explicated as a basic form of interrelationship constituted independently of the framework of given locality. So far, the naive concept of a local oscillation has been superceded by a different structural analysis of the pulse of radiation, specified by reference to the prelocal dimension of space formed by transition in the engenderment of place. Background explanations of radiation phenomena have utilized "horizontal" and "vertical" aspects of the structure.

Causal (or genetic) transition has been determined to be the elemental structure of physical events, so that the ultimate and irreducible natural existent has the structure of coherent differentiated development, of "growth from phase to phase."[12] Consider the generic instance of propagating radiation: not only the connections between individually identified phases of propagation in a particular transmission have this structure, but also the causal composition of the individual phase-events along the stratified structure of their "medium," which is identified as an immediate and total physical context of natural origination. There is horizontal and vertical genetic structure, corresponding respectively to transition as mediation in a regenerative series and to transition as the individual genetic structure or course of development of the phase-event. In this physical context, "arising out of" also means "arising" in and of itself, due to the fact that unitary transition or "passage" is always primary and contextual for any regenerative series; this is the root of the identity-difference relation in physical causality, whose complexity is one source of the philosophical problem of causation. In the case of a particular "photon" interacting with matter, the individualized aspect prevails in the effect, contributing the measured properties of frequency and energy indicating causal composition into and out of separately identifiable events.

If both the cycle of propagative succession and the fourth-dimensional composition of the phase-event are identified in terms of underlying genetic structure, this means that the causal composition of the phase-event in its individual identity, the analysis of which would terminate in the "quantum of action" measured by Planck's constant, can be conceived using the diagram of Chapter 6 (Figure 6.1) as a superposition "stacking" events along the vertical within the genetic background of a particular place; this superposition is an alignment corresponding to what I called the "formative trace" of an atom, but as applied to a part of empty space transmitting radiation (or to an occupied place simply *as* a place, if the transmission is penetrating matter without interacting). Causal structure in all roles and at all levels is asymmetrical event-contour. The events composing the radiation pulse in its propagative mode individually compose (as opposed to their composition in series) along transition in the presencing of space, forming head-to-tail alignment in this vertical superposition and so cumulatively forming the transition-structure of the composite event (it is tricky to keep in mind the complete integration of horizontal and vertical structure; it helps to remember that there is no purely linear structure in space and time here, but only gradient-developmental structure). The constitution of the phase-event ("photon") of light is a *coalignment or cumulation* of event-structure along the differentiation of stages of the field transition that is the electromagnetic aspect of the total field (discussed further below). The propagation arises and passes in the causal background by generating a particular coherent contour in and of prespatial ("field") process.

A way of stating the point of the previous section is to say that it is of the essence of a prelocal process that is has no *undifferentiated* continuity in its relations, against Salmon who suggested that undifferentiated continuity must identify the connectedness of any causal transmission. Instead, each phase-event of propagating radiation, for example, analyzes into a vertical series conforming to Whitehead's general idea of event-structure on the analogy of a "Chinese box," and just as in Whitehead's idea it neither has "atomic structure" (*discrete* composition) nor is itself *discrete* from others in its own propagative series. To have the relation of being in a particular propagative series is in part to share a level of inclusion with respect to a stratification of background events, and thus to be congruent in a sense specific to extension-as-inclusion. Note this well for purposes of Chapter 10.

One general fact to be accounted for about the vast spectrum of radiation is the variation among "particles" with respect to the measured quantity physicists call "mass." The variation in the measure of this variable seems simply to extend the continuum of variation in masses of ordinary objects *so far as the procedures of physics are concerned*. But I have denied that the term "particle" as used by physicists is accurately descriptive at all. The electron, for instance, which apparently itself localizes in some phase or aspect, does so *as a presencing event*; it is thus not a thing present in space, but a pattern of activity, as is demonstrable when it takes the form of empty space propagation and displays a reiterative structure. If one thus rejects the whole picture of the "building blocks" of atomic matter, then what does "mass" *mean* as applied to such "particles"? Presumably the extension of this measurement parameter in application to radiation is simply another indicator of continuity in the matter-field differentiation: "mass" identifies a continuum along which certain causal properties exhibited in the measurement of matter shade off into that which in distinction from matter is called the field, wherein the measured properties, antecedent to actual measurement, are carried by different forms of physical transition.

The "vertical" event by which radiation interacts with an atom of matter was said to align with, conform to, or retrace the genetic trace of the atom. When the measured property "mass" is attributed to what are, strictly speaking, elements of the field, this might refer to variations of the same *mode of genetic structure* exemplified in the genetic traces of atomic and aggregate matter, so that "mass" is really a certain continuous class of genetic configurations that includes the complex superpositions (as opposed to "subatomic particles") constitutive of atoms of substances. What is this mode of genetic structure? Dropping all artificializations, such as the pictorial representation of Figure 6.1, this originative transition toward discrete occupied locality is the very constitution of the prespatial extended reality of the emergent object; it must in fact amount to a *genetic field* pervading the local aggregate space and having the structure of prelocal *arrival in* the object (roughly speaking). This is a kind of background condition mediating propagation and interaction, which in its broadest extrapolation is a total event of presencing process. Enframed by a particular region of emergent, three-dimensional space, the genetic field—the primary graded structure of the field—is a kind of *convergence* taking place independently of all pregiven space, not passing through the local space as localized transition, yet pervasive in the sense that the fully constituted space is its irremovable manifest aspect, necessary for mental representation. Or perhaps the representation of given localized extension is not essential to the conception: recalling the account in the previous chapter of "prespace" as active space-*for*-space (pages 252–53), we can refresh our sense of what is meant by "presencing" for application to primordial convergence: it does not mean merely arriving or coming-to-be-present *at* a place, but rather coming-to-be *as* a place. (Localizing interaction retraces this particular transition.) Intuitively, the genetic structure of all entities with a positive "mass" might have an effective element of this particular form of genetic transition, the primordially convergent prespatial field, to varying levels of development, though in the case of the preatomic "particles" this structure would not terminate in fully formed locality, that is, aggregate mutual presence, but in a modification (or supervening structure) of the primary field, capable of local detection. The resistance and reactions of atoms and other "particles" to "collisions" with other "particles" would result from upsurgent presencing rather than from the inert occupation of space (it is probably safe to say that no knowledgeable person thinks these are "literally" tiny billiard ball events).

The "electron" or ultimate unit of electricity, for instance, has a measurable mass, as does the atom of chemical substance. Consider these remarks of Whitehead's:

> As long ago as 1847 Faraday in a paper in the Philosophical Magazine remarked that his theory of tubes of force implies that in a sense an electric charge is everywhere. The modification of the electromagnetic field at every point of space at each instant owing to the past history of each electron is another way of stating the same fact.[13]

Whitehead is referring to Faraday's reservedly promoted hypothesis that the ultimate entities of physical analysis, such as atoms of matter or what came to be called the electron, are not sharply defined, mutually impenetrable objects, but "centres of force" *constituted of,* not merely surrounded and permeated by, the associated force or field. Since the field is effective to no definite limit outwardly from the "central" atomic entity, this implies that

> matter fills all space, or, at least, all space to which gravitation extends (including the sun and its system); for gravitation is a property of matter dependent on a certain force, and it is this force which constitutes the matter. In that view matter is not merely mutually penetrable, but each atom extends, so to say, throughout the whole of the solar system, yet always retaining its own centre of force.[14]

In other words, the structure of physical being is such that a localized aspect grades into a nonlocalized aspect in a relation of ontological identity. This is a step away from envisioning the profound possibility of an integrated structure of presence and active presencing. The usual view would be that this passage quoted from Faraday presents a farfetched imaginative construction within a fatally unsophisticated concept of science. Here, on the contrary, it has the power to remind us that transition in presencing is at the same time a distancing, a "holding apart"; the "field" of an object refers to the fact that the constitution—which is to say the identity—of the particular localizing structure is bound up with the "alterity" of extended nature *in relation to* the terminal locality; in other words, the above mentioned relation of ontological identity contains the difference distinguishing a present (or presencing) thing and an absolutely dispersed field activity. This corresponds to the difference-structure in perception of "over there," "out there," or "here at hand." "But precisely on this point Faraday's and Whitehead's idea smacks of irrationality. The notion that particular physical events have 'internal' connections with the whole universe may be pleasingly 'holistic,' but how is it really possible?" This "how" is answered by prelocal genetic structure as the ontological basis of the field. In the causalist tradition of Faraday, Maxwell, Whitehead, and the present inquiry, the narrative concept of the field and the field-matter relation has developed with continuity, and it turns out that

for genuine comprehension one must leave the field free from conceptual reduction to the extent that it is *essentially a history or genesis*,
a transition pervading nature as a total concurrent event. This
upsurgent event is a context of localizing transitions that may or
may not terminate as or at sites in aggregate matter. It is not a mere
potential that at some point actualizes, but an inseparable mutual
resonance of "being" and "becoming" in and as primordially productive physical activity.

At the other end of the spectrum with regard to the parameter
"mass" is the "massless" photon, or unit of light, which from the
discussion so far would be described in its propagative mode as a
series of cumulative superpositions individually composed across
the strata of prelocal structure, thus reiterating transversely to this
structure, without the element of convergence toward locality. The
general basis for the difference of localizing and nonlocalizing
composite structure is the fact that events comprising the genetic
background of physical space are essentially generative and regenerative and at the same time have individual identity as situated
with respect to an event-totality via specific genetic traces. This is
why any prelocal entity or process has the modal possibilities of
propagation ("radiation") and convergence ("field"). A phase-event
of electromagnetic radiation interacts discretely with matter and
thus expresses its aspect of individual prelocal superposition without contributing an independent element of convergent structure,
while such an event in its propagative activity is wholly transverse
to localizing transition and thus "tangential" or essentially mediated in its relation to localized space.

The fundamental genetic constitution of matter, conceived as
transition in originative presencing, and the particular genetic con-
stitutions of substances corresponding to the particular event-composition of their atoms, is submitted as a starting point for the
explanation of fields of force, with the variation in constitutive genetic traces of substances accounting for traits of specific fields
other than universal gravitation. The latter might simply be the
primary genetic field, an original structure of generative presencing
in the background of aggregate objects and localities. Electromagnetic and nuclear fields would be the traces of structures arising in
the genesis of atomic ultimates, localizing structures that are not
local objects at a microcosmic dimension, but precursors to atomic

locality; it would be the atom which first actively constitutes discrete presence. Thus the different atomic and interatomic entities might be identified with stages of genetic process. The different "levels" of field activity (gravitational, electromagnetic, and nuclear) would on the one hand grade into one another and be mutually permeating, each bearing some trace of the activity of the others, though primarily they would "displace" one another as domains of fundamental genesis. This bears on the topic of the composition of forces, introduced in Chapter 4, and on the more specialized matter of the dramatically different measurably effective ranges of the forces (where the "microcosm" is normally conceived conveniently by following lines of convergence to smaller and smaller spaces). The basic fields would be integral regions of the process of presencing as the complete dynamic constitution of matter.

My overall proposal in this area is that the different interconnected modes of genetic transition and their complex causal composition, participating in a gradation of contexts successively broad and ultimately encompassing natural totality as absolute transition, can provide a first basis for explaining a plethora of facts, including those officially accounted for as "wave-particle duality" and the differentiation of "particles" in respect to mass. Specifically, the ambiguity of particular causal structures as essentially propagative and as events identified by unique relations with respect to contextual "upsurge" of events as a whole—the dual structure of physical causality—may give rise to a dual character of effect. Correspondingly, mass-masslessness might involve a variation in the relative status of two modes of superposition or composition in the genetic structure of the entities, modes that may actually identify a single continuum of possibilities, namely, primordially "convergent" presencing, and reiteration across prespatial structure. It is extremely important for appreciating these relations to keep in mind (a) the common physical context involved in the different modes of transition; and (b) the basic limits of representational thinking imposed by the whole procedure and indeed by the background narrative conception itself.

Some may feel a disturbing element of happenstance in these physical accounts; the complaint might be that explanation always occurs in a seemingly "weak" form: "it comes about that . . . "; "certain structures arise"; "it unfolds thus." This feeling may only be resistance to a necessary revision in concepts of understanding and explanation, away from tight and precise mechanisms and/or tidy ratiocinicity in functional correlation, away from lucid organic purposiveness. There is an aspect of necessity in the explanations, in the sense that the overall outline or background conception leaves no room for "things to happen some other way."

A speculation toward the further ontological exploration of fields is that bipolarities such as that of electric charge might be accounted for by the asymmetry of individual genetic structure, the unique "from . . . toward . . . " of transition-in-development; on this basis science might, for example, finally be able to move beyond representational models involving discrete charges and action at a distance (and beyond mere diagrams of events) in explaining the "cohesion" of atoms. It appears likely as noted above that following this explanatory strategy leads to ontological identification of the gravitational field as the primary genetic field of masses, a general active impetus of convergence to locality in their prespatial background. All of this must be regarded as cursory and provisional suggestion. Perhaps someone with more detailed knowledge of phenomena and empirical relations will take an interest in it.

Summary and Transition

If physics has come across genuine indeterminism, this does not represent a dead end for narrative physical explanation, because indeterminism can itself be explained as something occurring in a context of connected causal process having the narrative form suggested by Wittgenstein. The appropriate procedure is not to construct causes for the causeless features of events, but to contextualize interatomic indeterminism within a broader background explanation. This and other real problems of explanation in the "quantum" area are suitably accounted for by the conception of physical being under development (if it can be regarded as thinkable). The meaning of nonmechanistic physical propagation has been illuminated in a negative respect by contrasting its description with some spatial representations normally constructed for the basic analysis of propagations. Preparatory suggestions have been given toward future development of the theory of radiation into a causal account of the details of the empirical knowledge possessed by physics. This book is only making an argument for the truth of a background ontological explanation, and I believe these chapters by no means exhaust the problems to which it might be applied.

The final two chapters turn to some specific philosophical and explanatory problems of science which on the surface appear remote from the issues of this chapter, but which are quite amenable in their own ways to the Whiteheadian ontological suggestions toward background understanding for the domain of physics (indeed, they are the problems Whitehead was primarily concerned with in his work on physical science). Applying the conclusions reached so far to this "relativity/absoluteness" sphere of problems makes the important point that the range of explanatory application of the Whiteheadian ontological shift extends across the entire domain of physics.

9

Absolute Causal Reference

THIS CHAPTER ARGUES that the theory of nature developed in previous chapters in order to furnish explanations for an array of radiation phenomena also provides a resolution to a long-standing scientific/philosophical problem encountered by the basic theory of space. This problem is an ambiguity in the meaning of "system of spatial reference" applied across different kinds of motion. Unaccelerated motion in a straight line presupposes, it seems, a particular system of spatial positions against which the motion takes place, positions which the moving thing successively approaches, reaches, and leaves behind; certainly in any ordinary description of translatory motion some reference space such as the surface of the earth is assumed. But motion in a nonuniform or accelerated mode, as in the case of rotation, does not bear any essential reference to systems of spatial position given by local aggregate bodies. This has led to or contributed to disputes over the physical meaning of "space"; for example, this amorphous "absoluteness" of space evidenced by rotation and inertial forces has suggested that there might be such a thing as a single localized system of space extending throughout the universe, or in other words, the physical possibility of uniform motion irrespective of local bodies of reference. Another form the matter takes is the dispute whether space is "something in itself" or merely a "relation" between physical bodies.

My proposed explanation of the ambiguity of spatial reference is a straightforward ramification of the work of Chapter 7, in which the structure of finite totality was conceptualized and integrated into concrete physical explanations. Rather than being directly about radiation, this chapter points to a possible success at finally resolving an old philosophical problem of physics in order to add confirmation to the general cosmological story behind the theory of radiation. Perhaps the most important argument for the truth of the overall proposal is its range of explanatory power.

History of the Problem

Newton believed in the existence of an "absolute space." He had in mind a system of spatial positions, which by its nature cannot be itself in motion, and which is unitary throughout the universe. He saw no way of determining whether or not either of two relatively moving objects is at rest relative to this "absolute" system of space. But he did see definitive evidence for its existence in cases of rotation. This kind of motion does not appear to have any necessary reference to other objects, that is, the factuality of rotation does not depend on a particular spatial system (in practice specified by a material body) with respect to which the motion takes place, as when translatory motions are identified and measured using an earth-bound measure system in science and ordinary practical affairs. The rotating object moves about its own axis, which means, with respect to a spatial background permeating the object, but independently (it seems) from relations to other bodies.

The argument from rotation to the existence of absolute space was put in the form of an experiment by Newton as follows. A bucket of water is suspended by ropes. If the bucket is set spinning for a time, so that the water takes on the rotary motion, the surface of the water will be deformed in a certain way as a result of inertial forces. This deformation effect will take place when the water is spinning, regardless of whether the bucket itself is rotating along with the water or not; indeed, this effect does not appear to depend on any material facts in its surroundings. Newton extrapolated this result into the claim, which has a vague intuitive cogency but is impossible to test, that such "inertial effects" will occur whether or not there is any other matter in the universe, and that rotation is not a motion bearing an essential reference to any particular object or objects other than the universe as a spatial whole (conceived as a total locality). In my view, the philosophical problem here is that while motions of objects always have an aspect of *localized* transition, a purely kinematic aspect, they may also have a dynamic aspect involving a form of transition with a different spatial and temporal structure. Newton, however, filled the apparent need for some causal reference independent of local bodies of reference by postulating a special and unique localized space for the specifically dynamical reference, namely "absolute" space.

The hypothesis of "absolute space" played a prominent role in nineteenth-century physics, because the idea of a locality-system that would be the totality of space and thus comprise a unique system of reference independent of particular bodies or aggregations of bodies corresponded well with the notion of an all-pervading ether of special material which could serve as a medium for light waves (and for the processes underlying electricity and magnetism). This is why there were persistent attempts to measure by one means or another the "absolute" motion of the earth, that is, its motion with respect to the ether as the absolute locality-system. That such a measurement was possible in principle could be deduced (given these assumptions) from the fact that the measurement of the velocity of light was possible.

Ernst Mach was a major progenitor of twentieth-century scientific empiricism. As I have already discussed, in the era of ether physics and later he was arguing that science should rid itself of the idea that any postulated entities such as ether, fields, atoms, and absolute space have a status in scientific knowledge beyond that of useful, provisional models. He adopted the *relationist* view of space that had been argued against Newton by Leibniz. According to this view, nothing is meant by "space" other than the empirical fact of locality relations among material bodies, so that the reality of space depends entirely on the existence and mutual presence of matter. Thus space is regarded as a purely relational physical fact, while present material, a category valued for its positivity and unconcealed thinkability, can stand as the fundamental form of physical being. One source of the plausibility and scientific utility of the relationist claim is that in practice material spatial relations always lie behind the marking out and measurement of spaces and distances.

Mach gave a rejoinder to Newton's argument from the spinning bucket which many have considered definitive. A perfectly
good alternative exists, he said, to a Newtonian, quasisubstantive
"absolute" rest space as a reference system for rotation and its inertial effects, namely *the rest of the matter in the universe.* The axes of
this system can be defined through the horizon of observation established by the "fixed stars." When this suggestion is adopted,
rotation no longer falls outside the relationist theory of space.

Through the operationalist approach to physical theory,
Einstein *partially* carried out Mach's program of reducing space to
relations among material bodies. His special theory of relativity is a
quantitative model for the measurement of motion at all possible
velocities (exceeding the Newtonian model in its applicability)
which treats any translatory motion as purely relative to another
body. However, in the general theory, space (as spacetime) is identified mathematically with the gravitational field, an entity apart
from matter, which does not quite fit into the relationist program.
Whether the special theory implies an absolute system of reference
for acceleration and associated forces (as opposed to uniform motion), and whether the general theory provides such a system in the
form of a "global" or universal gravitational field, are matters of
controversy. Despite these ambiguities, it has been quasi-officially
settled upon that Einstein's relativity theory and Mach's answer to
the problem of "absolute" reference are mutually reinforcing theoretical solutions.

Whitehead found a kernel of soundness in Newton's argument, and he could not accept Mach's solution to the problem of
spatial reference for rotation. He thought that a philosophically superior post-Newtonian solution was yielded by the alternative to
Einstein's relativity models that he himself had developed; but this
solution was only explained in scattered and cursory passages.
Whitehead's "theory of relativity" has been widely acknowledged
to produce many if not all of the same predictions as Einstein's.[1]
The fact that Einstein's theory was adopted and celebrated by
physicists while Whitehead's theory was largely left alone has conventionally led to the automatic conclusion that the former has for
all time been scientifically certified while the latter bears the onus of
scientific rejection, but recent discussions have tended to loosen this
conclusion.[2] The next chapter helps to provide bearings for the

question of how to sort out and compare these two theories in regard to the various functions of explanation, useful modeling, and quantitative laws.

Whitehead's Solution

In *The Principle of Relativity* Whitehead expressed his affinity with Newton as against Mach and Einstein as follows (emphasis added):

> In conclusion I will for one moment draw your attention to rotation. The effects of rotation are among the most widespread phenomena of the apparent world, exemplified in the most gigantic nebulae and in the minutest molecules. The most obvious fact about rotational effects are their apparent disconnections from outlying phenomena. Rotation is the stronghold of those who believe that *in some sense* there is an absolute space to provide a framework of dynamical axes. Newton cited it in support of this doctrine. The Einstein theory in explaining gravitation has made rotation an entire mystery. Is the earth's relation to the stars the reason why it bulges at the equator? Are we to understand that if there were a larger proportion of run-away stars [as opposed to the "fixed" stars], the earth's polar and equatorial axes would be equal, and that the nebulae would lose their spiral form, and that the influence of the earth's rotation on meteorology would cease? Is it the influence of the stars which prevents the earth from falling into the sun?[3]

To guide ourselves into a closer look at the problem, we might go to the root of Whitehead's reasoning and reconstruct it as follows. What is meant by a reference system for rotational motion, or for any motion? Why is such a thing called for in each case? First of all, when the motion of an object, translatory or rotational, is spoken of, it is understood that there is a space through which the object moves. Without such a "reference" space in some sense, an object in motion is not distinguishable from a case of a motionless object. Consider first uniform translatory motion: a field of local space from which the object itself is physically differentiated or differentiable *belongs to its mode of transition as local motion.* Then there are motions of a certain class—"non-uniform" or accelerated motions, such as rotation—that are associated with "inertial" forces. An object does not rotate with respect to some particular other object, or with respect to itself, but does seem to require a reference "framework of axes" of some kind. Something specific to the *mode of transition* in the accelerated motion escapes dependence upon reference to *some particular local spatial system*; nevertheless the acceleration as such is a kind of motion, that is, a kind of spatial transition, one whose actuality and uniqueness is identified by the occurrence of forces. The first thing to notice about such a dynamic or inertially active mode of spatial transition is that it retains an overall aspect or pervasive component of local motion; yet equally pervasive is a component or aspect of physical transition that differentiates against its spatial background *in a distinctive sense*. From a perspective envisioning an ontological shift the problem is that the mode of spatial transition specific to the accelerational aspect of natural motion is not well understood (despite all calculative mastery), in that there is no conception of the physical background specifying this mode of differentiation; stand-in conceptions have involved absolute or total *localized* spatial systems. Whitehead's questions and the general concern of scientists with the problem of inertial space show that there is a parallel explanatory lacuna with *inertial* forces as with forces associated with the "fields" of objects. In both cases causal-ontological inquiry seeks to understand the physical context of forces—their causal space, so to speak—whether the fields or forces produce accelerations or the accelerations give rise to forces.

Whitehead could not accept the idea that the physical meaning of rotation or the reality of inertial forces depends on the "presence" of remote astronomical objects. In *The Concept of Nature*, he chides operationalist physics for providing no physical explanation of the dynamical properties of motion, and explains his objection to Mach's solution:

> It is difficult to take seriously the suggestion that these domestic phenomena on the earth are due to the influence of the fixed stars. I cannot persuade myself to believe that a little star in its twinkling turned round Foucault's pendulum in the Paris exhibition of 1861. Of course anything is believable when a definite physical connection has been demonstrated, for example the influence of sunspots. Here all demonstration is lacking in the form of any coherent theory.[4]

Wherever there are the characteristic motions that are called accelerations, a space of reference *in a sense apparently peculiar to motions of this kind* is required as a basis of differentiation for the motion, and to bring this to conceptual clarity would in the same stroke account for the existence of inertial forces. Today the ideas of an ether and an "absolute" localized system of space cannot be recovered. Even so, thinks Whitehead, how could such a causal property exhibited by the relation of the object to its own space have anything to do with the existence of the "fixed stars" and of other galaxies? What kind of efficacy or influence resulting in the inertial properties of bodies could possibly emanate from remote astronomical objects or from their total array? Or are these facts of inertia supposed to depend merely on cosmic coexistence? Though Mach gives a logical alternative to Newton's absolute space, is it a plausible or physically intelligible alternative? Attempts have been made[5] to defend Mach against Whitehead and others by arguing that the overall gravitational field generated by the most distant matter in the universe, calculated with the help of the general theory of relativity, is sufficient to account for local inertia. But since the gravitational field is just as much in darkness ontologically as it ever was, this amounts to a purely formal solution, an untestable law that inertia is in fact dependent on such a field. Also, this argument maintains the counterintuitive suggestion that rotational inertia depends on a *contingent* property of space. Such an attempt in the tradition of Einstein recognizes in its own way that the problem of absolute or inertial spatial reference is an aspect of the broader question of the physical field.

Continuing the passage first quoted, Whitehead describes his own account of absolute causal reference as follows:

> The theory of space and time given in this lecture, with its fundamental insistence on the bundle of time-systems with their permanent spaces, provides the necessary dynamical axes and thus accounts for these fundamental phenomena.[5]

The general notion of different time-systems, or more precisely the relativity of measured time, became a feature of physics during Whitehead's life. But Whitehead gave this physical idea a distinctive treatment, and the resources for physical ontology lie in the

distinctive features of Whitehead's physical theory, not in its com-
monalities with Einstein's. A fuller explication of his idea of a
"time-system" is reserved until the next chapter. But with a minimal exposition it can be shown that Whitehead's account of the axes
of inertial reference is structurally the same as one that emerges
from the results of the previous chapters, to be presented in the
section following. Considerable interpretive exertion is necessary
even for this, because as Whitehead scholar Robert M. Palter observes, Whitehead never concretely spelled out his solution to the
problem of absolute reference.[7]

A "time-system" with its "permanent space" is Whitehead's
physical analysis of a spatial system of reference; any actual spatial
system (such as an object or set of objects in a unitary state of
motion or rest) is in Whitehead's theory also a particular "time-
system," which means, it involves duration along a succession of
time which is actually unique to it for purposes of highly refined
measurement, though for nearly all practical purposes it can be
regarded as belonging to the same time-system as other potentially
utilizable spatial systems. According to the "relativistic" determinations in physics, two such spatial systems in relative motion diverge, become different, as time-systems by an infinitesimal amount
and in a graduated way as a function of their relative velocity. This
means that a given set of physical events, when "referred" to one
spatial system for measurement, will have certain exact relations of
time order and simultaneity, but when "referred" to the other system (moving relatively to the first), this set of time relations will be
minutely different, will be shifted in a systematic way, so that
events simultaneous "as referred" to system A may succeed one
another "as referred" to system B, or vice versa. Thus, Whitehead
accounts for one important observational basis of the new "relativistic" era of physics—the singular result of experiments of the
Michelson-Morley type together with its implication that exact
measured simultaneity will depend on states of motion—in these
brief remarks:

> [This result] is completely explained by the fact that, the space-
> system and the time-system which we are using [in performing
> the experiment] are in certain minute ways different from the
> space and the time relatively to the sun or relatively to any other
> body with respect to which it [our space- and time-system] is
> moving.[8]

The apparent enigma (from a naive, nonformalist standpoint—not to physicists) of different bases for time relations in a particular physical circumstance will have to be endured while it is explained how in the abstract these different systems of time figure in a theory of event relations, in which the meaning of Whitehead's reference axes for rotation can be uncovered. Meanwhile it should be noted that relativistic results of measurement certainly find a successful empirical model in Whitehead's time-systems account, even if this account were to prove unsuitable for causal ontology or even generally untenable philosophically. In Chapter 10 it will begin to appear that although the time-systems conception may be a perfectly good formal model, it cannot ultimately serve as physical explanation.

After the passage critical of Mach in *The Concept of Nature* Whitehead frames his own account of "absolute" reference in terms of the background structure of space and time:

> According to the theory of these lectures the axes to which motion is to be referred are axes at rest in the space of some time-system.[9]

"The space of [a] time-system" means for practical purposes any reference body at rest or in uniform motion. For Whitehead, uniform translatory motion *means* that one time-space, one object/space system, is differentiated from another, the "referential" system. But a time-space system must mean something else also if it is to account for absoluteness of space. He says that "axes at rest" in the time-system are what serve as a reference for dynamical properties of motion. The structure of absolute reference is described in the other passage as "the bundle of time-systems with their permanent spaces," which is clearly something more than a system of local axes. "Axes at rest in the space of a time-system" might seem to say that any local spatial system may be arbitrarily chosen as a reference, and that this fact is all that "absolute reference" amounts to. But this would be rather a denial that the problem exists than a solution, and the notion of a time-system is superfluous to this claim. Whitehead's solution must be sought in the fact that in his theory a space- and time-system does not reduce to a system of simple locality; this is only its manifest or "derivative" aspect. "Axes at rest" therefore cannot be understood in this context by recourse to Cartesian coordinate systems or simple spatial "reference frames."

The key, it seems to me, is the "fundamental insistence on the bundle of time-systems." For Whitehead, any three-dimensional space has an underlying constitution of a manifold of intersecting time-systems, as I show in detail in the next chapter. As a space and time-system it selects from a multiplicity of time-series intersecting in the constitutive physical events, each ordering the times and places of the events *in its particular way.* Such a confluence of time-systems underlies the manifest aspect of any local spatial system and belongs to its physical constitution; due to the fundamental status of events (as opposed to objects) as relata in spatial and temporal relations, multiple time-systems form a background structure to all contexts of matter in motion (as to all physical contexts). What the claim seems to be is that the background events of any physical circumstance, as ontologically prior to local material objects and three-dimensional space, retain an element of indeterminateness or amorphousness as regards time relations, and that this is what comprises the dynamical reference apart from particular bodies, for example, in the case of the earth's rotation. How might this background structure of space as postulated by Whitehead be understood more concretely?

Whitehead describes the subject matter of physical science as in any particular case an interconnected complex of events. Any such complex has the fundamental structure of a multiplicity of time-systems or time-successions ordering these events. Each of these is a derivative of the passage of nature as a whole, but also, each might in principle function as a reference space for motions; for practical purposes an object or set of objects in a particular state of motion defines such a time-system. Physical space in its full account is concrete, occurrent circumstance with relations of extensiveness in interconnection with a totality of events. The science of physics has disclosed that any limited or "finite" circumstance consists of a manifold of events related in diverse (exact) systems of time. A case of rotation, for instance, is embedded in a multileveled manifold of empty space events and interatomic events, and there are various possible systems of time-sequence by which the exact temporal assignment of each and every event in relation to others in the manifold can be specified. Thus the fundamental character of physical space, which is to say the structure of natural "process" permeating and, as I would say, engendering space, consists (for

Whitehead) of the fact that it is constructed and constituted by a convergence or confluence of intersecting time-systems. Every actual event occurs at, and in fact *is, some moment* in *each one* of these time successions; thus every time-system intersects every other in each actual event. The time-systems taken together form a structure of spatial systems (potentially, referential or "rest" spaces) intersecting in any physical circumstance as a complex of events; they represent a multiplicity of possible temporal analyses of this complex (it is helpful to keep in mind that the divergence of time sequence only occurs at a minute level of analysis). Any physical space *as constructed in this way out of time-systems* possesses the "dynamical axes" to which rotational inertia is referred.

For the pursuit of causal ontology, consideration of Whitehead's time-systems theory of inertial reference should emphasize the account of event-relations. Every event can receive its exact temporal assignment (in relation to a whole of other events) in a set or "bundle" of different ways, that is, for any concrete manifold of local events there are systematic differences in their overall ordering in time when these time relations are taken to ultimate precision. This unique "bundle" of time specifications possessed by any actual event situates the event in an absolute sense, that is, in reference to a totality of events, since no other event bears this unique "bundle" of temporal relations toward other events. Its participation in a diversity of temporal orderings supplies any event with a unique set of relations to nature as a whole:

> . . . nothing in nature could be what it is except as an ingredient in nature as it is. The whole which is present for discrimination is posited in sense-awareness as necessary for the discriminated parts. An isolated event is not an event, because every event is a factor in a larger whole and is significant of that whole. There can be no time apart from space; and no space apart from time; and no space and no time apart from the passage of the events of nature.[10]

An increment of the rotational motion of an object is such an event, marked by an internal reference to a whole. Its "absolute" mode of differentiation comes from the fact that its spatial context is constituted of a systematic confluence of temporal successions in the relational structure situating it among events. As a relational entity, any physical event is therefore fundamentally conditioned by a manifold of ordered successions of time in which it participates, and is identified by differentiation against this fundamental background structure.

If this interpretation is correct, then for Whitehead the primordial structure of events given by differentiation of time-systems becomes effective as a referential background with accelerated motion, whereas uniform motion is a transition confined to "derivative" localized spaces, just as the motionless occupation of space is confined as transition to a particular temporal derivative of events. If there were an ideal uniform motion in nature, this spatial transition would reveal only a simple differentiation of two time-systems in the purely manifest aspect of events. It would show no properties or effects whose explanation lies outside of this description: local spatial transition. Where there is deviation from uniform motion (where inertial forces arise), on the other hand, the mode of transition is differentiation against an inherent temporal complexity in background events, and the forces are effects arising from this mode of transition composing the motion.

Consistently with the overall interpretation of Whitehead, his solution to absoluteness of space is interpreted to involve neither a local nor a total system of given space, but the depth-structure of time and space as coemergent and in this sense involving a background totality or field of events.

Flat Transition and Structured Transition

Now I will give my own explanation of inertial reference in the developing concepts of physical ontology, showing how its claims are closely analogous to Whitehead's and may be considered a reformulation of Whitehead's ideas on this matter.

Mach's solution is rejected as being a direct appeal to the object-picture of natural totality, which has been found unacceptable because (a) it inherently excludes the genetic structure which is a brute scientific fact displayed in the field of cosmic observation, and (b) it is part of a conception of natural being in which locality is fundamental, and this conception sharply limits possibilities for the explanation of radiation phenomena.

I have proposed a background explanation in which physical space can be said to have two fundamental aspects, connected in a relation of ontological identity that is the structure of physical being. These can be designated the "product" and the "process" aspects of space. To speak of "the space of three dimensions" (represented abstractly as a coordinate grid), or "space occupied by matter positioned within it," is to speak about the product aspect. The process aspect is the "prespace" of transition in presencing. This twofold structure of space yields a simple basis for explaining the complexity of space as a reference for motions and concomitant forces: the evidential and speculative bases for the relationist view come from the product aspect of space, and Newton's and Whitehead's arguments for "absoluteness," or reference independent of particular systems of locality, are satisfied by the process aspect.

On my theory, the very concepts of spatial grid and of occupied and unoccupied parts of space belong to a common concept of natural being under a reduction to pure product. The truth of relationism is that space-as-product has in fact no reality independent from matter and its locality relations; space *in this aspect* necessarily arises whenever and wherever matter arises, since matter is concrete, systematic locality. Insofar as matter comes into being, local (relational) space is given along with it. That natural science has advanced as far as it has while limiting its concept of physical being to terminal and manifest aspects of space and time, and that it continues to employ this concept during a period in which physical explanation has been radically suspended in physics, is the background of the contemporary views of space promoted by Mach and Einstein. Space is purely relational insofar as it is localized actuality, to which scientific thought (leaving out advocates of event-ontology) has so far been confined.

The spatial context of any physical circumstance in its process aspect, on the other hand, supplies every element of the circumstance with a unique relation to a totality of events, due to the fact that each component event comprising this aspect is internally organized and constituted through interconnection with this transitional whole. The structuring of a physical context through its event composition, fully analyzed and specified, is thus none other than the structure of transition in a concurrent totality. The physical circumstance, for example, a case of rotation, is an event situated in primordial relationships independently of its manifest physical context of simple and vague material objects and their aggregate activity, the elements of its derivative spatial and temporal properties. Absolute causal reference results from the fact that physical events are ontologically prior to objects in that they determine all relations with respect to natural totality as the constituents of a suspended structure of "passage." Physical absoluteness points toward the root questions of ontology: "being," "nothing," "engenderment."

To clarify the role of space-as-process in providing an "absolute" system of reference, recall first the basic explanation of the atom in the last chapter: along the structure of the prelocal dimension a multilayered, convergent superposition is systematically composed, bringing about a *locality* and a primitive, transitional *object*. To say "bringing about" implies that local space cannot be given in and for the description of this genetic aspect of the circumstance. Locality is *something brought about* in its fundamental cosmological status. No series of points in nearby space extending in any particular direction or directions could be singled out for the representation of "convergent" prelocality, whether this refers to the genetic field of an object or a localizing interaction. Despite the complexity of the route of convergent presencing in terms of its own differentiating structure, all of its contributing processes equally and uniformly permeate the locality in a prespatial sense. Considered in its complete structure, this primal convergent aspect of the field is a connection of the atom with nature as a whole, a connection uniquely specific to this atom and which thus situates it in nature, *as a manifestation of events*, in a manner independent of particular spatial reference systems.

Not only the individual atom, but every individual physical object, is fundamentally an *outcome* of nature. This does not mean it is an isolated and free-floating fragment of an ancient explosion, but rather, that it is grounded or rooted in a permanent background of engendering process which in a terminal phase actively constitutes the object. Such process consists of routes of non-localized transition terminating in locality in its aggregate detail. Since these causal routes comprise the "process" by which nature arrives at and in determinate locality, this overall four-dimensional system—the object with regard to its pervasive genetic structure—lies outside description in terms of any local reference space. This systematic structure of prelocal process, the sum of genetic traces of the object connecting into a whole, is what can be called in a useful but potentially misleading way an absolute "system of reference."

When an object rotates, the physical contrast between (a) its aspect of an occupied volume and (b) the structure of engenderment process permeating its locality has come into play and become effective in properties of the object. The continuous connection between the object, its primordial genetic field, and engendering totality identifies the sense in which rotation is an absolute motion. Given the present background hypothesis, if one becomes puzzled over the spatial reference for rotation, one need only remember that unaccelerated motion is a mode of transition confined to flat, featureless continuity, whereas accelerated motion discloses (I claim) a referential context of four-dimensional differentiation in genetic process. The notion of an "absolute" localized spatial system is an empty construction. Accelerated motion is a combination of localized transition and transition that differentiates "prelocally," that is, in the mode of primordial event-structure. As with other scientific problems studied in this work, this problem as it has traditionally been discussed has artificial aspects and genuine aspects, the artificial in this case being the assumption that any referential structure for spatial transitions must be localized structure.

Conclusion

This chapter has shown how the strategy of physical ontology being developed to explain experimental findings of physics has the range to provide a fresh solution to the long-standing problem in the theory of space represented by rotation and other accelerated motion. The alternative contemporary options are either to leave absolute reference out of account as a question without practical significance, or to proffer the ad hoc solution proposed by Mach. I propose a neo-Whiteheadian solution in which an engenderment or presencing structure of space *replaces* both the postulate of an ether and that conception of the whole which is actually common to Newton and Mach (and pervades contemporary cosmology), the conception of a total aggregate locality. The event-ontological analysis of the question views the types of motion as different modes of physical transition. Acceleration is graded or structured, so that its special mode of differentiation is that of prelocal structure, not that of "flat" or "linear" localized transition. "Multiple time-systems," the centerpiece of Whitehead's theory of relativity, is for Whitehead also the basis for absolute or nonlocal reference structure. In my explanation of inertial reference what I call the genetic background of space is analogous to what Whitehead calls a "bundle of time-systems," in that both specify the relation connecting local events with an event-totality. This reinforces the feeling that a correspondence exists between Whitehead's derivation of particular time- and space-systems from a spatializing convergence of multiple time-systems, and what has been postulated here as a genetic and prespatial dimension of time.

10

Velocity *C* and Emergent Extension

THIS CHAPTER DOES NOT attempt a general exposition of the encounter with "relativistic" measurement phenomena on the part of experimental and theoretical physics. In keeping with Part Two as a whole, its goal is to outline a whole new way of inquiring into the matter, namely, for once as if this group of observations might be physically *explained*. This approach challenges a set of deep prejudices.

Relativity: Theories, Models, Explanations

The topic of this chapter is the fact of nature revealed by the Michelson-Morley and similar experiments, that the measured velocity of light shows no variance depending on the state of motion of the measuring apparatus along the direction of transmission. This is interesting because the natural assumption would be that the magnitude of a relative velocity is always the sum or difference of component velocities, for instance, if two cars are proceeding in the same direction on the freeway, one at sixty miles an hour and the other at sixty-five miles an hour, the velocity of the second relative to the first is five miles an hour. The primary aim here is to make the argument that the strategy of physical ontology developed in previous chapters can furnish an explanation why the velocity of light does not conform to this natural assumption, an explanation complementing the successful mathematical treatments that have been devised. The connection between this explanation and Whitehead's alternative (to Einstein's) method of revising the background concepts of space and time in physical science is also discussed.

The expository procedure with regard to Whitehead is to give the discussion of his theory the emphasis given by Dean R. Fowler,[1] as opposed to that given in most other studies. Fowler promotes Whitehead by distinguishing, as I do, between predictive and explanatory science, though he also speaks of this in a traditional way as a distinction between the "physical" and the "philosophical,"[2] which tends to make "physical explanation" an oxymoron. The more conventional approach to the study of Whitehead is to compare, in regard to their predictive successes, the quantitative laws of his theory with those of Einstein, whereas Fowler and I would seek out and explicate what we regard as the distinctly explanatory content of Whitehead's theory. My version of this approach is to focus on those aspects of the theory which in some way are engaged in putting together elements of narrative physical explanation—space, time, objects, and events—however rudimentary and/or abstract this textual material may be. This is because, as should be clear by now, I feel that explanations in natural science are narrative accounts of generative occurrences in nature and not predictive algorithms. The unconventional choice of focus is allied with Whitehead's own antioperationalist outlook, as shown in his criticisms of Einstein. My own "theory" which emerges from the discussion, presented first on its own and then in the context of an interpretation of Whitehead, itself challenges the current circumscription of "physical theory" by pressing toward concrete narrative explanation while regarding the abstract and formal models of existing theory—Whiteheadian or Einsteinian—as derivative and inessential (for this purpose) aspects of their inventors' ideas.

Some readers may be bothered by the fact that no parallel explication of whatever constitutes physical description in Einstein's theory is presented here for purposes of comparison. But because of the radically different functions that theory in science can have, the simple question, Who is right, Einstein or Whitehead? is probably pointless. The point here is to shed some light on Whitehead's explanatory approach, arguing that whatever its achievements and limitations, it was set on the correct course of transformation in physical ontology.

The most important advice to the reader is to maintain complete innocence from the presuppositions wrought by our habituation to the Mach-Einstein pragmatic and operationalist legacy about what is a proper procedure toward a physical theory. Due to the conventional point of view on this matter, bearing this legacy, the first response of the informed reader to the stated purpose of this chapter may well be to complain that Einstein already explained the Michelson-Morley result—or at least he discarded the assumption that rendered it perplexing, namely, the assumption of an ethereal medium at rest with respect to the universe . . . or did he *define* the problem out of existence? In any case (the thought will be), this is no longer a living problem of science. Or possibly, as some would argue, a choice exists between Einstein's and Whitehead's very different theories accounting for the relativistic facts, since they appear to make at least very nearly the same predictions[3] (the very criterion by which Einstein's theory is celebrated and adulated as exemplary science). Perhaps there is really a choice of three explanations, counting the Fitzgerald-Lorentz automatic contraction theory, that is, perhaps explanation takes place in and with any of a number of empirically adequate and empirically equivalent models. Are we not obliged to accept the determinations of specialized theoretical science as we find them?

This train of thought is completely unaware of a deep muddle and confusion concerning the different things that can come to be called "theory" in science. With all the intellectual energies exerted in the philosophy of science, there is a continuing failure to sort out cases in which a theory is a predictive artifice and cases in which it is a physical explanation, and cases in which there are elements of both as in a convenient and transitory physical model (e.g., electricity as a fluid). The features of the present era of physics took philosophers and physicists alike by surprise, I think, because it had not been fully realized how far an ingenious predictive model can go with its successes and uses, its charming interpretive elaborations, indeed the breadth of "confirmation" with which it can be consecrated, while it still may not satisfy the basic criteria for being a genuine explanation of the facts of nature of which it is a model.

The magical efficacy of inventive algorithms, and their range of usefulness over the various levels and kinds of prediction, have been widely misinterpreted as having consequences for the very meaning of explanation in natural science, which has contributed to the establishment of scientific nihilism.

In Chapter 5, the section entitled "Explanation as a Theme: A Mark of Basic Acausalism," I pointed to one of the ironic consequences of the way in which scientific explanation is usually discussed as a philosophical issue. There I showed that the traditional way of thematizing this issue in philosophy of science is a mark of an empiricist predisposition. Now, if one were to adequately discuss and compare existing theories on the Michelson-Morley result, it would require involvement in precisely the issue of the meaning of "explanation." In this case the theories are a confusing mixture of narrative or quasi-narrative and empirical/mathematical aspects, whereas quantum theory, which Salmon conventionally presumes is a case of explanation, consists only of quantitative laws that are sometimes accompanied by useful pictorial models—its knowledge-character is relatively clear.

Concerning the knowledge-character of Einstein's mathematical models for relativistic phenomena, however, there are some genuine difficulties and pitfalls for reflection. The conventional view is as follows: These models comprise an exemplary end product of science, a comprehensive theory (leaving aside certain apparent conflicts with quantum theory), amounting to a full and possibly final accounting of a certain subdomain of physics. But I submit that this view needs to confront some truths. For one thing, as I have already mentioned, Einstein's two-part theory in no way answers the question, What is radiation?—even though electromagnetic transmission has a role at the center of both the special and the general theories of space and time (as it must)—and this question is not answered elsewhere in physics, but has only sunk into oblivion and confusion.

There is a widespread idea, found for example in P. W. Bridgman,[4] that the realities explored by physics (forming the fundamental constitution of physical reality in general) are somehow *described* by the empirical relations signified in the algorithms of prediction, so that the mode of reality of radiation (e.g.) simply conforms to the given character of these quantitative relations; in other words, that somehow it is in and through existing empirical models that such realities are properly *understood*. This ontological recourse to the algorithm leads to a variety of Platonistic anti-naturalisms, philosophical rationalizations of the notion that the empirical and quantitative model in its different and interchangeable images can substitute for, or at least suggest a (radical) substitute for, traditional background concepts of the physical, which in any case are completely confounded by certain recently disclosed physical realities. But from a point of view expecting and attempting an advance in physical ontology this is an extreme and ultimately unsuccessful resort to forcing today's examples of physical theory into the category of explanations. My view therefore is that ontological question marks in the domain of physics are certainly not limited to one or two startling "quantum" facts. The casual substitution of formulas for physical stories (which formulas may then have to be "interpreted") appears to me to be about as strong a tendency in the case of nonuniform Riemannian space-time and its account of gravity as it is in the case of the Schroedinger equation. (Physical ontology so far has every reason to think that the physical reality that Einstein's general theory manages to express mathematically as "curvature of spacetime" is actually not physically described by this formulation at all, but instead must be understood in terms of the genetic/temporal structure of emergent space in a particular context.)

My approach to today's problems of physical explanation is not to present a theory about the meaning of "explanation" in science, but to provide some examples of explanation in and for the cases at hand. In this chapter, in addition to seeking an explanation of the invariance of the measured velocity of light in the framework of the ontological shift, I intend to show how this particular explanation is a way of elaborating what I have called Whitehead's guiding idea as he applied it in his theory of relativity. I have interpreted this guiding idea in terms of narrative causal explanation as follows: not only objects, but also ordinary space and time relations, are engendered realities only to be fully and fundamentally understood by reference to the engendering physical processes. The mathematical aspects and derivatives of Whitehead's theory are not my concern, which is not to diminish the importance of an exploration of the relationships among the various theories and modes of theory in this field of inquiry.

This is surely some of the rockiest and craggiest ground for the philosopher to explore in all of Part Two. The goal is to get a few glimpses of yet another perspective on a vista that is becoming familiar.

The Nature of the Problem

The reason for the general, unspoken attitude that the Michelson-Morley result is not a living scientific problem today is the fact that it is successfully incorporated into any of a series of quantitative models; but for physical ontology this fact leaves the question of explanation untouched, just as Newton's law of gravity left the question of the cause of gravity a matter for speculative hypothesis. Once the prejudice toward the finality of the Einsteinian relativity models has been successfully swept aside along with the general inclination to think that mathematical models can satisfy questions of explanation, it is no longer by any means obvious that the problem of explaining the Michelson-Morley result goes away once the ether/absolute space background picture is given up. It is intriguing to common sense independently of this older framework that the measured velocity of a particular transmission is not increased or diminished by the amount of independent velocity along the direction of transmission possessed by the material system against which it is measured. It would ordinarily be expected that when two separate velocities are measured relatively to one another the result would be an arithmetic sum or difference of the individual velocities. There is no reason yet to suppose that the failure of this assumption in the case of light is a genuine logical paradox, as if physical discoveries could actually suspend logic in application to themselves, which is far more problematic than current well-worn and facile tendencies of thought would suggest. After all, we have not even begun to think through how the case is affected when the idea of a background reference system of localized space common to the light transmission process and the motion of the apparatus is removed entirely from the picture, which one can safely assume is the starting point according to physics.

The successes of relativity theory in the course of abandoning the ether do not mean that our very wonderment at this fact about the velocity of light involves the assumption of an ether or absolute spatial locality. That the velocity of light relative to a particular body does not depend on the state of motion of the body along the direction of the transmission does point to the issue of the reference spaces for motions, and raises other questions about measurement. In the ordinary examples that provide the initial models for comparisons of motions and propagations, a common background reference of local space for the two velocities is assumed, as with two passing cars that have the common reference of the highway and the earth. But for the case at hand the key to an explanation appears to be the fact that the common referential space for the two processes being compared is not to be conceived in this way, as was the traditional notion of a total system of space modeled on the abstract idea of a local reference space. Wonderment consists precisely in this, that the common spatial context for the material and radiation processes *has to be understood in some other way.* Here one is asking for nothing less than a story about nature that would make full explanatory sense of this interesting fact by surrounding and contextualizing it, a story that fits with the rest of knowledge and comes under the reach of physical concepts. But this question never surfaces given the prevailing meaning of physical theory and its successes in this area.

Albert Michelson's apparatus was designed to detect the motion of the earth through the ether. The principle behind its conception was that the velocity of light over a fixed distance ought to measure slightly differently when two cases are compared: one when the course of the transmission is aligned in the direction of the earth's motion around the sun, and the other when it is aligned in the opposite direction. The expectation was that if the earth is really moving in its solar orbit through the etherial substance that is the medium of the propagation, which would give a fixed, fully localized reference for both the propagation and the earth's motion, then in one case the laboratory on the earth would advance slightly toward the wave front of light even during the minute interval of time the transmission would take, and in the other case recede from it slightly during this interval. Under these assumptions the light should arrive at the detection end of the experiment at minutely different times (in the chronology of the measurement) in the two cases.

Michelson achieved the required sensitivity (over the doubts of theorists) by making use of the phenomenon called "interference." He reasoned that if two beams were made to produce an interference pattern, then if one of the two wave fronts producing this combined effect were to be caused to arrive slightly before the other, the visible pattern would be shifted. To utilize this principle, one of the transmissions to be superposed would always traverse the apparatus at right angles to the earth's motion around the sun, making it "neutral" regarding this choice of direction. Along the other arm of the instrument the reading was taken at twelve-hour intervals, so that in one case the transmission was aimed along the earth's motion, and in the other it was aimed against this motion, and the two resulting superposition patterns could be compared. The results of this and other experiments of this type were that no difference is observed. The upshot is that there is no such thing as *relative* local velocity of an object with respect to a beam of light; one cannot gain measurable ground, for instance, on a wave-front of light regardless of the velocity one attains.

Michelson was apparently disappointed that his painstaking efforts measured no motion of the earth. He resolved to make the best of it by promoting the technique developed for the experiment as a way of giving a precise definition to the standard units of length in terms of a string of light wavelengths. Because the measured velocity of light proved to be so solidly invariant (in a vacuum), a fixed number of phases or wavelengths of a given frequency will intervene between any two points along a perfectly or ideally rigid object. This standard of length would be independent of temperature fluctuations and other changes in materials. That light wavelengths can serve as a standard of ideally rigid length is in fact one way of looking at the results of the experiment, and it is a convenient focus for the explanation I will be giving.

In prerelativity physics, the only explanation for the experimental result was the hypothesis known as the Fitzgerald-Lorentz contraction, according to which all moving objects contract minutely along the direction of their motion through the ether—Lorentz theorized that this was the result of the interaction of the ether with the atomic structure of the material—so that the absence of the expected effect of such motion would be explained by a compensatory contraction of the experimental apparatus. This allows the stationary ether to be retained; and Lorentz rendered the proposal mathematically in some equations for electromagnetic phenomena under conditions of "absolute" motion. Then Einstein came along and incorporated the experimental result into a novel "relativistic" theory of space and time (rendering the ether "superfluous" and doing away with "absolute" motion and rest), and Whitehead followed with his alternative relativistic theory of space and time—both of whom incorporated a minute contraction of measured length with motion into their theories, universalizing the Lorentz equations of motion beyond their original context of electromagnetic theory. A fact underlying these determinations is that exact measured simultaneity is affected by states of motion.

The explanation to be offered in this chapter is unlike any of these theoretical formulations in one respect: it does not consider it a primary goal to produce a set of applicable mathematical formulas. I have been arguing (against the standard doctrine) that the concept of explanation in natural science is only satisfied by physical stories of the genesis of phenomena, and that where there is only the law or laws (as has traditionally been the case with forces of attraction and repulsion, and came to be the case with all radiation phenomena), an explanation is missing. My explanation of the invariance of the measured velocity of light under varying kinematic conditions proceeds by enriching the description of the experiment, supplying the apparatus and the light transmission with a common context of background events. It does not subject ordinary concepts of space and time to any theoretical alteration, because it is only a theory of space and time in the sense that it utilizes the notion that these are physically emergent modes of extension, describing their underlying constitution in terms of "genetic process" in the sense of transition in presencing.

Furnishing the Physical Context

Recall that in the Introduction to Part Two I suggested that a contextualization of nonlocality might dissipate its seemingly extracausal character while stopping short of explaining any specific nonlocal phenomenon. In this section the problem at issue is similarly set in the context of our background hypothesis, making an initial and very general sense of the facts, and this result only evolves into a detailed explanation in the subsequent section. The key to understanding this peculiar phenomenon in the measurement of light is to realize that all elements of the simple spatial and temporal framework applied in the activity of measurement are emergent in the sense that the circumstance is pervaded by prelocal transition connecting all local aggregates and their motions into natural totality as a whole-in-process. The capacity of this hypothesis, supposing it is acceptable, to explain the phenomenon might already be intuitively appreciated; the remainder of the chapter is given to an attempt to think further about this explanatory possibility.

It has been proposed that the wavelike structure of propagating radiation involves a certain mode of differentiation through stages of a process that lies outside description in terms of changes of position in local spatial systems, a mode described broadly as genetic transition in the context of emergent space and time. The "shape" or structure of the wave-pulse is not formed by an evolving distribution of material or "field" variables through a local spatial system, but is composed of a prelocal transition-structure of events extending into an engendering dimension of physical reality, a dimension that always and everywhere effectively underlies the concrete actuality of things and localities. This means that when space and time relations in physical description are enframed by pregiven spatial extension and uniform duration, a pervasive depth-dimension of pulsating activity, the physical context of light propagation ("ether of events"), is omitted in its entirety. The contour, or mode of genetic reiteration, of the "wave" can only be analyzed as a secondary formation within this background structure of engendering-presencing activity, a particular reiterated pattern of superposition of its elements. These elements are intercomposing genetic transitions, ultimately composing into a whole as a unitary structure of "passage." This structure of physical being is not further analyzable.

This description provides one fundamental aspect of the explanatory narrative for the experiment at issue: the context of the measured propagation. The other aspect of the experimental circumstance is that of the apparatus itself, a material object and a spatial system of measurement coinciding with a part of the earth's surface, in a particular state of motion with respect to its material environment beyond the earth. The relationship born by this material complex, including its state of motion through a broader space of reference, to the engendering background activity forming the context of the radiation is that of a *terminal stratum* to its antecedent gradation of strata. Events that comprise this terminal object/space system—local motions—are structurally different from events composing the antecedent stratification, in that the former have a "flat" spatial and temporal constitution. But despite this difference in event-character, the narrative of engenderment implies that there is systematic, transitional connection between all local motions and prelocal background events. This difference itself is the primary instance of natural genetic transition.

Specific connections between the two aspects manifest themselves in the interaction events, emission and detection, at either end of the measurement of velocity. Though we have not examined the case of emission, it has been shown that regarding the interaction at the site of detection, the reconstruction of some features of background events can result in a sense-making story where one is sorely needed (Chapter 6, "The Transition in Interaction"). In this story radiant transmissions interact with the localized aspect of space in being absorbed by matter, and in doing so trace routes of engenderment specific to atomic loci within matter. These genetic routes of interaction, though not traceable through local space, follow permanent connections which permeate the depth dimension along with the radiation "fields" themselves, in the mode of presencing "convergence" to terminal locality rather than the mode of advancing succession across localities.

In measuring the velocity of light a condition of congruence is set up between a stretch of transmission joining two localized interactions and the length of material between the sites of interaction. The point of origin (emission) of the transmission and the site of its detection are both positions along this material length. The succession in the interactions is clocked. In the experiment at issue, the two interactions in sequence mark out a stretch of space, and the second (detection) occurs simultaneously with that of a separate (neutral) transmission; this is the crucial comparison to be ascertained. The actual quantity of material length is a matter of practicality and conventional units; and of course a particular uniform succession of time is in play for the measurement.

But for the ontological account pursued in this inquiry the structure of light transmission physically exemplifies a meaning of "extension in a physical process" that is different from and bears no simple correspondence to extension in local space. In fourth-dimensional terms, then, this form of propagation by its own activity spans the stretch between interactions through an independent, prespatial mode of extension. The physical congruence in the experiment is thus between, on the one hand, segments of a local spatial and temporal system, and on the other, units of a kind of extensiveness only explicable in terms of genetic-transitional process, as understood by following the ontological shift. The latter is a *meaning of extensiveness in a process* which in the proposed theory of radiation is *independent of given spatial extensiveness functioning in the comparison of lengths of material,* due to its prelocal context and character of activity. Intuitively this is a potent clue to the understanding of invariant velocity.

If this much is true, that the *form of extension* structuring the transmission activity itself is different than that marked out by the successive interactions taken for themselves as positions along a moving grid space, then this case of measurement is already not a simple coincidence between spans of space and time occupied by two side-by-side physical systems, like two cars that come to line up front-to-rear on the freeway. The latter understanding of the situation, that which gave rise to the belief that the "absolute" motion of the earth could be detected using light beams, is based on understanding both parties to the comparison in a common framework of simple and uniform space and time relations. But according to our explanations so far the physical character at least of spatial relations in fact differs across this particular condition of congruence. Through there is no disconnection, a structural contrast permeates the relevant aspects of the measurement circumstance. One side is a flat dimensionality relating positions in space, combined with simple duration established by elements of pure continuity or as Whitehead said, "persistence"; the other is primordial genetic process constituting all fundamental physical detail of the circumstance and extending prespatially, which means, tracing the physical engenderment *of all localized elements of measurement.*

The two basic features of this contextualization of the phenomenon, then, are (a) the thorough interconnection between these structurally contrasting aspects of the circumstance, and (b) their fundamental differentiation as modes of extensiveness. To describe (b) more exactly, the transmission physically spans the interval between interactions not by advancing through a given spatial distance in a way that could be filled in as a series of continuous or discrete positions, but instead purely and simply by *serial composition of its locality-independent differentiating structure.* This congruence across fourth-dimensional gradient difference does not utilize or make reference to any material or ideal localized space, neither one that is on hand for practical purposes nor one that is "at absolute rest." While a prelocal process as described in this work can by a series of interactions mark out spatial lengths in a particular measure system, it does not as a propagation traverse intermediate points in this or any other measure system. Similarly here: because prelocal process *is not motion through a reference space,* the expectation that a coextensiveness established between such a process and a certain span of measurement space over an interval of time will depend on the state of local motion of this measurement space is not fulfilled. This is an extension of the negative analysis of the course of propagation in the fourth section of Chapter 8.

The explanation of the phenomenon in question thus depends on the complementarity of two physical factors: *interconnectedness* and *differentiation.* By differentiation I mean the independence of the physical context of the transmission from local parameters of space and time due to its physical-ontological priority. By interconnectedness I mean that the background structure forming the context of the propagation is also the field of connections between the material aspect of the experiment, with all of its detailed localized structure, and nature as an event-totality. One might even put it this way: there is no such thing as movement of the apparatus against this field of connections as if it were an "absolute" rest space, since this would have to mean that this physical system could remove itself from its own genetic physical constitution and situate itself differently, adopting, as it were, a different causal past. An increment of earth-laboratory motion has a genetic field uniquely mediating its connection with a totality of events. This transition comprising the total field of local events is also the context that structures the extensiveness of the propagation being measured; specifically, as stated in the explanation of localizing interaction, the propagation crosses space by aligning transversely to the genetic routes of local space and matter. It is because any element of localized space is permeated by all phases of its own "becoming" that distances in this space can find an invariant measure—recall the use to which Michelson put his discovery—in such transverse systems of periodic events participating in the extensive structure of this "becoming."

Even the kinematic relation between two object-spaces discriminated by a state of motion is not *simply* this relative motion, devoid of any further physical reference, but is an engendered fact with determinate and effective relations to a whole of engendering process; this is the background explanation of the relativistic laws of motion. All relational facts about the earth-laboratory system share with the light transmission this background reference, which has a fundamental character contrasting with that of flat continuity. The basic intuition here developed as to the physical reason for the invariant velocity phenomenon is that if one begins by thinking of light transmission as fully analogous for measurement purposes to local motion, one will be puzzled by the facts, because its natural context (and accordingly its spatial and temporal character as a form of physical occurrence) is different. Here one does not merely give up absolute space and the ether, but also the automatic practical representation of transmission as local transition. To have the true concept of transition for these cases one must refer to the fact that the measurement circumstance by virtue of its own concrete physicality is embedded in and permeated by a systematic engenderment or presencing process, and the light transmission is a transversal feature *of this process.* In this conception engendering totality supplies the "causal absoluteness" or nonlocal referential context for events, as opposed to an "ether" of omnipresent locality. Unconventionally enough, I submit that explanations in the "relativity" domain derive from the same source as explanations in the "quantum" domain, because the characteristics of both domains arise from different aspects of a premechanistic and prelocal dimension of the physical.

The contextualizing explanation for the measurement phenomenon in question is that the measured transmission is structured causally through a system of relationships independent of localized frameworks of measurement. This system is identified only as the "passage of nature" in the sense of transition in the emergence of spatial and temporal contexts. Any experimental circumstance is perpetually engendered as a terminus of a totality of origination; the Michelson-Morley result is thus another manifestation of *causal absoluteness* in nature.

The final step in this inquiry is to carry out a confluence between the foregoing in its aspiration toward further specifics as explanation and an interpretation of Whitehead's work in this area. Whitehead's conclusions were first established in his alternative interpretation of relativity physics in the 1920s, and were restated in successive works.

Whitehead's Theory

It is important to keep in mind that my project in this section is neither a thorough nor a definitive analysis of Whitehead's physical theory; instead it is a selective interpretive explication in the service of my attempt to reformulate Whiteheadian explanation as causal explanation, the broader aim being to further physical ontology with the help of constructive solutions and intuitions found in Whitehead's work. It traces in greater detail the indebtedness of the present book to Whitehead, and in the process the explanation in the previous section gains by comparison and contrast with what can be called its original formulation. In addition, this section draws attention to the need for further study of Whitehead with regard to detailed matters of physical theory, with the help of competent expositions that have been done.[5] What follows, in contrast to such future scholarly endeavor, blazes an interpretive path to Whitehead's explanation of velocity c. "Interpretation" here means that physical ontology has developed to the point of pursuing explanations in the area of relativity, not as a separate issue from the theory of radiation, but as part of that theory.

Whitehead's "theory of relativity" can be sorted out into the two by now familiar aspects, quantitative formulas (laws) and narrative explanation. The former has been well studied already, and as I have said is outside my concern. It is useful to further distinguish two modes in which the narrative aspect, such as it is, is presented by Whitehead, though these are really integral to one another in Whitehead's theory. One is an explanation of the general facts of space and motion through a theory of intersecting "time-systems," which I began to explicate in Chapters 7 and 9. The other is the "event ontology," which is well developed by Whitehead in abstract terms but not in full-fledged narrative form. My interpretation continues to fill out this narrative through an elaboration of its basic claim that any actual spaces and times of observation are derived from relationships among primordial physical events conditioning the context of observation, relationships which, together with the events themselves, elude all description in which systems of spatial locality and uniform succession in time are pre-established. My procedure is to explicitly reassert narrative description nevertheless, moving outside of mechanism through a shift to different elemental terms.

One difficulty in Whitehead's texts that renders his own narrative explanation problematic is a subtle ambiguousness in central concepts between "time" and "process." As explained in Chapter 7, there is a close relationship between this ambiguity and my own causal narrative interpretation of the theory, namely, that the ambiguity yields to conceptual coherence (though a certain enigmatic quality is never lost) when "process" is interpreted as natural engenderment. My interpretation thus supplies Whiteheadian theory of space and time with a new coherence, besides reformulating the theory as narrative causal explanation, radicalizing the realism and naturalism to which Whitehead aspired. In the process the "time-systems" account begins to lose its essential role. (Whitehead himself eventually resolved the ambiguity in a metaphysical way while continuing to speak of "time-systems.")

Whitehead's explanation of the kinematic invariance of velocity c must be presented in two parts. The first phase is to continue to explicate the theory of space and motion in terms of the relations among "time-systems," found in the set of middle-period works that were directly concerned with physical theory in the area of relativity. Once this is accomplished, the status of velocity c in this theory can be explained. If it becomes difficult to tell whether my own theory or Whitehead's is being discussed, this is the interpretation groping for genuine dialogue with Whitehead's writings from its own developed and developing standpoint.

I have already cited the following succinct explanation of the phenomenon under discussion, which presents us again with the task of elaborating the developed theory that is summarized in it:

[The Michelson-Morley result] is completely explained by the fact that, the space-sytem and the time-system which we are using are in certain minute ways different from the space and the time relatively to the sun or relatively to any other body with respect to which it [our space- and time-system] is moving.[6]

As a practical matter, in any ordinary or scientific measurement a spatial reference system is preestablished or selected; in science experiments it is normally that of a laboratory and measuring apparatus, fixed with respect to the earth. The system selected is differentiated from systems (objects) relatively to which it is in motion, such as the sun. In reference to these other systems, the spatial configurations, at least, of events under consideration will be entirely different than they are as referred to the apparatus utilized as a rest space. But Whitehead's claim goes beyond this simple difference of possibilities for a *spatial* system of reference. He extends the concept of a system of reference to *time,* so that the exact time relations among events also have a certain kind of variance as a function of the choice of reference space. Despite its philosophical novelty, this notion of different reference systems for the assignment of time, which for Whitehead appears to mean that any local manifold of detailed physical events is ordered by exact time-sequence in a multiplicity of ways, is a fairly straightforward intellectual response to one well-known result of relativity physics, that exact measured simultaneity is relative to states of motion. A physicist would probably not object to characterizing the emergence of relativity physics as the discovery of differences of time-system, though what Whitehead does with this idea is idiosyncratic.

On Whitehead's view any relative motion of two objects is *explained and analyzed as* a differentiation of two time- (and space-) systems, so that two objects relatively at rest would by virtue of that fact form a single time-system:

Every enduring object is to be conceived as at rest in its own proper space, and in motion throughout any space defined in a way which is not that inherent in its peculiar endurance. If two objects are mutually at rest, they are utilising the same meanings of space and of time for the purposes of expressing their endurance; if in relative motion, the spaces and times differ.[7]

A time-system is precisely the "particular endurance" of an object and its approximate spatial region of occupation (this calls to mind

Salmon's classification of a material object at rest as a "process"). But this simple unitarity as a spatial system of reference does not give a complete spatial and temporal account of such a physical locality, because any actual space is formed out of the relationships among different time-systems:

> Position in space is merely the expression of diversity of relations [of an event] to alternative time-systems. Order in space is merely the reflection into the space of one time-system of the time-orders of alternative time-systems.
>
> A plane in space expresses the quality of the locus of intersection of a moment of the time-system in question (call it "time-system A") with a moment of another time-system (time-system B).
>
> The parallelism of planes in the space of time-system A means that these planes result from the intersections of moments of A with moments of one other time-system B.[8]

Elsewhere:

> ... a series of parallel planes in the space of our time is merely the series of intersections with a series of moments of another time-system. Thus the order of the parallel planes is merely the time order of the moments of this other system.[9]

And again:

> Our whole geometry is merely the expression of the ways in which different events are implicated in different time-systems.[10]

To summarize Whitehead's formal account of the primary relations among time, space, and motion: Positional order in space—locality difference—is itself the result, or the "expression," of the convergence or intersection (in events) of diverse systems of exact temporal relations among events. The derivative or manifest aspect of the natural convergence of time-systems is a stratified structure of aggregate material motions (e.g., earth, solar system, galaxy, and so on). Any particular motion identifies a distinct time-succession, unique in the way it precisely orders events, disclosed within the manifest aspect of the circumstance. A time-series can be represented as series of parallel planes of simultaneity, each of which is an instant in and for a particular system of three-dimensionally extended event-space, which thus amounts to a "space- and time-system." The relative motion of two objects, a kind of disjunction of their spaces, means just that these time-series cut across each other, or differentiate, in a manner specific to the direction and velocity of the motion. Any actual motion exhibited when two spatial systems differentiate is the "expression" of a divergence between temporal series (in the sense of the durational aspects of the spatial systems), so that any condition of mutual rest means that a single basis of exact simultaneity/time order relations can be applied in all measurement.

It is helpful in thinking about the time-systems theory to refer to the terms of the other narrative mode, bringing in some of the elaboration I have given it. Think of a case of relative motion not simply as a kinematic relation between two objects, but with its full physical context: a many-aspected flux of activity including observable motion, vibrations of molecules, interatomic events, processes constituting the surrounding space—of which the simpler description covers only the manifest and "derivative" aspect with its "flat" spatial and temporal character. Then according to Whitehead this ongoing manifold of events in the background of any instance of ordinary motion has for its measurable temporal relations two different systems of planes of simultaneity ("slices" of time) drawn through it in two (minutely) diverging ways—and in fact, taking into account all systems of reference given by particular object-motions in the broader environment, in a plurality or "bundle" of different ways. What would ordinarily be called a difference of spatial (kinematic) perspective means not only that relative motion exists as a manifest feature of these events, but also means that events exactly simultaneous in the spatial system of one object will, in the spatial system of the other object, be to some degree "shifted" apart in time.

It would be incautious to conclude that for Whitehead time itself is peculiarly internal to and unique to individual objects and localities as particular spatial systems. The particular time-system identified with an object and its "proper space" is a time-sequence in a whole of events extending beyond the object. That is, Whitehead claims to understand the word "time" in a common-sense way, as a particular ordering applying to the whole universe: "The flow of time means the succession of moments, and this succession includes the whole of nature."[11] A system of temporal parallels or time-slices through events—and a multiplicity of such systems are possible for any physical circumstance—is a *way in which* "nature is stratified by time."[12] The suggestion of a relativism of time in the time-systems account is discussed further below.

Intuitively, if the idea of alternate systems of simultaneity, that is, of a variable basis in all systems of natural events for the assignment of simultaneity, could prove scientifically plausible and philosophically acceptable, such a description of physical reality could easily do the job of accounting for any phenomena of measurement involving ultrahigh velocities which perplex our ordinary concepts of space, time, and motion. To alter a fundamental assumption that normally operates everywhere in scientific measurement, namely, that the assignment of simultaneity is unproblematic, simply a matter of observation and of the sense that is made of the synchrony of clocks—such a loosening of this bedrock assumption certainly would provide the flexibility to account for any systematic deviations in spatial and temporal variables from the predictions of previous understandings. And as I have indicated, in a certain way Whitehead's idea seems a quite natural response to the relativistic developments in physics, especially from an individual of mathematical ingenuity who had mastered the knowledge in physics and mathematics and was quite independent-minded. One reason for the lack of acceptance of Whitehead's physical ideas seems to be the novelty of his basic approach to such matters, which had its antecedents largely in his own earlier thought; if so this seems ironic in view of the astonishing and perplexing character of discoveries and ideas throughout physics today.

Theories of relativity are designed to account for a systematic departure in measurements from expectations that follow from the application of a simple temporal and spatial conception of "motion" to all modes of spatial transition, to all motions and propagations. This departure in measured results from the simpler expectations correlates with magnitude of velocity and reaches its limit case in the velocity of light. One requirement of an explanatory model is that it account for the fact that this velocity measures as an unvarying quantity regardless of the kinematic conditions of the measurement. I have yet to explain just how Whitehead's theory yields an explanation of this basic finding. Before beginning the discussion of how the fact of kinematically invariant measured velocity is accounted for by Whitehead, I will first consider a different and simpler example of Whitehead's explanatory model in action, as an exercize for what is to come.

A well-known phenomenon predicted by Einstein and confirmed by experiment is what is usually called "time dilation." As a result of being accelerated to a high velocity a clock will lag behind another clock that has remained at rest. On Whitehead's theory, what happens would be described in the following way: There are two perfectly synchronized systems of events embedded in the one nature (as totality of ongoingness), which are the myriad events composing and surrounding each of the clocks—call the systems (a) and (b)—and subjecting one of these to extreme acceleration will cause the temporal analyses of these systems to diverge in some degree. "Temporal analysis" framed abstractly involves a picture of a succession of *time-slices* that intersect all instantaneous parts of events occurring at particular instants, forming a discrete series comprising the extension in duration of a *particular* time sequence (note that an individual time-slice through events is itself extended *spatially* and in three dimensions). If two instantaneous events, let us say, are exactly simultaneous in clock (a) remaining at rest, their corresponding events in clock (b) will be successive in time sequence (a), not due to any rearrangement of particular events relative to the rest as if its mechanism were disturbed, but due to divergence between the two *systems* of exact time order. For simultaneous parts of events of (b) to be shifted apart in time when referred to system (a) would amount to a systematic (comparative) retardation of the system as a local complex of events, since any given process would simply take more time to occur. (This "temporal perspectives" model is difficult and perhaps impossible to make concrete. I am fairly sure that one can and should regard it as a possibly useful abstract approach and need not, fortunately, retain it for the purpose of physical explanation.)

But, it seems, the strange talk of "time dilation" which has helped to give Einstein's theory the mystique of inscrutability need not arise on Whitehead's account, where the effect of the acceleration is merely the unfolding of different systems of exact simultaneity within otherwise determinate sets of events. On this account there is only a shift in the way events in a particular local event-system (clock [b]) are distributed in time (compared to event-system [a]); this difference across time-space frames amounts to a systematic slowing of the processes composing clock (b), not somehow an increase in the duration of duration, a "slowing (or stretching) of time." The latter notion is a product of the reapplication of traditional physical concepts under convenient modifications in the context of a certain mathematical model. But for a philosophical appreciation of the problem, one can dispense with this conceptual Procrustean bed for time: the case is exactly as if all the machinery of the clock *slowed down uniformly.* This is slightly different from "this clock runs slow," which indicates that a distortion has occurred somewhere in the clock mechanism. Nevertheless a systematic, uniform retardation of the clock while it is working perfectly and its mechanism is unaltered, though in a certain way out of the ordinary, is a perfectly understandable and coherent possibility. One should not elect to describe a famous imaginary illustration in terms that are fundamentally quixotic conceptually, "time slows down (or stretches out) for the space traveller and stays the same for the earthbound person," when it might be enough to say that the events making up the traveler's body and surrounding (colocated) events occur *as a system* at a slowed rate by comparison with the other individual. Whitehead said,

> My own explanation is that there is a universe, of which both the traveller and the chronologer have diverse experiences dominated by the diverse histories of their bodies as elements in that universe. The real diversity of relations of their bodies to the universe is the cause of their discordance in time-reckoning.[13]

This eminently sensible remark, which contains the essence of Whiteheadian ontology, does not of course dispose of the issue of "time dilation." The real physical/philosophical questions here loom large but are generally ignored or dismissed. Strikingly, there is still general ambivalence in discussions of relativity as to whether the effect going by this name is due to the acceleration of one of the clocks or is due to the purely relative motion of the clocks. The latter, which is most often conveniently adopted (following Einstein), actually makes little sense: why then would one rather than the other of the two clocks lag behind when they are brought to rest? This option would have to be subject to the complete arbitrariness corresponding to a pure relativity of motion; this seems thoroughly nonsensical. Whitehead's suggestion was commodious to thinking compared to this logical snarl (which, understandably, is for the most part deftly evaded). Forgetting about prejudices introduced by the choice of theoretical procedure, there may be ways of formulating and appreciating these discoveries that are relatively unremarkable. We could say, for instance, that with this phenomenon nature is reminding us that the events of a running clock accelerated to a high velocity are under distinctly and systematically altered physical conditions compared with those of an identical running clock at rest.

In view of the findings of the previous chapters, one avenue of approach to Whitehead's explanation of velocity c suggests itself at this point, even before consulting Whitehead's statements specifically about this. This will eventually lead into Whitehead's own brief statement of his solution.

Speaking formally and abstractly, for two different events to be perfectly simultaneous means that they are differentiated from one another only in the spatial mode of extension and not in the temporal mode. In Whitehead's theory, the spatial mode of differentiation among events—extension wholly within a plane of simultaneity—is that by which a particular time-system is discriminated. "Multiple time-systems" apparently corresponds to the diversity of material objects in motion and their systems of space that comprise the extended environment of any physical circumstance. It should be noted that Whitehead did not subscribe to the operationalist reduction of motion to pure relativity. For him the reference of theoretical terms such as "multiple time-systems" was always actual perspectival nature locally and in totality, which in the present context means a series of levels of aggregate bodies in periodic motion extending indefinitely. Nevertheless, according to his basic account it is *in and with relatively moving spatial systems* that diversity of time-systems occurs. (This discloses a problem for the formal time-systems account: the multiplicity of time-systems is somehow both antecedent to and confined to "derivative" space and time.) Therefore, any set of events that has in its physical structure a multiplicity of exact temporal orderings has internal structure or extension of a *purely spatial* sort, corresponding to any particular exact temporal ordering.

Such a set of events must then be, extraordinarily enough, a *complex of motions*, with each participating spatial system discriminating a particular time-system.

But this study has been developing the idea of *prelocal* events, which implies that certain sequences of events (e.g., light) are genetically antecedent to given space, and hence in a way transcend and escape the purely spatial differentiation, thus representing, as extended process, an independent mode of extension. Such a sequence could involve no differentiating structure that identifies and discriminates particular time-systems, because it would not have the kind of extension (as event-differentiation) that defines planes of simultaneity: purely spatial extension. Because the structure of transition in the sequence would involve no simple spatial extension (even though it *results in* a certain traversing of space), the events of propagation would not identify a particular linear temporal series in simultaneously extended space, hence would not discriminate the time-system of any particular body of reference. In other words, this case of a velocity would not have the spatial and temporal properties of a locally moving object, because its structure of transition would be independent of (though interacting with) the reference systems that come into play with local motion. This could account for the fact that differences of relative motion do not affect measured space and time properties of the sequence, such as its measured velocity of propagation. There is a problem with this as interpretation of Whitehead, since it would appear to place light transmission outside the application of the formalism of multiple time-systems, which is presented as describing a general feature of nature. The present reformulation avoids this problem by saying that linear space and time are preformed in the context of radiation events.

Is the interpretation on the right track as to the way in which invariant velocity enters into Whitehead's time-systems model? If so it would fit with my claim that propagating radiation is a sequence of events having its full and detailed structure of differentiation and composition in a form which does not incorporate the kind of ordered relations represented in coordinate grids, since it is constituted of a prelocal physical gradient. It remains to be seen whether this hunch is confirmed by further examination of Whitehead's theory.

If my gradual reconstruction of Whitehead's explanation of the Michelson-Morley result is correct so far, it represents a highly furthered interpretation of Whitehead's own writings on the topic, which always tend toward condensation into abstract and formal conceptions. But the exact place of velocity c in his model of space and time is articulated by Whitehead, and I think what he says corroborates my reading of the theory. His claim here is best approached by continuing to interpret the theory as genetic explanation of physical observations in terms of "events," where these are stipulated to have a structure that does not utilize the traditional elemental concepts of physical objects and their relations or abstract models of such relations. Rather than making quotations from Whitehead premier exhibits and examining them from all scholarly angles, they will be embedded in my own interpretive exploration in order to confirm that this interpretation is on track.

The basis of Whitehead's theory about velocity c (on my reading) is a genetic account of space and of motion and other propagation, an account explaining the known "relativity" phenomena, which runs as follows: objects, localities, and motions are engendered in the course of, and genetically analyzed into, events of physical "passage" or "process," and this means that particular localized spaces or spatial systems, each identifying by its uniform "endurance" a particular simple linear temporal ongoingness, are likewise engendered and physically analyzed. On this view there are engendering events, composing into a pervasive background process, which involve in their happening a sense of physical "extension" whose meaning is independent of either of the two distinct types, spatial and temporal extension, with their distinct modes of undifferentiated continuity. Extendedness as a property of primordial process consists of the relations of component events to a contextual totality or "field," with each component event having a unique internal connection with the structure of the whole. These are the fundamental space and time relations.

As Whitehead expressed it in the later work, *Process and Reality: An Essay in Cosmology,* "The extensiveness of space is really the spatialization of extension; and the extensiveness of time is really the temporalization of extension."[14] I would emphasize the active content of this statement: spatial*ization,* temporal*ization.* I am interpreting his claim as follows: Given any actual three-dimensionally extended space with temporal duration, there is a dimension of processes permeating the physical context of this spatial system, a dimension constituted of a structure of total "passage," of a field-totality, quite independent of the positional definitions, the local volumetric confines, and the spans of simple duration comprising this system. At stages prior in the genetic structure to the differentiation of space from time there is *extension only in the sense of the stages through which certain events, composed through primordial structure, arise and pass.* This "extension" has for its sole applicable concept genetic-transitional differentiation. The two species of flat extension, spatial and temporal, are fully differentiated only at the *terminus* of a fundamental transition, such a terminus being the aspect of given locality in a physical circumstance. As a result the interrelations within the primordial productive activity have a form, a systematic order, and an identity independent from any reference to local spaces and times. In other words, there are background processes genetically prior to local distances and durations employed in measurement, the "extension" in these processes being constituted of a form of transition bearing no reference to these measuring intervals, though the processes permeate their product spaces and times. Quite naturally, extension in product is "derivative" from extension in process.

The derivation of space and time from primordial activity—the spatialization and temporalization of extension—would mean (as long as one does not make the mistake of consigning the "derivative" to the "purely conceptual") that any physical reality consists of two aspects: (a) ordinary space and time as distinct forms of extension, and (b) the gradient of physical strata genetically antecedent to this distinction, a gradient extended in the mode of "passage of the whole." There will be certain events that belong inherently, in their structure, to this genetic transition or gradient. Suppose these are propagative alignments *across* it, that is, transversal to it, according to the four-dimensional schema that was already worked out for the basic theory of propagation and interaction. Such propagation would then itself have two (perfectly compatible) aspects: one aspect would be described in a manner in which the sharp distinction of distance from temporal duration is not applied, and the other would be described as a crossing of space through a clockable series of interactions with matter. Combining these aspects means that the propagation occurring along this boundary or transition will lie alongside, or be in a sense tangential to, the spaces of scientific measurement without being a localized transition mediating between the interactions. A conjecture based on some otherwise inexplicable evidence was earlier made (Chapter 6, penultimate section) that interactions at sites marking a successive linear order will take place through orthogonal shifts of alignment. Also, it was proposed that the phase-structure of propagating radiation has no further physical or spatio-temporal analysis than as a particular coherent contour in the events forming its natural context; though there is pervasive compositional structure in the form of causal differentiation in unitary and serial modes, there is no analysis of phase-transitions by means of entities of a different spatial and temporal character, such as localized objects, point-masses, or points of space. One can expect, then, that primordial "extension" is or has its own referential basis for *congruence of its unit parts*, for instance, the ordered phases composing prelocal transmission. In Chapter 8 (page 304–5) I specified this congruence in a particular propagation as "sharing a level of inclusion" (of background events). Though such extension is attributed to occurrence interactive with matter and hence susceptible to measurement, that is, the occurrence marks spans in measurement space and time, its activity is not local motion and bears in itself no special reference to any particular "rest space."

A tracing of the genesis of space and time along the course of their differentiation, investigating the process of their reciprocal unfolding, would of course encounter a physical condition underlying any and all flat terminal spaces and (their) uniform periods of duration. Events comprising this primordial physical condition would be structured and interrelated in a manner contrasting fundamentally with observed and measured events of motion or qualitative change in objects through time. In the genesis of ordinary space and time, specifically spatial intervals would differentiate from specifically temporal intervals as both emerge from extension in primordial process. This reciprocal emergence would involve structure in the mutual background of the time and space of particular local systems, so that a certain quantity of spatial extension would emerge from a given span of primordial extensive structure, and from *that same* portion of extensive structure a certain quantity of simple duration would emerge as the durational aspect of this stretch of emergent space. This would mean that actual spatial and temporal intervals in the same physical context have a basis for being *compared against one another* in respect to primordial (preemergent) extension; there would exist, for example, a relation of congruence between spatial and temporal stretches as "measured" in units of background structure from which they coemerge. This comparison refers back to extension in a genetically prior context.

Now, successive phases of propagation are congruent to one another in this very sense, namely, that they involve the contoured extension in prelocal events occurring at a particular level of composition. This gives the seriality of transmission a common basis of extensive relations with the primoridal event-congruence between coemergent intervals of space and of time. According to this, electromagnetic propagation compared against a particular material system would in fact measure or mark out this congruence in background events transitional to the system; the high magnitude of velocity *c* would mean that from a practical standpoint congruence in underlying genetic structure occurs between what emerge as extremely large units of distance and extremely small units of duration. The special property of velocity *c* in experiments of the Michelson-Morley type would result from the way the prelocal structure of space and time relations is reflected in phenomena of scientific measurement. Now here is the statement of Whitehead's for which I have been preparing:

> . . . the existence of *c* with its peculiar properties really means that the space-units and time-units are comparable; namely, there is a natural relation between them to be expressed by taking *c* to be unity. Either the time-unit would then be inconveniently small or the space-unit inconveniently large; but this inconvenience does not alter the fact that congruence between time and space is definable.[15]

On my interpretation, then, the Whiteheadian claim about the special status of velocity *c* is as follows: This velocity, this particular ratio of distance to time units, represents the proportions in which a given span of physical space and a given span of physical time in a circumstance of measurement are in their genetic physical constitution—that is, in terms of background events—perfectly congruent, since these are the respective magnitudes in which these derivative forms of extension *emerge together and mutually differentiate* from a given quantity of extension—given by levels of "inclusion" or composition—in events physically prior to their differentiation. Thus despite the high magnitude of this ratio of distance to time in terms of the spaces and times convenient for measurement, in terms of the structures of extension in event-systems participating in primordial "passage," this ratio is in fact *unity*. Light has this measured velocity because its form of activity follows a particular systematic congruence in events that belongs to the structure of engendering-presencing, that is, to the unfolding differentiation of space and time; its phases incorporate stretches of this unique relation. Velocity c (expressed in convenient, arbitrary units) signifies a manifestation in temporally extended three-dimensional space of a fundamental relationship between space and time which results from their common origination. It is a manifestation of the fact that the quantities of extension in temporal intervals can be compared with those of associated spatial intervals by reference to the events in their common causal substructure.

The time-systems account, to whatever extent it is valid or valuable, can be brought into the picture as follows. Light transmission has as its context a condition in the "process" of nature which differs in its basic physical character from the "derivative" relations that conveniently analyze object-motions. Motions of material objects in their context of broader motions are aggregate events that for practical purposes differentiate in a purely spatial dimension, which means they have an approximate manifest extensiveness that can be reconstructed as "planes" of simultaneity which group into a single series of parallels. On the new relativistic conception "velocity" *as applied to object motions* means that a measurable divergence occurs between this "stratification" of time-space given by the object itself and that given by the measuring or reference object. But a light transmission process does not differentiate internally in a purely spatial sense, so that there is no unique stratification of time conditioning its propagative succession. It does not uniquely identify a particular space- and time-system, hence does not discriminate local motion or rest. In my terms, light follows the fourth-dimensional periphery of any grid of spatial position, and does not transit successive local positions, though it is a measurable crossing of space simply by virtue of succession in spatially separate interactions. Its extensive structure as physical happening refers only to structures of individuation lying in the genesis of space, and that now means structures that pervade and precede the active differentiation of space from time. Therefore its transition can involve no privileged reference to any particular rest space or any particular moving space; it does not differentiate time-space systems.

This interpretation of Whitehead corroborates the claim in the present theory of radiation that propagation is aligned across a primordial physical transition or gradient. The two accounts concur that light transmission, manifested as a velocity in local spatial systems, is a propagative succession in the structure of events which lies along a transition joining the local space and its prelocal physical context, and therefore it does not proceed *through* space conceived as a localized reference system; as I would phrase it, light propagation occurs at the periphery of the *being* of space. The magnitude of velocity points to a "natural relation" in which spaces and times are congruent in terms of extension in prelocal events.

(That the velocity of light signifies a primordial relation may shed some light on the traditional philosophical puzzle about perception, still endemic and oppressive throughout professional thinking [its latest symptom is the "metaphysical realism-idealism" debate]. The puzzle arises as soon as one thinks of perception as occurring somewhere near the end of a process beginning at the object and ending up in our brain; we "really see" some terminus of this process, not an object. Standard epistemology begins with the idea that to think scientifically about perception is to think about light propagations and organic stimuli, the physical conditions of perception approximately understood, which is to turn away from any concrete occasion of perception and its actual structure. No organic stimulus is finally a perceived object. Even once this is acknowledged, however, if the causal transmission underlying perception is conceived according to the simple notion of an influence crossing a given spatial distance in a certain amount of time, this tends to place an irremovable temporal wedge between the object and our seeing it. But if light and other primordial causality is understood more completely as constituting the medium of prespatial connectedness, then the question of perception may lose this aporia as well. Then perhaps the discussion of mediating causality will not so readily give rise to the conclusion that to see an object is to "see" an image or remnant of the object or to process a piece of information. But naturalistic explanations cannot address the intentional structure of perception,[16] e.g., seeing *something*, feeling, sensing, or intuiting *something* [similarly, there is no actual determination of natural science affecting autonomy of action or moral responsibility].)

But does not Whitehead's formal theory, with its distinct and separate time-systems, after all amount to just that philosophically undesirable result which Merleau-Ponty in a discussion of relativity called a "radical plurality of times,"[17] a pure relativism of the measure-system, as though a unified ongoingness of nature did not encompass all possible measure-systems? Whitehead stresses the *convergence* of the group of time-systems in events as a unitary and fundamental "passage of nature." But suppose the time-systems account does unavoidably describe a "radical plurality of times"; how this reflects on the Whiteheadian approach might depend on how one views the status and function of different formal presentations of the theory. "Time-systems" could be viewed as a useful formalizing commentary on the fact of the relativity of measured simultaneity and other relativistic phenomena: "It is *as if* nature were always and everywhere 'stratified' by time in multiple ways." This function of the time-systems account would be comparable to the true function of Einstein's models, as I see it: the special theory treats physical circumstances *as if* simultaneity were only definable by light "signals" and *as if* the motion of objects were always purely relative, and the general theory treats the spatio-temporal basis for measurement *as if* it were not uniform and Euclidean but nonuniform and Riemannian. This "instrumentalist" view of today's theories (they do not explain; explanations lie elsewhere) combined with my approach to "causal realism" (physical ontology) might if necessary avoid the "radical plurality of times" on Whitehead's behalf by assigning to the idea of multiple time-systems the status of an abstract construction useful for generating laws and perhaps for aiding in the pursuit of explanations, and at the same time holding that Whitehead's theory would amount to explanation only insofar as it can be fully articulated in its aspect of a narrative account of a structure of concretely understood physical events in a *genetic* four-dimensional framework (not an Einsteinian "spacetime," at least not according to its popularized conception). In Whitehead's theory, the explanatory context for relativistic phenomena is multiple actual time-space systems inter-relating events, whereas in the proposal developing here it is transition in the emergence of time-space systems, an unanalyzable physical structure.

Also on the matter of the specter of a radical relativism of time, my sense is that what is called the relativity of simultaneity does not have the shattering consequences for commonsense under-standing of the world as one might by tempted to suppose. It certainly does not threaten the general temporal coherence of the world. The divergence of simultaneity is only observable under extreme conditions and concerns some minute detail in physical events. One should keep in mind that the *measurement* of time nec-essarily refers to particular locality systems for marking off periods of time. To pursue the matter of time relativity, I speculate, would be to inquire into the effect of forcing time-reckoning, by proce-dures of measurement, into the relational context of discrete locality at its furthest practical refinement. There may be nothing here be-yond the enigma of the limited scope of such relational contexts in nature. If a conclusion of physical ontology is that there is structure in events that is not mappable on coordinate grids, this might have already suggested to us that simultaneity is not in all cases resolv-able by reference to points of space. *Measurable* time is indetermi-nate for any physical process in which locality is indeterminate. But the relativity of simultaneity does present a serious problem for a conception of time as entirely separable from natural processes that may or may not be occurring "in time"; this conception operates, for instance, when cosmological science endeavors to fix the date of birth of the universe.

An effort to think further about time relativity would begin with the idea of time as emergent in the sense outlined in Chapter 7, especially the idea that time *as simple duration* is not ontologically prior to space, since ordinary time and space are cotermini of ge-netic transition. Other guiding insights for a physical-ontological approach would include (a) that while theoretical models covering this area that are currently in use may be perfectly satisfactory from a practical, technical, and predictive standpoint, they do not adequately address the questions from the standpoint of natural philosophy;[18] and (b) that if attempts to describe and/or explain phenomena in the measurement of time under relativistic conditions (that is, involving velocities significant compared to the velocity of light) are enframed with the simple conception of space as a local "reference frame" or coordinate grid that is contained in the ordinary notion of "velocity," they will fail to make narrative sense of the phenomena.

Whitehead's account of the properties of velocity c can be interpreted in its basic nontechnical content as follows: A measure system (laboratory in a state of motion) is an event-complex or process with a unique sequence of temporal succession which is one derivative of primordial and universal genetic process. All particular systems of spatial and temporal ordering, such as this simple "process" of material in motion or the light transmission itself, have as a common explanatory context an event-totality in unitary transition. The process of propagation is situated mediately between the material process and its total originative field, that is, it lies across an absolute structure. My own version of the explanation is not directly about time, but about the participation of interstructured genetic successions in local complexes and in a total genetic gradient. Thus the interpretation develops entirely as an extension and expansion of the event-ontological mode of presentation, without commitment to the technical-formal constructions.

Through the stages of his thought about relativity and into his later cosmology, Whitehead moved toward narrative theory and away from abstract construction in terms of planes of simultaneity; but at the same time he moved away from physical science (or as he might have put it, the interpretation of physical science) and into metaphysical explanation. But the later work did come to emphasize genetic process,[19] continuously with the stages of his work on physical science, though framed in ontotheological and teleological terms (with his own conception of God and creation, of course). In *Process and Reality* the operative elements of ultimate explanation are individual units of activity by which the world is reproduced and undergoes "creative advance," with a theory of time appropriate to this metaphysical program. It seems to me that this synoptic work has to be understood in one of its major aspects as continuing to elaborate the theory of nature. But along with this expectation I have strong reservations about the teleological elements of this work, and some concerns about the biological metaphors of the later period.

Whether from the standpoint of the present work an interpretation of Whitehead's mature cosmology can be given which would contribute to the ongoing discussion and debate about it is a question I leave open. The basic relation of Whitehead's own earlier work up to and including the theory of nature to his later philosophical "system" is a matter of controversy.[20] I would so far contribute the following suggestions. It was his early critique of "simple location" that allowed him to proceed beyond the preconception that locality must be pregiven in any physical description while maintaining a physical theory that was to some degree ontic and narrative. Later ideas regarding "process" are (at least partly) traceable to the theory of time in the physical theory as it developed into a metaphysical narrative of genetic process. The difficult "epochal" or "atomic" theory of time, which is a theory of "endurance as re-creation," came about in the course of turning to metaphysics for explanations while following out the theoretical construct of different time-systems. Without coming to a final judgment on it, from the present point of view the complex reasoning behind it seems flawed, though it is an attempt to formulate what I have called a genetic dimension of time.[21] Whitehead was convinced by a Zeno-type argument that the "becoming" of extensiveness cannot itself be extensive;[22] for his metaphysics such becoming is instead the becoming-actual of pure potentials (rather like Platonistic forms or universals). If any "becoming" has extension, he thought, its temporal aspect must be pictured as follows: the event of "becoming" has a simple temporal extension such that it takes up a uniform interval of time contiguous with intervals before and after it, leading to irrationality.[23] This assumes that a flat continuity of time is necessarily presupposed in and for the temporal aspect of extendedness in genetic structure, which is exactly analogous to the materialist assumption regarding events and space that he critiqued in his earlier work. Because of what he saw as the irrational consequences of continuous time in a metaphysics of becoming, that is, where actuality is a terminus of becoming, he offered an alternative conception of the relation of primordial genetic process to time, namely, a kind of metaphysical atomism, not about atoms of dis-crete material, but about "process": there are a series of discrete recurrences ("durations") of the being-becoming of extensiveness in events, the "sheer succession" of which constitutes a time-system. I submit the way of extrapolating his middle period work presented in Chapter 7 as a third and ultimately simpler alternative that main-tains an extensiveness in and of "becoming" understood as physical activity of presencing. The upshot of that account is that physical time in one aspect shares a common structure with the *differentiated continuity* that is the "passage" or "process" of nature-in-totality, so that time in its full conception incorporates genetic structure as well as the derivative element of simple duration arising along with spatial extension. This allows theory of nature to be self-contained, rather than deferring to a metaphysical background explanation involving successive re-creation.

One source of the atomic theory of time appears to have been that Whitehead was under the standard misapprehension that the "quantum" discoveries required some basis for fundamental dis-creteness/discontinuity in nature,[24] despite his long-standing com-mitment to an event-continuum underlying objects; reinforcing this may have been the conceptual pull of his terminology of "scientific objects" referring to electrons, protons, and the like, which would correspond to the general problems of thought that result from speaking of "fundamental particles" in reference to what are actu-ally reiterative structures in events. Whitehead was not fully cogni-zant of the radical technicalization of physics that had taken place in his time and the concomitant demand on philosophy to assume full autonomy in questions of physical explanation. Fundamental atomicity need not be resorted to insofar as there is the explanatory possibility that physical reality is essentially generation and regen-eration in a structured continuum. Overall, Whitehead's mature cosmology is antithetical to the point of view of physical ontology in that it assumes a split in the roles of science and philosophy, though his conception of this split is not that of his (and our) con-temporaries. According to *Process and Reality,* science discloses and correlates empirical fact, while *explanation* is the business of philoso-phy as metaphysics, whose task is to harmonize science and reli-gion.[25] The question of the continuity or discontinuity of time might benefit from some diligent scholarly and philosophical controversy. A forthcoming book includes discussion of the later Whitehead in relation to Nietzsche, Heidegger, and physical ontology.

As anticipated in Chapter 6, my neo-Whiteheadian approach to physical explanation would supply structural content to the bare abstractions "event" and "extension" that occurred in the middle period work, in the form of differentiation through engendering transition which was disclosed in its concrete physical meaning in Chapter 7. "Inclusion" has been interpreted as composition through specific modes of genetic interstructuring in four dimensions. Certainly there are more connections to be drawn with the details of Whitehead's abstract mode of presentation. It strongly appears that Whitehead himself was striving to fill in these early constructions through a narrative of interconnection and transition in his later period.

Conclusions

This chapter together with the previous has shown that the explanatory range of the theory of radiation developed in earlier chapters in connection with the "quantum" area of physics extends also to the "relativity" area; in other words, that the theory covers the ontological question, What is radiation? in its major ramifications, even though the answer is incomplete in the sense that a vast dimension of detail regarding differentiation of species of radiation and the genesis of the ordered species of atomic matter remains to be explored, which only means that physical ontology has been occupied with developing a new approach to these fundamental aspects of the physical in recognizing and filling the need for a first, ultimately basic ontological understanding.

The overall physical fact indicated by the laws of measurement in the area known under the rubric of "relativity" is that physical becoming is inherent to physical being, and in particular, that temporal and spatial extension are emergent. Due to the fact that being—the actuality of events, objects, places, and times—is integrated with process of becoming in a single twofold structure, there are two distinct aspects to any circumstance of physical measurement: a manifest aspect that can be understood by means of the simple framework of spatial grids and occupying objects, and another which consists of a prelocal dimension of engendering–presencing process. The diligent and ingenious study by physics of quantitative detail in motions and propagations eventually came across consequences of the latter for measurement in given space and time. Wherever there is simple mutual reference between distinct local systems of space, there is also a common and independent causal reference in a total contextual structure of origination. The Michelson-Morley experiment was carried out under the assumption that "the universe" is a total (though perhaps infinite) locality, namely, the spatial system identified by the ether, or in other words, by "the condition of absolute rest." The failure to detect the absolute motion of the earth is due to the fact that "the universe" does not conform to this concept (which is not altered essentially by moving to a relationist doctrine of space), but is rather a total structured event, essentially transitional. The kinematic invariance of the measured velocity of light, along with other properties of radiation, points to transition in a genetic structure forming a causal background to all physical space.

Notes

Part One—Introduction

1. Biologist Stephen Jay Gould defends the scientific status of "historical explanation" in *Wonderful Life: The Burgess Shale and the Nature of History* (New York: W. W. Norton, 1989), 277–90.
2. Martin Heidegger, *Nietzsche*, vol. 2, trans. David Farrell Krell (New York: Harper and Row, 1991), 113.
3. Friedrich Nietzsche, *The Will To Power*, trans. Walter Kaufmann and R. J. Hollingdale (New York: Random House, 1967), 5–82.

Chapter 1

1. Galileo Galilei, *Two New Sciences*, trans. Stillman Drake (University of Wisconsin Press, 1974), 157–58.
2. Galileo Galilei, "Letters On Sunspots," *Discoveries and Opinions of Galileo*, trans. Stillman Drake (New York: Anchor-Doubleday, 1957), 123.
3. Pierre Gassendi, *De Motu*, in Craig B. Bruch, ed. and trans., *The Selected Works of Pierre Gassendi* (New York: Johnson Reprint Corp., 1972), 133–36.
4. Gassendi, *Exercises against the Aristotelians*, in Bruch, Selected Works, 107.
5. René Descartes, *Principles of Philosophy*, trans. Valentine Rodger Miller and Reese F. Miller (Dordrecht: D. Reidel, 1983), 105.
6. *Isaac Newton's Papers and Letters on Natural Philosophy* (Cambridge: Harvard University Press, 1958), 302–3.
7. H. G. Alexander, ed., *The Leibniz-Clarke Correspondence* (Manchester: Manchester University Press, 1956), 53.
8. Cited in L. Pearce Williams, *Michael Faraday* (New York: Basic Books, 1964), 268.
9. Ibid., 453–59.

10. James Clerk Maxwell, "On Physical Lines of Force," *The Scientific Papers of James Clerk Maxwell*, vol. 1 (Oxford: Cambridge University Press, 1890), 488.
11. Maxwell, "On Faraday's Lines of Force," *Scientific Papers*, vol. 1, 159.
12. Maxwell, "Faraday," *Scientific Papers*, vol. 2, 789.
13. Maxwell, "Elements of Natural Philosophy," *Scientific Papers*, vol. 2, 324–25.
14. Maxwell, "Faraday," *Scientific Papers*, vol. 2, 360.
15. Maxwell, "On Faraday's Lines of Force," *Scientific Papers*, vol. 1, 156.
16. Maxwell, "Elements of Natural Philosophy," *Scientific Papers*, vol. 2, 328.
17. Maxwell, "Faraday," *Scientific Papers*, vol. 2, 360.
18. Maxwell, "Elements of Natural Philosophy," *Scientific Papers*, vol. 2, 327.
19. Maxwell, *Scientific Papers*, vol. 2, 316, 487.
20. "Let nobody doubt that whoever stands that much in *need* of the cult of surfaces must at some time have reached *beneath* them with disastrous results [and become as] burnt children . . . " (Friedrich Nietzsche, *Beyond Good and Evil*, trans. Walter Kaufmann [New York: Random House, 1966], 71).
21. Richard P. Feynman, *QED: The Strange Theory of Light and Matter* (Princeton University Press, 1985), 82.
22. Ibid., 36.
23. Alfred North Whitehead, *Science and the Modern World* (New York: Macmillan, 1925), chap. 8, "The Quantum Theory."
24. Whitehead, *An Enquiry into the Principles of Natural Knowledge* (New York: Dover, 1982), 25–26; also Whitehead, "Einstein's Theory," in *Science and Philosophy* (New York: Philosophical Library, 1948), 311.
25. Whitehead, *The Principle of Relativity with Applications to Physical Science* (Cambridge: Cambridge University Press, 1922), 36.
26. See Robert M. Palter, *Whitehead's Philosophy of Science* (Chicago: University of Chicago Press, 1960), appendix 4.
27. George R. Lucas, Jr., *The Rehabilitation of Whitehead* (Albany: SUNY Press, 1989), chap. 10.
28. Albert Einstein and Leopold Infeld, *The Evolution of Physics* (New York: Simon and Schuster, 1950), 258.
29. Einstein's introductory essay, "Maxwell's Influence on the Development of the Conception of Physical Reality," in James Clerk Maxwell, *A Dynamic Theory of the Electromagnetic Field*, ed. T. F. Torrance (Edinburgh: Scottish Academic Press, 1982).
30. Whitehead, *Principle of Relativity*, 39.

31. Suzanne Bachelard, "The Specificity of Mathematical Physics," *Phenomenology and the Natural Sciences* (Evanston: Northwestern University Press, 1970), 431.

32. Max Born, *Natural Philosophy of Cause and Chance* (Oxford: Clarendon Press, 1949), 105.

33. Ibid., 125.

34. Ibid., 104.

35. See the discussion by physicist Henry P. Stapp on Copenhagenism and anti-Copenhagenism in "Light as Foundation of Being," *Quantum Implications: Essays in Honor of David Bohm*, ed. B. J. Hiley and David Peat (New York and London: Routledge and Kegan Paul, Ltd., 1987), 257.

36. See Nick Herbert, *Quantum Reality* (New York: Anchor-Doubleday, 1985).

37. For insight about prevalent misconceptions of democracy I am indebted to my father, Robert A. Athearn; see "The Constitution, Taste, and 'Violence of Factions,' " *Dissent* (Spring 1989): 263–65.

38. P. W. Bridgman, *The Logic of Modern Physics* (New York: Macmillan, 1927).

39. Ibid., 164–71.

40. Karl Popper, *The Logic of Scientific Discovery* (New York: Basic Books, 1959.)

41. *The Philosophy of Karl Popper,* book 1, ed. Paul Arthur Schilpp (La Salle: Open Court, 1974), 28.

42. Popper, "Natural Selection and its Scientific Status," *Popper Selections,* ed. David Miller (Princeton University Press, 1965), 242.

43. Ibid., 244–46.

44. Thomas Kuhn, *The Structure of Scientific Revolutions,* 2d ed. (Chicago: University of Chicago Press, 1970), 1.

45. Ibid., 3.

46. Ibid., ix.

47. "Philosophers of science have repeatedly demonstrated that more than one theoretical construction can always be placed upon a given collection of data" (Kuhn, *Scientific Revolutions,* 76).

48. Popper, *Logic,* 317, 280.

49. For an excellent discussion see Vincent Descombes, "The Interpretative Text," *Gadamer and Hermeneutics,* ed. Hugh J. Silverman (New York and London: Routledge, 1991), 247–68.

Chapter 2

1. Philipp Frank, *Philosophy of Science: The Link between Science and Philosophy* (Englewood Cliffs: Prentice-Hall, 1957), 133.

2.   Otto Neurath, Rudolf Carnap, and Charles W. Morris, eds., *International Encyclopedia of Unified Science*, vol. 1 (Chicago: University of Chicago Press, 1955), esp. Carnap, 60–62.

3.   David Hume, *A Treatise of Human Nature* (Oxford: Oxford University Press, 1978), 166–68.

4.   Lawrence Sklar, *Philosophy and Spacetime Physics* (Berkeley: University of California Press, 1985), 4.

5.   Rudolph Carnap, *Foundations of Logic and Mathematics*, Neurath, Carnap, and Morris, *Encyclopedia*, 210–11.

6.   Nancy Cartwright, *How the Laws of Physics Lie* (Oxford: Clarendon Press, 1983), 4.

7.   Ian Hacking, *Representing and Intervening* (Cambridge: Cambridge University Press, 1983), 262–75.

8.   Carl G. Hempel and Paul Oppenheim, "The Logic of Explanation," *Philosophy of Science* 15 (1948).

9.   Ibid., 135.

10.   Ibid., 135–36.

11.   William Dray, *Laws and Explanation in History* (Oxford: Oxford University Press, 1957), 66–79.

12.   Carl G. Hempel, *Aspects of Scientific Explanation and Other Essays in the Philosophy of Science* (New York: The Free Press, 1965), 336.

13.   Hilary Putnam, *Philosophical Papers Volume 3: Realism and Reason* (Cambridge: Cambridge University Press, 1983), 297.

14.   Hempel, *Aspects*, 335–36.

15.   Ibid., 336.

16.   Ibid., 337.

17.   John Dewey, *How We Think* (Boston: D. C. Heath, 1933), 90.

18.   Rudolph Carnap, *Philosophical Foundations of Physics* (New York: Basic Books, 1966), 6–7.

19.   Paul Oppenheim and Hilary Putnam, "Unity of Science as a Working Hypothesis," *Minnesota Studies in the Philosophy of Science*, vol. 2, (Minneapolis: University of Minnesota Press, 1958), 3–36.

20.   Karl Popper, *Objective Knowledge, An Evolutionary Approach* (Oxford: Clarendon Press, 1972), chap. 5, "The Aims of Science."

21.   See for example Abner Shimony, "The Reality of the Quantum World," *Scientific American* (January 1988): 46–53. The same may be said of general cosmology from the viewpoint of astrophysics; see Wallace Tucker and Karen Tucker, *The Dark Matter* (New York: William Morrow, 1988).

Chapter 3

1.   David Hume, *A Treatise of Human Nature* (Oxford: Oxford University Press, 1978), 167.

2. Ernst Mach, *History and Root of the Principle of Conservation of Energy*, trans. Philip E. B. Jourdain, M.A. (Chicago: The Open Court, 1911), esp. p. 57.

3. A good account of this history can be found in Wesley Salmon (see note 12, chap. 5).

4. Herbert Feigl, "Notes On Causality," *Readings in the Philosophy of Science*, ed., Herbert Feigl and May Brodeck (New York: Appleton-Century-Crofts, 1953), 408.

5. Moritz Schlick, "Causality in Everyday Life and in Recent Science," *Readings in Philosophical Analysis*, ed. Herbert Feigl and Wilfrid Sellars (New York: Appleton-Century-Crofts, 1949), 515–33.

6. Ibid., 516.

7. Ibid., 516.

8. Ibid., 520.

9. Ibid., 520–21.

10. Ibid., 521.

11. Ibid., 521.

12. Ibid., 522.

13. Hume, *Treatise*, 170.

14. Hilary Putnam, *Realism with a Human Face*, ed. James Conant (Cambridge: Harvard University Press, 1990), 85.

15. Even Milič Čapek, who favors an ontology for fundamental physics along Whiteheadian lines, adopts this view explicitly as indicating the limits of the Whiteheadian approach. *The New Aspects of Time, Its Continuity and Novelties: Selected Papers in the Philosophy of Science*, ed. Robert S. Cohen, *Boston Studies in the Philosophy of Science*, vol. 125 (Dordrecht: Kluwer Academic Publishers, 1991), 166.

16. Schlick, "Causality," 530.

17. Ibid., 530.

18. Ibid., 532.

19. Ibid., 530.

20. Wittgenstein tried to present this same suggestion in a different philosophical context. The passage occurs as entries 903–6 in *Remarks on the Philosophy of Psychology*, ed. G. E. M. Anscombe and G. H. von Wright, vol. 1 (Chicago: University of Chicago Press, 1980) (see Part II, Chapter 8).

21. Schlick, "Causality," 531.

22. Ibid., 533.

23. Ibid., 532.

24. See Richard K. Scheer, "Wittgenstein's Indeterminism," *Philosophy* 66, 255 (January 1991): 5–23.

25. Frank, *Philosophy of Science*.

26. Frank, *Foundations of Physics*, in Neurath, Carnap, and Morris, eds., *Encyclopedia*, vol. 1, pt. 2, 478–79.

27. Frank, *Philosophy of Science,* 287.
28. Ibid., 348.
29. R. E. Hobart, "Hume without Skepticism," *Mind* 39 (1930): 273–301.
30. Alfred North Whitehead, *Process and Reality: An Essay in Cosmology* (New York: Macmillan, 1929), 265–66.
31. Hobart, "Hume," 280.
32. Ibid., 281.
33. Carnap, *Philosophical Foundations of Physics* (New York: Basic Books, 1966), 189.
34. For some additional arguments favorable to these claims see G. E. M. Anscombe, "Causality and Determination," *Collected Papers,* vol. 2, *Metaphysics and the Philosophy of Mind* (Minneapolis: University of Minnesota Press, 1981), 133–47.

Chapter 4

1. Nancy Cartwright, *How the Laws of Physics Lie* (Oxford: Clarendon Press, 1983).
2. Ian Hacking, *Representing and Intervening* (Cambridge: Cambridge University Press, 1983).
3. Cartwright, *How the Laws of Physics Lie,* 128.
4. Ibid., 60.
5. Ibid., 59.
6. Rom Harré, *Varieties of Realism: A Rationale for the Natural Sciences* (Cambridge: Basil Blackwell, 1986), 287.
7. Lewis Creary, "Causal Explanation and the Reality of Natural Component Forces," *Pacific Philosophical Quarterly* 62 (1981): 153.
8. Cartwright, *How the Laws of Physics Lie,* 61.
9. Ibid., 61.
10. Ibid., 62.
11. Ibid., 67.
12. Ibid., 66–67.
13. Harré, *Varieties of Realism,* 291.
14. Werner Heisenberg, *Physics and Philosophy* (New York: Harper and Row, 1958), chap. 3; and Henry Stapp, "Quantum Theory and the Physicist's Conception of Nature: Philosophical Implications of Bell's Theorem," invited talk at conference "The World View of Contemporary Physics: Is There a Need for a New Metaphysics?" Colorado State University, Fort Collins, Colorado, September 25–28, 1986.
15. Harré, *Varieties of Realism,* 282.
16. Ibid., 291.

17. Ibid., 291.

18. John Dupré and Nancy Cartwright, "Probability and Causality: Why Hume and Indeterminism Don't Mix," *Nous* 22 (1988): 521–36.

19. Ibid., 521.

20. Ibid., 521.

21. Ibid., 521.

22. See Norman Malcolm, "Scientific Materialism and the Identity Theory," *Dialogue* 3, 2 (1964): 115–25.

23. The encounter with indeterminism is one motivation for the use of the terms "tendency" and "*potentia*" by physicists. An additional motivation is that measured properties such as position and momentum do not appear to preexist the measurement events, so that there is some reason to talk about these properties as latent or potential in what is measured.

24. Maurice Merleau-Ponty, *In Praise of Philosophy and Other Essays* (Evanston: Northwestern University Press, 1963), 154.

25. Bas Van Fraassen, *The Scientific Image* (Oxford: Clarendon Press, 1980), 17.

26. Hacking, Representing and Intervening, 262–75.

27. Ibid., 27.

28. Ibid., 262–66.

29. Cartwright, *How the Laws of Physics Lie*, 92.

30. Ibid., 99.

31. Alfred North Whitehead, *The Principle of Relativity with Applications to Physical Science* (Cambridge: Cambridge University Press, 1922), 17–19.

32. Ibid., 26.

33. Stephen P. Maran, "The Mysteries of Miranda," *Natural History* 97 (1988): 74.

34. Rom Harré, "Modes of Explanation," *Contemporary Science and Natural Explanation: Contemporary Conceptions of Causality*, ed. Dean J. Hilton (New York: New York University Press, 1988), 142.

Chapter 5

1. Wesley Salmon, *Scientific Explanation and the Causal Structure of the World* (Princeton: Princeton University Press, 1984).

2. Ibid., 138–39.

3. Ibid., 139.

4. Ibid., 141–42.

5. Ibid., 142.

6. Ibid., 143–44.

7. Ibid., 209.
8. Ibid., 139–40.
9. Ibid., 182–83.
10. Ibid., 156–57.
11. Ibid., 148.
12. Ibid., 214–20.
13. Ibid., 279.
14. Ibid., 202.
15. Ibid., 203.
16. Ibid., 203.
17. Rom Harré, *Varieties of Realism: A Rationale for the Natural Science* (Cambridge: Basil Blackwell, 1986), 264, 304, 308.
18. Ibid., 315–16. For my critiques of Bohm see Chapter 8, pages 281–83.

Part Two—Introduction

1. Maurice Merleau-Ponty, *The Visible and the Invisible/Working Notes* (Evanston: Northwestern University Press, 1988), 16.
2. Ibid.
3. Ibid., 274.
4. Maurice Merleau-Ponty, *In Praise of Philosophy and Other Essays* (Evanston: Northwestern University Press, 1963), 153.

Chapter 6

1. Alfred North Whitehead, *An Enquiry into the Principles of Natural Knowledge* (New York: Dover, 1952), 26.
2. Alfred North Whitehead, *Science and the Modern World* (New York: Macmillan, 1925), 48–52.
3. W. H. Watson, *Understanding Physics Today* (Cambridge: Cambridge University Press, 1967), 132.
4. Ibid., 151, 167.
5. Whitehead, *Enquiry*, chap. 7, entries 25.5–25.9.
6. R. T. Herbert, "The Relativity of Simultaneity," *Philosophy* 63, 242 (October 1987): 463; and G. E. M. Anscombe, *Collected Papers*, vol. 2, *Metaphysics and the Philosophy of Mind* (Minneapolis: University of Minnesota Press, 1981), 154–56. Herbert characterizes Anscombe's view as follows: "It seems that for her only an event 'which was the first,' an event occurring not only before all events but evidently also before the existence of any objects, brings about the existence of a place." But Herbert's arguments to this point in the article tend to show that an ordinary physical

event, such as a flash of light, "though not the first event, may bring about the existence of a place." The results of Part Two indicate that both are right.

7.   See Milič Čapek, "Particles or Events," *The New Aspects of Time, Its Continuity and Novelties: Selected Papers in the Philosophy of Science,* ed. Robert S. Cohen, *Boston Studies in the Philosophy of Science,* vol. 125 (Dordrecht: Kluwer Academic Publishers, 1991), 167–218; also David Ray Griffin, ed., *Physics and the Ultimate Significance of Time: Bohm, Prigogine, and Process Philosophy* (Albany: SUNY Press, 1986).

8.   Martin Heidegger, *Basic Problems of Phenomenology,* trans. Albert Hofstadter (Indianapolis: Indiana University Press, 1982); and Heidegger, *The Essence of Reasons,* trans. Terrence Malick (Evanston: Northwestern University Press, 1969).

9.   Jorge Luis Nobo, "Whitehead's Principle of Process," *Process Studies* 4, 4 (Winter 1974): 275–84; also George R. Lucas, Jr., *The Rehabilitation of Whitehead* (Albany: SUNY Press, 1989), 167–74.

10.   Maurice Merleau-Ponty, The *Visible and the Invisible/Working Notes* (Evanston: Northwestern University Press, 1988), 184, 211.

Chapter 7

1.   Alfred North Whitehead, *The Concept of Nature* (Cambridge: Cambridge University Press, 1971).

2.   Alfred North Whitehead, "Einstein's Theory," in *Science and Philosophy* (New York: Philosophical Library, 1948), 303.

3.   Whitehead, *Concept of Nature,* 51.

4.   Alfred North Whitehead, *The Principle of Relativity with Applications to Physical Science* (Cambridge: Cambridge University Press, 1922), 15.

5.   Stephen Hawking, *A Brief History of Time: From The Big Bang to Black Holes* (New York: Bantam, 1988), 9, 46, 173.

6.   Whitehead, *Concept of Nature,* 52.

7.   Ibid., 57.

8.   Ibid., 53.

9.   Ibid., 66.

10.   Ibid., 56: "A duration *can be* all nature present as the immediate fact posited by sense awareness" (emphasis added).

11.   Ibid., 52–53.

12.   Ibid., 53.

13.   Ibid., 53. The phrase "slab of nature" also occurs in *An Enquiry into the Principles of Natural Knowledge* (New York: Dover, 1982), 69.

14.   Whitehead, *Concept of Nature,* 54.

15.   Ibid., 178.

16. Ibid., 58–59.
17. Martin Heidegger, *Basic Problems of Phenomenology*, trans. Albert Hofstadler (Indianapolis: Indiana University Press, 1982), 231–74.
18. Ibid., 248–49, 251.
19. Ibid., 249.
20. Ibid., 249.
21. Whitehead, *Concept of Nature*, 59.
22. Ibid., 73.
23. Martin Heidegger, "Time and Being," *On Time and Being*, trans. Joan Stambaugh (New York: Harper and Row, 1972).
24. Stanley Rosen, *Nihilism: A Philosophical Essay* (New Haven and London: Yale University Press, 1969), 126–27. My comment should not be taken as an endorsement of this book as a whole.
25. Alfred North Whitehead, *Process and Reality: An Essay in Cosmology* (New York: Macmillan, 1929), 197.
26. Whitehead, *Concept of Nature*, 68–69.
27. Maurice Merleau-Ponty, "An Unpublished Text by Maurice Merleau-Ponty: A Prospectus of His Work," *The Primacy of Perception and Other Essays*, pt. 1, chap. 1, ed. James M. Edie, trans. Arleen B. Dallery (Evanston: Northwestern University Press, 1964), 5–6.
28. Maurice Merleau-Ponty, "The Relations of the Soul and Body and the Problem of Perceptual Consciousness," *The Structure of Behavior*, trans. Alden L. Fisher (Boston: Beacon Press, 1963), 186.
29. Whitehead, *Enquiry*, 14.
30. Frank Ebersole, "How Philosophers See Stars," *Things We Know* (Eugene: University of Oregon Books, 1967), 54.
31. Whitehead, *Enquiry*, 14.
32. Maurice Merleau-Ponty, *The Visible and the Invisible/Working Notes* (Evanston: Northwestern University Press, 1988), 229–30. For additional discussion see Merleau-Ponty, *The Structure of Behavior*, 217–18.
33. Ebersole, "How Philosphers See Stars," 62–63.
34. Whitehead, *Concept of Nature*, chap. 2; *Principle of Relativity*, 72. Ludwig Wittgenstein, *Philosophical Investigations*, trans. G. E. M. Anscombe, 3d ed. (New York: Macmillan, 1953), 193–206. J. L. Austin, *Sense and Sensibilia* (Oxford: Oxford University Press, 1962). Ebersole, *Things We Know*. Maurice Merleau-Ponty, *Visible*, 14–27, 156; *The Structure of Behavior*, 185–244; *Phenomenology of Perception*, trans. Colin Smith (New York: Humanities Press, 1962).
35. Maurice Merleau-Ponty, "The Concept of Nature, I," *In Praise of Philosophy and Other Essays* (Evanston: Northwestern University Press, 1963).
36. Ibid., 151.

37. Ibid., 134.
38. Merleau-Ponty, *Visible*, 211.
39. Ibid., 151.
40. Merleau-Ponty, *Visible*, 211.
41. Merleau-Ponty, *In Praise*, 138–39.
42. Ibid., 137.
43. Merleau-Ponty, *Visible*, 238.
44. For a recent historical discussion of evolutionary cosmologies see George R. Lucas, Jr., *The Rehabilitation of Whitehead* (Albany: SUNY Press, 1989), chap. 4.
45. Merleau-Ponty, *In Praise*, 155.
46. Whitehead (*Enquiry*, chap. 5, entries 15.4–15.9) speaks of the "self-identity" of objects, as opposed to events, and explains this by saying that an object is what it is independently of its relations to the rest of nature, which are purely "external" and contingent to it; in other words, it has identity apart from the differentiation in "passage" that is essential to the identity of an event as having a place and a time. By "objects have no parts" he means that an object as opposed to an event has no inherent differentiating structure.
47. Merleau-Ponty, *In Praise*, 155.
48. Ibid., 134.
49. Heidegger, "Time and Being," 1–24.
50. Merleau-Ponty says (*In Praise*, 130) that to "leave nature in silence" as a region of ontology leaves general ontology with an untenable conception of spirit, namely one that is "immaterialist."
51. Friedrich Nietzsche, *The Will To Power*, trans. Walter Kaufmann and R. J. Hollingdale (New York: Random House, 1967), 262.

Chapter 8

1. For example, see *Microphysical Reality and Quantum Formalism*, 2 vols., ed. A. van der Merwe, F. Selleri, and G. Tarozzi (Dordrecht: Kluwer Academic Publishers, 1988).
2. David Bohm, *Causality and Chance in Modern Physics* (London: Routledge and Kegan Paul, 1957), 120.
3. David Bohm, *Wholeness and the Implicate Order* (New York: Routledge and Kegan Paul, 1980).
4. Ludwig Wittgenstein, *Remarks on the Philosophy of Psychology*, ed. G. E. M. Anscombe and G. H. von Wright, vol. 1 (Chicago: University of Chicago Press, 1980), entries 903–6.
5. Nancy Cartwright, *How the Laws of Physics Lie* (Oxford: Clarendon Press, 1983), 19.

6.   See the dispute between John W. Cook and Norman Malcolm over Wittgenstein's *On Certainty*. Cook, "Notes on Wittgenstein's *On Certainty*," *Philosophical Investigations* vol. 3, no. 4 (Fall 1980): 15–37; Malcolm, "Misunderstanding Wittgenstein," *Philosophical Investigations* vol. 4, no. 2 (Spring 1981): 61–70.

7.   Maurice Merleau-Ponty, *In Praise of Philosophy and Other Essays* (Evanston: Northwestern University Press, 1963), 152–55.

8.   Ibid., 150.

9.   W. H. Watson, *Understanding Physics Today* (Cambridge: Cambridge University Press, 1967), 153.

10.   See note 14, Chapter 4.

11.   I would like to thank Henry P. Stapp for bringing this problem to my attention.

12.   Alfred North Whitehead, *Process and Reality: An Essay in Cosmology* (New York: Macmillan, 1929), 433.

13.   Alfred North Whitehead, *The Concept of Nature* (Cambridge: Cambridge University Press, 1971), 146.

14.   Michael Faraday, "A Speculation Touching Electric Conduction and the Nature of Matter," *Experimental Researches in Electricity*, vol. 2 (New York: Dover, 1965), 293. This line of Faraday's thought contains an intimation of the wavelike constitution of matter. An associated speculation was that the lines of force, rather than the ether, might serve as a "medium" for light propagation. Faraday, "Thoughts on Ray-Vibrations," *Experimental Researches in Chemistry and Physics* (London: Taylor and Francis, Inc., 1991), 366–72.

Chapter 9

1.   Notably by Ernest Nagel out of no particular favoritism toward Whitehead. Nagel, *The Structure of Science* (Indianapolis: Hackett, 1979), 273.

2.   See especially George R. Lucas, *The Rehabilitation of Whitehead* (Albany: SUNY Press, 1989), chap. 10.

3.   Alfred North Whitehead, *The Principle of Relativity with Applications to Physical Science* (Cambridge: Cambridge University Press, 1922), 87–88.

4.   Alfred North Whitehead, *The Concept of Nature* (Cambridge: Cambridge University Press, 1971), 138.

5.   Whitehead, *Principle of Relativity*, 88.

6.   For example, D. W. Sciama, *The Unity of the Universe* (New York: Anchor-Doubleday, 1961), 83–129.

7. Robert M. Palter, *Whitehead's Philosophy of Science* (Chicago: University of Chicago Press, 1960), 212.
8. Whitehead, *Concept of Nature,* 179.
9. Ibid., 138.
10. Ibid., 141–42.

Chapter 10

1. Dean R. Fowler, "Whitehead's Theory of Relativity," *Process Studies* 5, 3 (Fall 1975): 159–73; and Fowler, "Disconfirmation of Whitehead's Relativity Theory: A Critical Reply," *Process Studies* 4, 4 (Winter 1974): 288–90.
2. Fowler, "Critical Reply," on the issues between Einstein and Whitehead.
3. See note 26, Chapter 1.
4. P. W. Bridgman, *A Sophisticate's Primer of Relativity* (Middletown: Wesleyan University Press, 1983), 151–65.
5. For example, Fowler, "Whitehead's Theory of Relativity"; and Robert M. Palter, *Whitehead's Philosophy of Science* (Chicago: University of Chicago Press, 1960).
6. Alfred North Whitehead, *The Concept of Nature* (Cambridge: Cambridge University Press, 1971), 179.
7. Alfred North Whitehead, *Science and the Modern World* (New York: Macmillan, 1925), 120.
8. Alfred North Whitehead, *The Principle of Relativity with Applications to Physical Science* (Cambridge: Cambridge University Press, 1922), 8–9.
9. Ibid., 60.
10. Ibid.
11. Ibid., 7.
12. Ibid., 54.
13. Whitehead, contribution to a symposium on "The Problem of Simultaneity," *Aristotelian Society Supplementary Vol. 3,* "Relativity, Logic and Mysticism" (London: Williams and Norgate, 1923), 35.
14. Alfred North Whitehead, *Process and Reality: An Essay in Cosmology* (New York: Macmillan, 1929), 442.
15. Whitehead, *Enquiry,* 164.
16. See Maurice Merleau-Ponty, "Science Presupposes the Perceptual Faith and Does Not Elucidate It," *The Visible and the Invisible/Working Notes* (Evanston: Northwestern University Press, 1988), 14–27.
17. Maurice Merleau-Ponty, *In Praise of Philosophy and Other Essays* (Evanston: Northwestern University Press, 1963), 154.

18. For some considerations that lend support to this claim See R. T. Herbert, "The Relativity of Simultaneity," *Philosophy* 63, 242 (October 1987): 455–71.

19. Nobo (see note 9, Chapter 6) provides a helpful clarification of this aspect of *Process and Reality.*

20. George R. Lucas, Jr., *The Rehabilitation of Whitehead* (Albany: SUNY Press, 1989), 174–79.

21. Whitehead, *Process and Reality*, 433–34, 442.

22. Ibid., 53.

23. Ibid., 105–7; Whitehead, *Science and the Modern World*, 125–27.

24. Whitehead, *Science and the Modern World*, 135; *Concept of Nature* 162.

25. Whitehead, *Process and Reality*, 20–24. But perhaps one should not regard this as a dogmatism on his part: "Mankind never quite knows what it is after."

Index

absoluteness, physical
  classical concept of, 314–15, 327
  general "process" concept of, 274
  Mach on, 315–16
  Newton on, 314–15
  ontological account of, 324–28
  and photon emission, 301–2
  problem of for understanding space,
    313–28
acausalism, 5–6, 8, 13
  two forms of, 62–63
Anscombe, G. E. M., 207, 372, 374–75
anthropomorphism. *See also* teleology
  114–15, 127, 128, 132
antirealism, scientific, 63–67, 88–90, 134,
  138–46
Archimedes' principle (as example), 73–75,
  80–81
Athearn, Robert A., 369
atom
  constitution of, 294–95, 305–10, 326
  genetic trace of, 222–23, 305–6
  ontological status of, 294, 308–9, 326
  thought-pictures of, 275, 294–95
  Whitehead on, 203–4
atomic interaction, 214–24, 289, 290, 339

being, physical. *See also* ontology: physical
  and becoming, 219, 293, 307–8, 364
  and the field-matter relation, 307
  and light propagation, 299, 358
  Merleau-Ponty on, 292–93
  new concept of, 270
  and physical absoluteness, 326
  primary structure of, 253, 293
  question of, 180, 220
  time-causality relation and, 252

Bell, John S., theorem of, (mentioned) 39, 177
Big Bang theory, 231–32, 254–55
Bohm, David, 173, 281–83
Bohr, Niels, 38, 129, 168, 296
Born, Max, 38–39
Bridgman, P. W., 11, 38, 44–45, 332
Brownian motion, 294

Čapek, Milič, 371, 375
Cartwright, Nancy, 66–67, 120–25, 129–35,
  139, 141–46, 160, 281, 291
causalism. *See also* causal realism
  before Newton, 17–19
  as coinciding with physical ontology, 13,
    15–16
  in early physical research, 29–30
  in Faraday, 22–23
  introduced, 13
  in Maxwell, 23–29
  in Newton and Clarke, 19–21
  ongoing tradition of, 307–8
  suppression of, 58, 60. *See also*
    mechanism, breakdown of
  in Whitehead, 31–35
causal explanation. *See* explanation: causal
causal realism, 13–14, 55, 66–67, 119–73, 291
"causality, law of," 110–11
causality, physical
  basic structure of, 96, 172, 303–4, 309
  expanded conception of, 178, 293–94
  and forces, 146–50
  and locality, 180–81, 281, 284
  serial versus nonserial, 239–42, 286–89,
    291–92, 309
  technical meaning of in physics, 52, 108
  and time, 219–20, 262–63
  ultimate nature of, 224, 274–75, 288–89

Index

causation. *See also* causality, physical
connection in, 92–93, 96, 101–5, 209–13
for Salmon, 162–63
in explanation of light, 209–13
in interacting causes, 93–95, 112–14,
121–25, 134–35, 160
for neo-Human outlook, 52, 91
as primordial relation, 105
problem of in philosophy, 13, 85–117,
165–66, 205, 210–11, 242
mechanistic models and, 85–87, 117,
210–11
reduction of to seriality, 105, 159, 205
transitional structure of, 96, 116
chaos, philosophical problem of, 291
charge, speculation regarding, 310
Clarke, Samuel, 20–21, 127, 191–92
Cook, John W., 378
confirmationism, 45–47
contiguity. *See* discreteness/contiguity
picture
convergence argument
from Perrin experiments, 166, 187
in physical ontology, 188, 314
Copenhagen interpretation, 12, 37–38, 88,
145, 295–96
critiques of by physicists, 40–41, 168
cosmology, evolutionary (philosophy), 270–
71
cosmology, scientific
conceptual problems of, 180–81, 231–32,
254–55, 277
space and time models in, 231–32, 254,
261, 277
Creary, Lewis, 122–24

Darwin, Charles. *See* natural selection,
theory of
democracy, 41–42, 249, 369
Descartes, René, 18–19, 32. *See also*
ontology: Cartesian
determinism
biocultural, 42
ontological approach to, 111, 286–89
scope of in nature, 106–7, 293
difference/differentiation. *See also*
transition; extension
absence of in a "process," 161, 163–64
in causal structure, 96, 101–5, 116, 165–66,
209–13, 239–42, 283, 293–94, 340–42

in events, 165–66, 206, 208
forms of, 85–87
in "passage" (Whitehead), 234
in perception, 262, 307
in prelocal structure, 219
and whole-part relation (Bohm), 282–83
difference, identity and. *See* identity and
difference
discreteness/contiguity picture, 85–87,
98–105, 116, 199, 210–11, 263
double-slit experiment, 198, 223
Dupré, John, 129–32

Ebersole, Frank, 376
Einstein, 11, 35–37, 39, 177, 178, 206–7, 216
and Whitehead. *See* Whitehead: and
Einstein
Einstein-Podolsky-Rosen problem, 39, 177–
79. *See also* nonlocality
electricity, 187
electron, 305–8
empiricism. *See also* neo-Humean outlook
in Mach, 89, 315
objection of to physical ontology, 185–87
in Ostwald, 89
in Van Fraassen, 145
energy, 197, 297, 298
epistemology. *See also* empiricism; neo-
Humean outlook
as ally of nihilism, 10–11
and causal theory of perception, 358–59
freeing science from, 183
positivist, 87–89
as skeptical antirealism, 62–63
as theory of science, 1
and "time lag" problem, 256
ether
contemporary issue of, 202–3
different roles of, 36
downfall of, 30, 186, 191, 196
for Faraday and Maxwell, 23–25, 195–97
for Mach, 35
replacement for, 151, 218, 253
"ether of events" (Whitehead), 32, 201–9,
272
Everett, Hugh (Everett's hypothesis), 284
explanation. *See also* understanding
and description, 67–68, 87, 182
Einstein on, 36
grounds for correctness of, 189

for later Whitehead, 363
law-based model of. *See* question of in philosophy
Maxwell on, 23–29
narrative (causal, historical, physical), 4–8, 59, 71, 105, 110, 172, 177, 184, 205–6. *See also* science: as natural history
question of in philosophy, 10, 13, 57–84, 168, 189, 332–33
"scientific," 52, 67–81, 167
for Salmon, 166–68
theory of, 81–84, 153–54, 166–68, 332–33
and understanding, 84
extension. *See also* difference/differentiation
contrasting forms of, 219, 239–42, 340–41, 353–57
in perceptual depth, 255–56, 261–63
Whitehead on, 32–33, 204–5, 234–35, 228–31, 353–54, 362–63

falsifiability, doctrine of. *See* confirmationism
Faraday, Michael, 22–23
advanced speculations of, 299, 306–8, 378
causalism of, 22–23
and Maxwell, 23–29, 30
and Whitehead, 34, 306–8
Feigl, Herbert, 90, 112
Feynman, Richard, 30–31, 53
field, physical. *See also* forces
as constituting matter
for Einstein, 36, 216
for Faraday, 306–8
contemporary recognition about, 18, 202–3
Einstein on, 36, 216
Faraday on, 23
and inertial forces, 318, 319
as latent causal factor, 39, 135–38, 148
new conception of, 215–24, 296–97
physical explanation of, 149–50, 215, 273, 305–10
as physically fundamental, 11
scientific approaches to, 4–5
and space, 18, 149
as subject of present study, 2
Whitehead on, 32–34
field-matter differentiation, 294, 298–99

Fitzgerald-Lorentz contraction theory, 331, 336–37
forces. *See also* fields
of attraction and repulsion, 2, 196–97
empiricist concept of, 122
history of inquiry into, 19–35
as latent properties of bodies, 125–29
possibility of inquiry into, 146–50
inertial, 314, 318, 319, 324
Fowler, Dean R., 330, 379
Frank, Philipp, 58–59, 60, 110–11
"free will," problem of, 133, 359
Fresnel, Augustin Jean, 183, 195–96

"Galilean science," 17
Galileo, 16–17, 21
Gassendi, Pierre, 17, 21
Gould, Stephen J., 367
gravity, force of
causal question about, 148–49
Clarke on, 20–21
Newton on, 6, 19–20, 29
ontological explanation of, 308–9, 310
as radiation, 124–25

Hacking, Ian, 66–67, 139–46, 281, 291
Harré, Rom, 121, 122, 125–29, 148–49, 160, 170, 172–73
Heidegger, Martin, 7, 180, 219, 244–45, 248–49, 364
Heisenberg, Werner, 128, 129
Hempel, Carl G., 68–81, 153
Herbert, Robert T., 207, 374–75, 380
Hobart, R. E., 112–14, 116
Hume, David
on causation, 88, 100–101, 103, 112–113
Salmon on, 162–66
as source of neo-Humean outlook, 13, 55

identity and difference
and the matter-field relation, 298–99, 306–8
and physical origination, 273
problem of for understanding causation, 165–66, 212, 303
and temporal passage, 262
indeterminism
and atomic interaction, 214, 285, 289–94
"capacity" interpretation of, 131–32
causal explanation of, 138, 223, 289–94

implications for physical causality, 107
and intentionality, 132–33, 286–89
philosophical possibility of, 286–89
positivist interpretation of, 105–11
"productionist" interpretation of, 111, 132–34
informationism, 283–84
instrumentalism
and antirealism, 61–67
in Cartwright, 121
question of, 2
in Kuhn, 48
in physical ontology, 64–65, 281, 359
as spur to positivism, 79–80, 84, 87

Kant, Immanuel, 52, 183
Kuhn, Thomas, 11, 38, 47–49, 50, 52

laws of nature
as content of scientific explanation, 13, 57–84
as mode of theoretical knowledge, 4–5, 130, 147, 331–32
Lenard, Phillip, 197
light
causal structure of, 164, 194–95, 299–302, 303–5, 308, 309, 338–39
and causalism, 33, 140, 191–93
microphenomena of, 30–31, 187, 195, 197–201
ontological question of, 191, 299
as problem of explanation, 183–84, 329–65
sources of wave theory of, 186
theories of, 20, 194–97
logico-empiricism, 13, 52–53, 55, 57–61, 168, 283
Lucas, George R., 377

Mach, Ernst, 35, 89, 145
Maclaurin, Colin, 29
Malcolm, Norman, 378
mass (physics), 305–9
matter wave, 285, 298–99
Maxwell, James Clerk
electromagnetic theory of, 24, 183, 195–97
on physical explanation, 22, 23–29, 30, 197
and Whitehead, 34
measurement, impact of on space and time concepts, 86, 360

mechanism
alternatives to, 20–21, 23, 209–13
breakdown of, 30–31, 136–37, 151, 180–83, 192, 197–98
conceptual foundations of, 21, 23, 32, 85–86, 179, 180–82, 206
cosmological background of, 265–67
critique of in Whitehead, 32–35, 227, 233, 265
and neo-Humean outlook, 85–87
possible meanings of, 58–59
and problem of causation, 85–87, 210–11
Merleau-Ponty, Maruice
commonality with Whitehead on perception, 251, 259
on microcosmic entities, 136, 292–93
on ontology of nature, 226, 264–72, 377
on philosophy-science relation, 189
as resource for physical ontology, 185
on time relativity, 359
metaphysics
as distinct from ontology, 7, 362–63
as framework of modern science, 275–77
reaction against, 50–51
in Whitehead's mature thought, 361–64
Michelson, Albert (Michelson-Morley experiment), 320, 329, 334–37, 365
Millikan experiment, 66
mind-brain identity thesis, 133, 286–87
mysticism, naturalistic, 276–77
mysticism, quantum. *See* quantum mysticism

Nagel, Ernest, 378
natural history. *See* science: as natural history
natural selection, theory of, 46–47
necessity
in concept of causation, 92–93, 95–96, 105, 109–10, 113
in ontological explanations, 309–10
neo-Humean outlook. *See also* philosophy of science; empiricism
on causation, 52, 85–87, 91, 115–17
on explanation, 52–53, 185
origins of, 5, 168
types of, 13
Newton, Isaac, 19–20, 29, 314–15
New Age, the, 42
Nietzsche, Friedrich, 12, 277, 364, 368

nihilism
  as general cultural condition, 39–40,
    41–44, 49–52, 189
  general model of, 12, 49
  "scientific," 8–14, 182–83, 276–77
  designation explained, 12
Nobo, Jorge Luis, 380
nonlocality
  anticipated by Whitehead, 202
  approach of physical ontology toward,
    181–82, 285, 295, 301–2
  Bohm's response to, 282–83
  standard conceptual responses to, 179
  encounter with by physics, 39, 176–79
  projected strategy concerning, 14
nothing (or nonbeing), 268, 269, 273, 326

observation
  as influencing experiments, 43
  for critical realists, 141–43
  versus perception, 89, 91, 140–41, 146
  for positivism/empiricism, 87, 140–41
  for Whitehead, 33, 141–43, 193, 228–31
ontic-ontological distinction, 180–81, 275
ontology
  Cartesian (traditional, classical) 32,
    180–81, 265–68, 292–93
  crisis of in physics, 13, 175–76, 179–81,
    182–84, 232
  event (or process), 7, 208–9, 215
    in Salmon, 155–66
    in Whitehead, 32, 184, 192, 201–9, 215,
      344
  physical
    central question of, 220
    contrasted with metaphysics, 7
    contrasted with physics, 53–54,
      177–82, 190
    introduced, 6–8, 11
    professional attitudes toward, 51–52
    shift in, 181, 188, 264–72, 275,
      296–97, 340
    strategy of, 7, 31–35, 149–50, 179–82,
      192–93, 251–52
    tasks of, 34–35, 183–84, 190
    and Whitehead, 34–35, 184
operationalism
  as methodology of physics, 35, 39, 206–7,
    331
  philosophical, 45, 332–33
Oppenheim, Paul, 68–72, 153

Ostwald, Wilhelm, 89

Palter, Robert M., 320
particle-wave duality, 5–6, 31, 179, 186–87,
    195, 223, 224, 285–86, 309
perception. *See also* observation: versus
    perception
  for Merleau-Ponty, 251, 259
  traditional philosophical theories of,
    210–11, 259–60, 358–59
  for Whitehead, 228–31, 250–51
Perrin, Jean, 89, 166
philosophy (general)
  achievements of, 49, 183
  cultural status of, 8, 259–60
  natural, 3, 6–7, 11, 54, 175–76
  relation to science, 3, 6–7, 15, 51–52, 54,
    96, 227–28, 363
  Wittgensteinian, 50–51
philosophy of science
  attitude toward physical ontology of,
    51–52
  conventional constraints on, 1, 119–20
  cultural influence of, 8, 15
  influence of acausalist physics on, 44–49,
    175
  nihilist dilemma of, 10
  on arbitrariness of theories, 48, 369
  as philosophically retrograde, 182–83
  traditional aims of, 1–2
  Whitehead's criticism of, 250
physics. *See also* quantum mysteries;
    quantum theory; Einstein
  modern history of, 4–5, 15–39, 60
  influence of philosophy on, 183
  mode of theory of, 4–5, 9, 11–12, 15–16,
    37, 55, 178, 189, 279–81, 293, 331–33,
    359
  ontological crisis of. *See* ontology: crisis
    of in physics
  as paradigm of science, 1, 12, 188–89
  quest for "particles" in, 302
  space and time models in, 261, 293, 333,
    359
  specialized character of, 2
  thought-pictures used in, 200, 294–95
Planck, Max (Planck's constant), 166, 199,
    304
Podolsky, Boris. *See* Einstein-Podolsky-
    Rosen problem
Popper, Karl, 11, 38, 39, 45–47, 48–49, 83

positivism, 13, 45
  and antirealism, 65
  on causal relation, 90–105, 112–14
  conception of science, 87–90
  distinguished from logico-empiricism,
    52–53
  on indeterminism, 105–11
"postmodernism," 49–50
prediction
  as general function of theory, 4–5. *See also*
    physics: mode of theory of
  probabilistic, 105–11, 135–37, 169–72
prelocal structure (or process)
  as constituting interatomic reality,
    305–310
  as context of propagation, 221–23,
    290–92, 337–43, 352
  and the field-matter relation, 299
  introduced, 215–23
  relation of to time, 225–26, 271–72
"probability wave," 38–39, 43, 111, 149,
    171–72, 214
Putnam, Hilary, 74, 105

quantum discontinuity, 199–200, 209, 211,
    214
quantum indefiniteness, 136–37, 285,
    295–96
quantum mysteries. *See also* nonlocality;
    mechanism: breakdown of 167, 182,
    214–24, 284–86, 289–99
quantum mysticism, 39–41, 43–44
quantum physicist-ontologists, 7–8, 30, 54,
    135–38, 166–68, 188, 279–86
quantum theory
  positivist arguments from, 105–11
  "realist" interpretations of, 279–86
  as source of nihilism, 37–38

radiation. *See* light
realism, scientific. *See also* causal realism
  in Einstein, 11
  "modest," 126–29, 134, 172
  about "particles," 138–46
  in quantum physics, 279–84
  resurgence of, 65, 138–46, 279
  about theories, 121
Reichenbach, Hans, 159
relativity, theory of, Einstein's
  knowledge-character of, 12, 37, 178,
    332–33, 359

nihilist impact of, 11–12
Rosen, Nathan. *See* Einstein-Podolsky-
    Rosen problem
Rosen, Stanley, 248, 376
Russell, Bertrand, 159

Salmon, Wesley, 153–73, 239, 283
Schlick, Moritz, 90–110, 116
science
  concept of
    in Maxwell, 23–29
    for natural philosophy, 2, 4, 15, 227,
      281, 330
    in philosophy of science, 11–12, 44–49
    positivist, 60
  distorted contemporary relationsh with,
    41–42, 183
  as natural history, 6, 46–47, 48, 50, 59, 82,
    367
  relation to philosophy. *See* philosophy:
    relation to science
  social, 41
  theory of. *See* epistemology; philosophy
    of science
  unity of, 60–61, 68, 82–83, 87
scientific worldview, 42, 275–76
skepticism. *See* epistemology
Sklar, Lawrence, 64–65
space
  as coemergent with time, 252–53, 353–57
  Descartes-Gassendi dispute over, 18
  events (or processes) and, 86, 148, 204
  and light propagation, 299–302
  as medium of forces, 22–23, 30, 225
  and motion, 313–28
  two aspects of, 217–22, 252–53
spacetime (or space-time), 86, 216, 226
Stapp, Henry, 128, 135–38, 369, 372, 378
Stein, Howard, 167, 214

teleology, 20–21, 127, 269–70. *See also*
    anthropomorphism
"theoretical entities," 2, 66, 121, 124, 140–46
theory reduction, 82
time
  as coemergent with space, 252–53, 353–
    57, 360
  in conception of processes, 86
  "dilation" phenomenon of, 349–51
  metaphysical concepts of, 238, 252, 362–64

problem of for process ontology, 219–20, 227, 249–50, 252–53, 273
problems of in relativity physics, 349–51, 359–61
relation to causality, 246–47, 251–52, 262–63
two aspects of, 237–53
Whitehead on, 226–53, 319–24, 343–60, 362–64
"time lag" problem, 253–63
totality, natural
Bohm on, 282–83
classical concept of, 235, 254–55, 277, 314–15, 327, 365
and physical absoluteness, 323–24, 324–27, 364–65
"process" conception of, 228–50, 269, 274, 326
transition. *See also* difference/ differentiation
as character of immediate experience (Whitehead), 250–51
and physical absoluteness, 324–28, 342
as structure of causation, 96, 116
as ultimate physical structure, 33, 215–16, 234–36, 239–41, 271–72, 274–75

understanding, 43–44, 50, 84. *See also* explanation

Van Fraassen, Bas, 138–46
verificationism. *See* confirmationism

Watson, W. H., 203, 291, 295–96
wave-particle duality. *See* particle-wave duality
Whitehead, Alfred North
approaches to interpretation of, 330
on "causal efficacy," 112
causalism of, 31–35
and Einstein, 31, 35, 178, 201, 226, 231, 316, 330–31
on "ether of events," 201–9, 221, 304–5
on event-object distinction, 377
and Faraday, 28, 306–8
later thought of, 361–64
on perception, 228–31, 232–35, 250–51, 255, 258, 259
on physical absoluteness, 316–24, 328
on physical observation, 33, 142–43, 193
on physical fields, 31–35
relativity theory of, 227, 319–24, 328, 343–60
as resource for physical ontology, 7, 184–85
on the science-philosophy relation, 227–28, 363, 380
theory of nature of, 32–34, 207–8, 215, 225, 227–28, 273–74
on time in relation to "process," 226–53
Wittgenstein, Ludwig
contemporary legacy of, 12, 50–51, 207, 259
on indeterminism, 286–89, 290, 371

Young, Thomas, 186, 188, 194, 198